Landraces - Traditional Variety and Natural Breed

Edited by Amr Elkelish

Published in London, United Kingdom

IntechOpen

Supporting open minds since 2005

Landraces - Traditional Variety and Natural Breed
http://dx.doi.org/10.5772/intechopen.91545
Edited by Amr Elkelish

Contributors
Ponraj Perumal, Arun Kumar De, Debasis Bhattacharya, Aurora Díaz, Inés Medina-Lozano, András Gáspárdy, José Luis Spinoso Castillo, Esteban Escamilla Prado, Victorino Morales Ramos, Víctor Heber Aguilar Rincón, Gabino García De Los Santos, Tarsicio Corona Torres, Cosmas Chikezie Ogbu, Devender Sharma, Rajesh K Khulbe, Ramesh S Pal, Jeevan Bettanaika, Lakshmi Kant, Monali Bhaskar Wakchaure, Mohammad Siddiqui, Akshay Sonawane, Radu Liviu Sumalan, Sorin Ciulca, Renata Maria Sumalan, Sorina Popescu, Bonnie Furman, Arshiya Noorani, Chikelu Mba, Suman lata, Ramesh Kumar Yadav, B. S Tomar, Sajid Rashid Ahmad, Sana Ashraf, Humaira Nawaz

First published in London, United Kingdom, 2021 by IntechOpen
IntechOpen is the global imprint of INTECHOPEN LIMITED, registered in England and Wales, registration number: 11086078, 5 Princes Gate Court, London, SW7 2QJ, United Kingdom
Printed in Croatia

British Library Cataloguing-in-Publication Data
A catalogue record for this book is available from the British Library

Additional hard and PDF copies can be obtained from orders@intechopen.com

Landraces - Traditional Variety and Natural Breed
Edited by Amr Elkelish
p. cm.
Print ISBN 978-1-83968-717-4
Online ISBN 978-1-83968-718-1
eBook (PDF) ISBN 978-1-83968-719-8

Meet the editor

Dr. Amr Elkelish received his Ph.D. in Plant Science from the Technical University of Munich, Germany. Currently, he is a visiting researcher at the Department of Plant Physiology, Matthias Schleiden Institute of Genetics, Bioinformatics and Molecular Botany, Friedrich Schiller University Jena, Germany. He is also an associate professor in the Botany Department, Faculty of Science, Suez Canal University, Ismailia, Egypt. Dr. Elkelish is an editorial board member of several reputed journals and a guest editor of Biology. He has published more than seventy papers in peer-reviewed journals. He is a certified associate trainer with the International Board of Certified Trainers (IBCT).

Contents

Preface

Conservation and genetic diversity in crops are essential elements of sustainable solutions for hunger, malnutrition, and livelihood improvement. Unsustainable use of natural resources, promotion of genetically uniform varieties over local varieties, introduction of alien invasive species, changing human consumption patterns, lack of or inappropriate legislation and policy, and climate change all threaten crop diversity.

Food safety has long been associated with abundant main crops producing cereal, roots and tubers, vegetables, and fruits to supply affordable nutrient energy sources. However, this image has changed as the concept of nutritional security has become an important component of food safety, and food diversity has become the fundamental component of human health.

In both extreme wild and trade forms, conceptualizations appear clear in the broad range of plant materials regarding domestication and/or reproduction. Wild plants are not domesticated or subject to artificial selection and reproduction processes. They do not show any typical crop features, such as uniform germination of seeds and fruit maturation. Commercial varieties are produced by a breeding program to improve certain features of the crop.

This book is a handbook of conservation and genetic diversity in plants and animals. A chapter on maize (*Zea mays ssp. mays*) summarizes its wild relatives/landraces and the genetic gain over time in biotic/abiotic, productivity, and nutritional quality traits. Maize is a crop of global significance, used as human food, animal feed, and in various industrial products. It is an essential source of calories and protein for livestock in developing countries.

One of the chapters discusses Coffee landraces. The popular drink is a primary contributor to annual revenue and employment on four continents and in many emerging nations. In the second half of the nineteenth century, the growth of the mass consumer market in the United States transformed it into an industrial product due to the acceleration in coffee production in Brazil.

Their domestication history has been a concern for our domesticated animals for many years. A major area of research is the considerably earlier development of domestic species. The Bovidae family (e.g., cattle, sheep, and caprine) has a less phylogenic nature. In Southeast Asia, about 50 million years ago, the now-extinct Hypertragulidae were the first identifiable primitive ancestor. Livestock agriculture is a profitable agricultural enterprise and a vital income-enhancement activity.

Amr Elkelish
Faculty of Science,
Suez Canal University,
Ismailia, Egypt

Section 1

Plant Landrace

Chapter 1

On-Farm Crop Diversity for Advancing Food Security and Nutrition

Bonnie Furman, Arshiya Noorani and Chikelu Mba

Abstract

In 2019, nearly 690 million people were hungry, indicating that the achievement of Zero Hunger by 2030 is *not* on-track. The enhanced conservation and use of crop diversity, which demonstrably improves farm productivity and hence food security and nutrition, could be one of the solutions to this problem. The broadening of the inter- and intra-specific diversity of crops contributes to dietary diversification and nutrition and improves the resilience of production systems to shocks, especially the biotic and abiotic stresses attributed to climate change. Examples of successful interventions that resulted in enhanced on-farm crop diversity are provided. Relevant tools and guidelines to strengthen national capacities for the enhanced on-farm management of plant genetic resources for food and agriculture are also highlighted. Guidance, based primarily on the *Second Global Plan of Action for Plant Genetic Resources for Food and Agriculture*, is presented to enable the conservation of farmers' varieties/landraces, their genetic improvement and seed delivery systems; promote their cultivation, consumption and marketing; develop and implement policies; foster partnerships and strengthen requisite institutional and human capacities. Finally, the case is made for research and development, including using modern techniques, to achieve these aims.

Keywords: plant genetic resources, farmers' varieties, landraces, conservation, sustainable use

1. Introduction

The most recent edition of the report on the *State of Food Security and Nutrition in the World* [1] contains very worrying statistics: nearly 690 million people are hungry, i.e. 8.9 percent of the world's population! This represents an increase of 10 million people in a single year and nearly 60 million in five years. In fact, in 2019, close to 750 million – about one in ten people in the world – were exposed to severe levels of food insecurity. Conversely, the incidence of overweight children and adult obesity continues to rise [1]. Thus, the world is not on track to achieve Sustainable Development Goal (SDG) 2: Zero Hunger by 2030 [2]. Should recent trends continue, the number of people affected by hunger will surpass 840 million by 2030. It is crucial, therefore, to find effective, sustainable solutions to address hunger. As implicit in the Agenda 2030 [2], the eradication of hunger and malnutrition must be achieved through sustainable means, especially those that preclude further damage to the environment.

The conservation and use of crop genetic diversity is a key component of sustainable solutions to hunger and malnutrition as well as improving livelihoods. Unfortunately, this crop diversity is threatened by such factors as urban encroachment on farmland, unsustainable use of natural resources, the promotion of genetically uniform varieties in replacement of local varieties, introduction of alien invasive species, changing patterns of human consumption, absence of, or inappropriate, legislation and policy, as well as climate changes [3]. The loss of this genetic diversity reduces the options for sustainably managing resilient agriculture [4] in the face of adverse environments and rapidly fluctuating meteorological conditions. As such, it is essential to strengthen their improvement and management on-farm and to enhance their documentation and complementary conservation *ex situ* to safeguard these valuable resources [5].

The Second Global Plan of Action for Plant Genetic Resources for Food and Agriculture (Second GPA) [5] is the internationally agreed framework for the conservation and sustainable use of the full range of plant genetic resources used for food and agriculture, including farmers' varieties/landraces managed on-farm. The actions which countries commit to take in order to achieve these aims are enunciated in the Second GPA in 18 thematic Priority Activities, several of which are specific to crop diversity managed on-farm. Developed as the global policy response to the gaps and needs identified in the *Second Report on the State of the World's Plant Genetic Resources for Food and Agriculture* [6], the Second GPA provides guidance on:

- promoting farmers' varieties/landraces, which is used as an indication of overall crop diversity in this chapter, through developing and strengthening national programmes;

- increasing regional and international cooperation, including research, education and training and enhanced institutional capacity for the conservation and use of plant genetic resources for food and agriculture (PGRFA); and,

- developing and implementing evidence-based policies to promote and improve the effectiveness of on-farm conservation, management, improvement and use.

This chapter highlights the importance of inter- and intra-specific crop diversity managed on-farm as a mechanism to address malnutrition and food insecurity, especially under worsening climate change scenarios. To promote the cultivation and use of the widest possible crop diversity, guidance, based on the relevant Priority Activities of the Second GPA, is provided. These encompass the actions necessary for the conservation and on-farm management of PGRFA; enhanced access to, and use of, local crop diversity – including through responsive seed systems; and genetic improvement as means to the sustainable use of crop diversity. Relevant enabling policy instruments and initiatives for the conservation and sustainable use of crop diversity, developed over the last 50 years, are also described.

2. Important elements of crop diversity conservation and use

With about 80% of all foods being plant-based, any effective solutions for the current trend of worsening food insecurity and malnutrition must address the shortcomings of crop production systems. Crop genetic diversity not only represents the basis of food and agricultural systems, it is also an enormous reservoir of useful genes and gene complexes that endow plants with coping mechanisms for evolution and habitat changes [7, 8]. The inter- and intra-specific variation of crops provides the basis for more productive and resilient production systems that are better able to cope with stresses such as drought or overgrazing [9]. This diversity also enhances the nutritional status of people [10–12]. Changes in land use, together with high rates of urbanization and emigration, displacement of traditional crops in favor of a few starchy staples, and abandonment of marginal agricultural areas, are posing an unprecedented threat to this diversity. Exacerbating this are the threats posed by climate change manifested through the increasing frequencies, distribution and intensities of extreme weather events.

2.1 The narrow genetic base of crop production systems

There are approximately 380,000 vascular plant species [13, 14], of which less than 30,000 (or barely 7%) have been consumed as food by humans [15]. Of these, some 6 000 (or 22% of edible plants) have been actively cultivated for human consumption [16, 17]. Despite this diversity, agricultural production systems depend on a narrow list of crop species. This is illustrated by the fact that less than 200 plants were the sources of global food production in 2019, with only nine of them (sugar cane, maize, rice, wheat, potatoes, soybeans, oil palm fruit, sugar beet and cassava) accounting for over 66 percent of all crop production and 53 percent of global average daily calories [3, 18] (See **Figure 1**).

Agricultural production systems, based on just a few crops, are more vulnerable to biotic and abiotic stresses, including incidences of extreme weather events which, in the past, have resulted in crop failures. Compounding this, many local crops and varieties are cultivated as small and isolated populations and thus tend to lose genetic diversity [19]. These small populations undergo limited geneflow and are subject to genetic drift, founder effects and inbreeding. This, seen ever more frequently due to progressive introduction of commercial varieties, changing climatic conditions, migration to urban areas and expansion of land use for infrastructure and social development, represents an unprecedented threat to local crop diversity [20].

In order to address the impact of the above on changes on diversity, it is essential to monitor farmers' varieties/landraces on-farm [3]. Understanding changes in genetic diversity over time entails the assessment of:

- species richness and evenness and associated environmental variables;
- population size and genetic structure of farmers' varieties/landraces; and,
- the impact of management or farming practices on populations.

Further, at the genetic level, diversity can be assessed using a range of modern genomics-based approaches, such as molecular markers to determine changes over time as well as phylogenetic analyses. An overview of these approaches can be found in Bruford et al. [21] and Dulloo et al. [22].

The Second GPA [5] provides guidance on developing and strengthening systems for monitoring and safeguarding genetic diversity and minimizing genetic erosion of plant genetic resources for food and agriculture. Priority Activity 16 of the framework highlights the importance of establishing and implementing monitoring mechanisms for the regular assessments of genetic erosion. Information from extension services, local non-governmental organizations, seed sector and farming communities can be linked to early warning systems at the national and higher

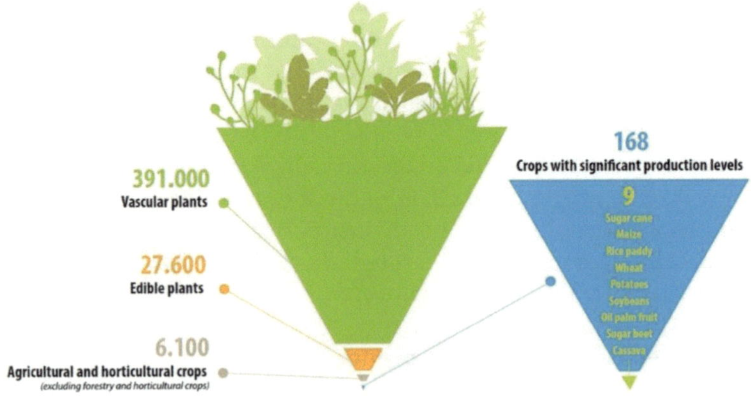

Figure 1.
The plant diversity 'funnel'. Humans rely on nine crops for most of their food while almost 400,000 higher plants have been described out of which a little less than 30,000 are edible.

levels. This Priority Activity also underscores the need to enhance the use of advanced methods, such as those based on information and communication technologies and molecular and spatial analytical tools, for monitoring the status of the most threatened diversity in crops.

2.2 Challenge of climate change

Crop production is affected by the consequences of climate change [23], such as increasing temperatures, changing precipitation patterns, higher concentration of carbon dioxide (CO_2) in the atmosphere and the occurrence of extreme weather events such as floods and drought conditions. Climate change is also affecting biotic factors such as emergence of new pests and diseases and change in the virulence of existing ones. While specific impacts in crop production vary by crop and the climate in which they are grown, there is a growing scientific consensus that increasing temperatures will be detrimental, especially in many developing tropical countries where food insecurity and malnutrition remain pervasive.

Temperature increase and prolonged drought affect a range of biological processes. For example, the physiological responses of plants to high temperature and/or drought conditions are translated into negative effects on growth rates, and therefore on yield. Substantial declines in yields of important crops have already been reported and are predicted to particularly affect those regions where food security is already a major concern [24, 25]. Fruit and vegetable crops are highly vulnerable to climate change during their reproductive stages and to more disease prevalence, and thus production of these crops is also expected to be affected [26]. A detailed study on data from 23 countries in different regions undertaken by Iizumi and Ramankuttym [27] identified temperature variation as a key constraint to maize, soybean, rice and wheat yields. The study showed that the year-to-year variations in yields of these crops from 1981 to 2010 significantly decreased by 19% to 33%.

Climate change also alters the quality of plant nutrients by affecting soil biology, physics and chemistry, and therefore impacts the availability of nutrients [28]. Food quality might similarly be negatively impacted. For example, temperature increases over the past decades in Japan have led to earlier blooming of apples, which in turn has impacted acidity, firmness and water content, and thereby reducing quality [29].

Climate change is expected to alter the range and severity of pest and disease incidence [25]. Predictive models forecast that there will be either increases or decreases of incidence, depending on the region and its climatic conditions; however, the mean probability of pest and disease incidence is expected to rise globally [30]. Quiroz et al. [31] report that climatic changes in the Andean region have led to an increase in pest and disease occurrence in potato cropping, which is driving farmers to shift their production to higher altitudes.

The effects of climate change on major crops are well studied, particularly at species level (i.e. [32–35]). The majority of studies focus mainly on the yield of a specific crop under climate change, yet there are fewer studies comparing the effects on climate change on different varieties of the same species. The use of inter- and intra-specific crop diversity is central to traditional risk management practices in many farming communities (e.g., [36–38]). Such practices will be even more essential as the effects of climate change become more frequent and profound. Many farmers' varieties/landraces are suited to local ecosystems, climatic conditions and farming practices, and have been shown to be more resilient to unpredictable and hardy conditions [8, 39–42].

The Second GPA [5] addresses climate change in most of its Priority Activities, which responds to concerns about the impact of climate change on agriculture. As mentioned above, climate change impacts farmers' varieties/landraces cultivated, with the result that farmers will need to have access to new germplasm. Priority Activity 2 of the Second GPA draws attention to the need for adapted crop varieties to cope with future environmental conditions. It recommends that a range of initiatives and practices should be employed to help farming communities benefit from local crop genetic diversity in their production systems.

2.3 Diversified diets and nutritional components

Plants are the basis of nutrition – whether directly or indirectly – providing key elements in the human diet. While it is clear that malnutrition overall is a major concern, the impact of malnutrition is disproportionately higher on women and children [1]. This can be addressed both through increasing the dietary diversity of the food consumed as well as increasing the quality of produce through breeding initiatives, such as biofortification, to develop nutrient-dense crop varieties.

In the last century, there have been major advances in food production, improving yields in many staple crops [43]. However, the focus of production has been on calorific intake – often negatively correlated to nutritional value in terms of protein content and quality [44–46].

In response to the above, systemic approaches to agriculture now include nutrition as a key component. This is essential for ensuring not only that sufficient calories are produced but that other key health requirements are addressed [43, 47]. In particular, there is a renewed interest in nutrient-rich neglected and underutilized species (NUS) [48–53]. While many of these species are environmentally resilient and cultivated in marginal areas as well as being rich in nutrients, bottlenecks for their increased production and consumption are common [16, 43, 54]. These include low yields, access to quality seeds and planting materials, low market demand and a lack of knowledge in their consumption. These issues, which occur along entire value chains, can be addressed through research and development (R&D) and coherent policy frameworks. In many cases however, financial resources are required to generate innovative solutions and build capacities for their implementation.

The Second GPA [5] provides guidance on promoting diversification of crop production; broadening crop diversity and promoting development and commercialization of all varieties, primarily farmers' varieties/landraces and underutilized species. Its Priority Activities 10 and 11 require that countries promote both the diversity of crops on-farm and the development and commercialization of the widest range of crops and their varieties, in particular farmers' varieties/landraces and NUS, respectively. Additionally, Priority Activity 11 highlights the need to develop and implement policies and incentives to create demands and the matching markets for the products of these crops.

Boxes 1 and **2** illustrate how local crops can be mainstreamed successfully, resulting in increased quality, availability and demand for these fruits and vegetables. The two examples presented, one in Micronesia and the other in Kenya, highlight the need for multisectoral approaches and strategies.

Vitamin A deficiency is one of the key causes of blindness in children [55]. This public health problem is prevalent in many countries, especially in Africa and South-East Asia [56]. One of the approaches for addressing the prevalence of Vitamin A deficiency has been to increase the nutritional diversity of local fruits and vegetables consumed.

Bananas are a key staple in many countries and one of the world's most popular fruits. Studies of different banana cultivars have revealed great differences in carotenoid content, from 5945 mgb-carotene/100 g in the some of the yellow/orange-fleshed Fe'i cultivars to 58mgb-carotene/100 g in the white-fleshed cultivar of the Cavendish subgroup [57, 58]. Fe'i banana (*Musa troglodytarum*) is indigenous to the islands of the Pacific (**Figure 2**) and is known to be rich in Vitamin A.

Figure 2.
Fe'i banana, showing the rich orange color of the fruit, an indicator of its high carotenoid content.

During the 1970s in the Federated States of Micronesia, diets based on non-local foods, together with an increase in consumption of refined white rice, flour, sugar, fatty meats and other processed foods [59], caused a serious Vitamin A deficiency [60]. In response, international agencies and local governments teamed up to promote the production and consumption of local banana cultivars, especially those identified as containing significant amounts of bio-available Vitamin A. The approaches included the development of policies promoting local cultivation, guidance on agronomic techniques, youth clubs, school activities and farmers' fairs. As a result of the various initiatives, the local production and consumption of the yellow/orange-fleshed banana variety, Karat, containing 2 230 µg/100 g of the provitamin A (50 times that found in white-fleshed bananas), was effectively promoted and these local nutritious bananas are now available in most markets. The success of this multisectoral approach – health, agriculture and education – is regarded as a model, linking dietary and agricultural diversity for healthy diets, to be replicated with other locally available, nutrient-dense crops in vulnerable populations.

Box 1.
Successes in mainstreaming local crops for better nutrition: Fe'i bananas in the Federated States of Micronesia.

There are many diverse species and varieties of indigenous leafy vegetables consumed locally in tropical sub-Saharan Africa. These include African nightshades (*Solanum scabrum*), leafy amaranth (*Amaranthus spp.*), spider plant (*Cleome gynandra*), cowpea (*Vigna unguiculata*), Ethiopian kale (*Brassica carinata*), mitoo (*Crotalaria ochroleuca* and *C. brevidens*), kahuhura (*Cucurbita ficifolia*), jute plant (*Corchorus olitorius*) and pumpkin leaves (*Cucurbita maxima* and *C. moschata*) [61]. The nutritional importance of African leafy vegetables (ALV) has been recognized by various experts over recent decades [62–65]. Yet, despite their nutritional advantage over many imported vegetables, levels of consumption had been decreasing in many countries, including Kenya [66].

One of the key reasons for the decline in the consumption of ALV includes migration to cities, causing a shift in production. With these changes, knowledge of the cultivation of ALV was also being lost, including, very importantly, methods of the production of quality seeds. Increasing the quality of seeds can increase yields. For instance, selecting those seeds with lower rates of dormancy results in higher germinability and hence, improved yields ultimately.

In this respect, African nightshades, for example, require the removal of the wet pulp that contains growth inhibitors, which affect germination rates [61]. Initiatives to improve ALV cultivation by disseminating this information, along with other techniques that enhance seed germination, to farmers through participatory methods were implemented successfully. The resulting uptake in the cultivation of quality ALV by smallholder farmers increased the production and quality of African nightshades in Kenya. Extension workers collaborated closely with researchers and international organizations to reconstruct a knowledge base, combining traditional and more technical information on these species.

Although these crops used to be considered a "poor man's food" until 15 years ago, due to, *inter alia*, improvements in seed quality, awareness raising and value chain interventions, ALV are now commonly found in Kenyan supermarkets [61, 63]. ALV, now gaining in popularity, as evidenced by seed companies' interest and the increase in area cultivated, are contributing to addressing malnutrition as well as to improving livelihoods [65].

Box 2.
Enhancing the quality of seeds to boost production: Seed dormancy in African leafy vegetables in Kenya.

3. Management of on-farm diversity

Enhanced crop diversity, including farmers' varieties/landraces, confers resilience on crop production and reduces vulnerability to shocks and are potential sources of traits for crop improvement, especially for developing varieties tolerant to biotic and abiotic stresses [3]. A significant amount of crop diversity, including farmers' varieties/landraces, is only maintained in farmer's fields, orchards or home gardens. Many farmers choose to cultivate farmers' varieties/landraces due to agronomic, culinary, or quality preferences [3, 40]. Much of this crop diversity also has locally important cultural values. The dynamic on-farm management of this diversity contributes to their continual evolution and adaptation due to farmers' selection and seed exchange systems [67].

In order to support countries in enhancing the diversity of crops and varieties which are cultivated by farmers, the *Voluntary Guidelines for the Conservation and Sustainable Use of Farmers' Varieties/Landraces* [3], were developed. They serve as reference material for preparing a National Plan for the Conservation and Sustainable Use of Farmers' Varieties/Landraces and are a useful tool for development practitioners, researchers, students and policymakers who work on the conservation and sustainable use of these valuable resources.

3.1 Germplasm conservation and on-farm management

The diversity of crops and varieties maintained on farmers' fields must also be backed up *ex situ*, to ensure their conservation in an effective, integrated and rational manner in case of loss on-farm. Conserving this diversity *ex situ* is additionally advantageous in that it can be assessed and made more readily available to researchers and plant breeders. Crop germplasm, a significant proportion of which are farmers' varieties/landraces, is conserved in more than 650 genebanks worldwide [68]. Complementary *ex situ* conservation of crop diversity is essential for safeguarding global food security for the present and future. The application of standards and procedures that ensure their continued survival and availability is therefore essential. The *Genebank Standards for Plant Genetic Resources for Food and Agriculture* [69] set the benchmark for current scientific and technical best practices, and support key international policy instruments for conserving crop germplasm in genebanks.

Ex situ conservation of plant genetic resources in genebanks and other facilities safeguards a large and important amount of resources that are vital to global food security [6]. Genebank conservation entails acquisition, storage, characterization, evaluation, regeneration, safety duplication and documentation of germplasm accessions [69, 70]. The methods used include the storage of orthodox seeds in seed genebanks and safeguarding species that produce nonorthodox seeds or are propagated vegetatively as live plants in field genebanks or as plantlets through *in vitro* culture or cryopreservation [69]. Genebanks serve the dual aims of the conservation of PGRFA and the provision of these genetic resources to plant breeders, researchers and other users.

Many collections, especially at the national level, remain vulnerable as they are exposed to natural disasters, including those caused by climate change, and manmade calamities such as civil unrest. These collections are similarly at risk due to avoidable adversities resulting from lack of funding and/or poor management. Well-managed genebanks both safeguard genetic diversity and make it available to breeders. As such, genebanks require adequate and continuous levels of sustainable funding.

In this context, Priority Activity 2 of the Second GPA [5] underscores the need for improved on-farm conservation and the management and use of farmers' varieties/landraces and underutilized crops. It also highlights the need to foster linkages between these activities and the conserving this diversity in genebanks. The Second GPA also recommends that governments consider how production, research, economic incentives and other policies impact the on-farm management and improvement of PGRFA. The actions that should be taken to enhance the *ex situ* conservation of germplasm are provided in the following Priority Activities of the Second GPA:

- Priority Activity 5 on the targeted collecting of germplasm;
- Priority Activity 6 on sustaining and expanding effective *ex situ* conservation of diverse germplasm; and
- Priority Activity 7 on regeneration and multiplication of *ex situ* accessions, including for distribution and safety duplication.

3.2 Enhancing access to, and use of, local crop diversity

The development of farmers' varieties/landraces is commonly undertaken through participatory plant breeding (PPB), which aims to bridge the formal and informal seed systems by supporting smallholder farmers and their collective efforts [71, 72]. PPB often uses demonstration plots in Farmers Field Schools [73] to increase farmers' awareness of the quality of varieties and seed produced, and to support adoption. Vernooy et al. [74] reported that PPB resulted in both the conservation of farmer-preferred landraces and the development of new PPB-developed varieties, as well as farmer-managed seed production and distribution (e.g., in China and Mexico). Community seedbanks played a crucial role in these activities through seed collection and distribution; seed production of improved local varieties; and education and awareness activities. Community seedbanks are informal, locally governed institutions whose core function is to preserve seeds for local use. They play an important role in increasing access to diverse and locally adapted crops and varieties [74, 75], especially farmers' varieties/landraces. These community-based endeavors also enhance related local knowledge and skills in the workflow for seed delivery, i.e. selection, treatment, storage, multiplication and distribution [3].

Community seedbanks can be an effective part of a comprehensive strategy for the conservation and sustainable use of crop diversity. Community-based small-scale seed initiatives, often linked to community seed banks, will play a vital role in the improvement of, and access to, quality declared seeds and planting materials, maintenance of crop diversity for food security, and positively contribute to the national breeding programs. For example, the formation of seed clubs in Vietnam enabled working with farmers to promote varietal selection through participatory plant breeding and the national varietal registration of these local varieties. This has enhanced farmers' access to the quality seeds and planting materials of preferred varieties [76] (see **Box 3**).

Enhanced farmers' access to quality seeds and planting materials of well-adapted crops and varieties is realized through the strengthening of community-level seed production with suitable quality assurance regimes, including protocols for quality declared seeds and quality declared planting materials. The *Quality Declared Seed System* [78] consists of guidelines and protocols that aim at assisting small-scale farmers, specialists in seed production, field agronomists and agricultural extension services in the production of quality seed. This system provides an alternative for seed quality assurance and is particularly useful for countries with limited resources

In Vietnam, the Southeast Asia Regional Initiatives for Community Empowerment (SEARICE) and the Mekong Delta Development Research Centre of Can Tho University (MDI-CTU) have been collaborating with communities on the formation of seed clubs to drive community-based conservation and sustainable use of plant genetic resources. These clubs enable local seed supply systems through seed conservation, exchange, and crop improvement activities. In particular, they facilitated:

- participatory variety rehabilitation, i.e. whereby the original characteristics of the farmers' variety/landrace is restored through selection;
- participatory plant breeding, where farmers collaborate in the process of crop varietal development and have opportunities to make decisions throughout; and
- participatory variety selection, which involves farmers growing and selecting varieties in their own fields, providing a way for breeders to learn which varieties perform well on-farm and are preferred by farmers.

These activities, which bridged the formal and informal seed systems [77], have resulted in the development of 360 farmers' varieties, five of which are nationally certified [76]. The formal registration of farmers' varieties, made possible by the policy and technical assistance provided by MDI-CTU and funding provided by SEARICE, paved the way for the eventual production of quality declared seeds – thereby enhancing the confidence of the farmers in the seeds. This approach to community empowerment has been fundamentally important in the improvement of access to and availability of seeds, maintenance of crop diversity for food security, and positively contribute to the national breeding program through the linkages established between the formal and informal seed sectors.

Box 3.
Seed clubs in Vietnam.

[79]. The system is less demanding than full seed quality control systems yet guarantees a satisfactory level of seed quality. Its partner publication, *Quality Declared Planting Material* [80], was prepared in collaboration with the International Potato Centre and follows the principles and approach of FAO's Quality Declared Seed System.

It is necessary to develop and implement national seed regulatory frameworks and to enable the participation of multiple actors, including farmers. This can be undertaken through cooperatives and small- and medium-scale seed enterprises, and the private sector, while supporting institutional and human capacities along the entire seed value chain. Areas of intervention typically include strengthening capacities for the production and processing of seeds and their quality assurance, packaging, storage and marketing. Priority Activity 11 of the Second GPA recommends that countries promote the "development and commercialization of all varieties, primarily farmers' varieties/landraces and underutilized species" [5]. Linked to this, Priority Activity 12 of the Second GPA focuses on supporting seed production and distribution. It underscores the importance of developing/reviewing seed regulatory frameworks that facilitate the development of seed systems and their harmonization at regional levels, taking into account the specificities of different seed systems [5].

To support practitioners along the entire seed value chain, the six-module *Seeds Toolkit* [81–86] is a resource to enhance knowledge and skills for delivering quality seeds and planting materials of well-adapted crop varieties to farmers. The modules are designed as practical guidance to assist in the implementation of the national seed strategies and capacity building activities, especially for small-scale farmers and small- and medium-scale entrepreneurs.

For policy specific guidance, stakeholders may refer to the *Voluntary Guide for National Seed Policy Formulation* [87]. This explains seed policies and how they differ from seed laws; describes the participatory process of seed policy formulation, the nature and layout of seed policy documents and their key elements; and addresses issues involved in their implementation.

3.3 Genetic improvement as means to sustainable use of on-farm crop diversity

A continuous stream of improved crop varieties that are adapted to particular agro-ecosystems and production systems is required for meeting the challenges posed by food insecurity and malnutrition, especially in the face of climate change. In this regard, Priority Activity 9 of the Second GPA recommends countries to support "plant breeding, genetic enhancement and base-broadening efforts" [5].

Crop breeders must aim to develop varieties that are productive, nutritious, resistant to biotic and abiotic stresses, and are well-adapted to target agroecologies and meet consumer preferences and market demands. Genetic diversity is an essential resource for breeders to improve new cultivars with desirable characteristics [88]. For crop diversity to be useful in addressing malnutrition and climate change through breeding, their characteristics need to be measured, evaluated and recorded in information systems that are available to all relevant stakeholders. The process of characterization entails the description of a minimum set of standard phenotypic, physiological and seed qualitative traits. The evaluation of PGRFA requires an analysis of agronomic data obtained through appropriately designed experimental trials. Both characterization and evaluation use crop descriptor lists that are available for a large number of crop species [89–91]. Additionally, to support standardizing the information, FAO and Bioversity International published passport descriptors that are widely used for the documentation and exchange of germplasm [92]. The FAO World Information and Early Warning System on PGRFA (WIEWS) [68] provides access to passport data of materials held in genebanks worldwide. Other global germplasm management systems, such as GRIN-Global [93] and GENESYS [94], document not only passport but also characterization and evaluation data in genebanks. GENESYS also includes information on the climate at the origin of accessions, and provide the option to search for accessions originating from similar climates. These systems provide plant breeders with a catalog of traits and germplasm for crop improvement.

Conventional plant breeding procedures can be time-consuming and expensive [95]. For example, the breeding, delivery and adoption of new maize varieties has taken up to 30 years [96]. Advances in biotechnology have substantially increased the efficiency for the identification of desirable traits for crop improvement and the knowledge of the genetic mechanisms that control the expression of traits of interest [97]. More targeted breeding can be undertaken as the links between traits and genes are better understood. This is especially important for those traits under polygenic control such as yield and those conveying heat, drought and other stress tolerances [98].

Crossing high-yielding varieties with lower-yielding but resilient local germplasm such as landraces can reduce genetic vulnerability [99] through the broadened genetic base of the improved varieties. This is achieved most effectively through pre-breeding, i.e. the generation of intermediate materials by crossing non-adapted germplasm that possess novel traits with standard breeding lines [5, 100]. A detailed step-by-step overview of pre-breeding procedures is provided in an e-learning course [101], developed under the auspices of the Global Partnership Initiative on Plant Breeding Capacity Building (GIPB). This course is made up of five modules covering the introduction to pre-breeding; genebank management relevant to pre-breeding; pre-breeding project management; creating and managing variation; and the distribution and use of the pre-bred materials and associated regulatory considerations.

In situations where sourcing heritable variations from existing germplasm is not possible or otherwise impractical, the induction of allelic variations through mutagenesis is a viable option [102]. Mutations can be induced by physical (i.e., gamma and x-ray technology) or chemical means [103] for a comprehensive review on this topic). DNA mutations tend to be chance events and therefore require that scientists generate massive numbers of putative mutants that are then subsequently screened for particular traits, a lengthy and costly process. However, advances in high throughput molecular genetics, cell biology and phenotyping techniques mitigate these constraints and facilitate the integration of induced mutations into improved crop varieties [103].

Morphological assessments using traditional phenotyping methods can be labor intensive, time consuming, subjective, and frequently destructive to plants. In fact, the access to large-scale phenotypic data has been one of the major bottlenecks hindering crop breeding [104]. High-throughput phenotyping (HTP) is a recently developed method that has potential to overcome this bottleneck and offers large-scale, accurate, rapid, and automatic data acquisition for crop improvement [105, 106]. A large number of advanced technologies [107, 108], including sensors, information technology and data extraction, combined with systems integration and reduced costs, means that morphology and physiology can be assessed non-destructively and repeatedly across entire populations throughout their development [104, 109]. Novel HTP approaches are necessary to advance the understanding of genotype-to-phenotype cause and effect relationships and therefore accelerate plant breeding [110, 111]. This can be of great importance for assessing the production and resilience traits of farmers' varieties/landraces.

Many traits have been mapped to specific genes and as a result, more analyses are being conducted per unit of time that allow for more specific mapping of traits. Quantitative trait loci (QTL) mapping results provide useful information to understand the genetic mechanisms of important traits and improve the efficiency of marker-assisted selection and genomics-assisted breeding [112, 113]. Taken together, existing genomics knowledge and tools may be used to overcome the constraints to the development of adapted varieties that combat malnutrition and climate change [114, 115].

Advances in phenotyping technology and methodologies for multi-population data analysis have made possible the mapping of QTL [116, 117]. In addition, DNA sequencing has become more rapid, more precise and less expensive [104, 110]; the genomes of most staple crops, and some minor ones, have been sequenced [118]. A recent initiative driven through the African Orphan Crops Consortium (AOCC) is applying genome-enabled methods to improve the production of 101 under-researched ('orphan') crops on the continent [119]. To date, eight genomes have been sequenced and published and another 26 are underway [120]. The ultimate goal of this initiative is to develop resilient, palatable and nutritious varieties of local crops for local peoples to consume and sell – thereby enhancing their nutritional status and livelihoods.

4. Existing policy frameworks

As means to enhance intra- and inter-specific on-farm crop diversity, diverse initiatives, policies and global frameworks have been developed and implemented. In recent years, focus has been on areas of synergies and streamlining efforts among the health, environmental and agricultural sectors (**Figure 3**). The number of policy and legal frameworks targeting crop diversity, reflects the growing global interest and concern and the commitment of countries for their conservation and sustainable use [51, 121].

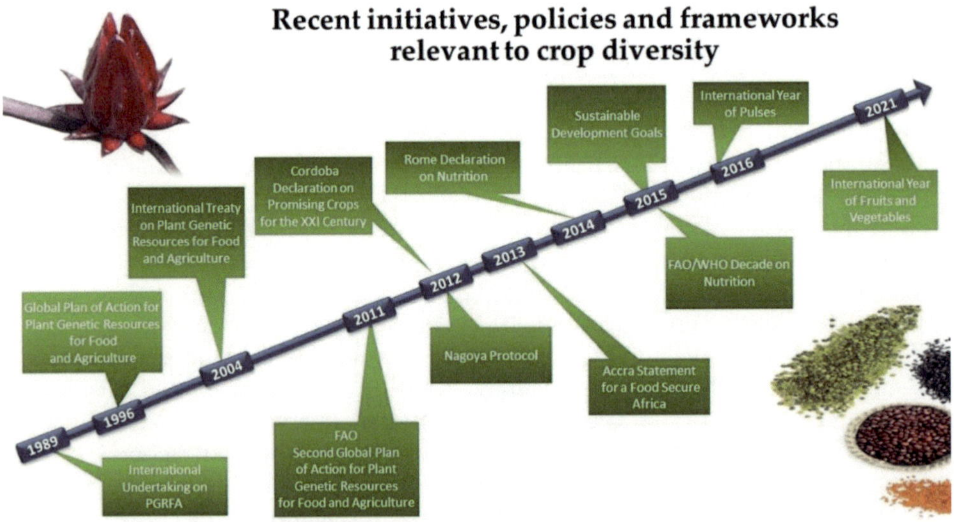

Figure 3.
Timeline showing the development of initiatives and frameworks important for the conservation and sustainable use of crop diversity (adapted with permission from [122]).

While crop diversity has been a key focus of many policy discussions since 1950 onwards [7], the International Undertaking on Plant Genetic Resources which was adopted by resolution 8/83 of the FAO Conference in 1983 was a watershed moment. The objective of this Undertaking was "to ensure that plant genetic resources of economic and/or social interest, particularly for agriculture, will be explored, preserved, evaluated and made available for plant breeding and scientific purposes" [123].

This laid the groundwork for the development of cornerstone frameworks for crop diversity, especially:

- the Global Plan of Action (GPA) for the Conservation and Sustainable Use of Plant Genetic Resources for Food and Agriculture (PGRFA) adopted by 150 countries in 1996 [124];

- the International Treaty on Plant Genetic Resources for Food and Agriculture (the Treaty) that entered into force in 2004, providing a legal framework whereby governments, farmers, research institutes and agro-industries can share and exchange PGRFA and benefits derived from their use [125];

- the Global Crop Diversity Trust, established in 2004 by FAO and Bioversity International on behalf of the CGIAR, to support the efficient and effective *ex situ* conservation of crop diversity over the long term [126];

- the Second GPA in 2011 [5];

- the Cordoba Declaration [127], which emphasized the importance of underutilized and promising crops at the international level;

- the Second International Conference on Nutrition (ICN2) held in Rome in 2014 [128], which showcased the profile of NUS and adopted the Rome Declaration on Nutrition after which 2015–2025 was declared the UN Decade of Action on Nutrition [129]; and

- adoption of the 2030 Agenda for Sustainable Development by 193 Member States of the United Nations [130].

5. Looking forward

Addressing livelihood options for smallholder farmers requires that the focus of R&D be broadened to include a much wider range of crop species and cropping systems. This diversity is essential for breeding new plant varieties that confer the ability to adapt to changing environments, including new pests and diseases and adverse climatic conditions, on cropping systems. Thousands of years of farming and targeted selection have resulted in an invaluable heritage of locally adapted varieties of major and minor crops [16, 127]. The greater the diversity, the greater the chance that at least some of the individuals will possess an allelic variant suited to changing environments, and will produce offspring with that variant [7].

5.1 Bridging conservation, sustainable use and the seed sectors

To achieve the most benefits from PGRFA while at the same time safeguarding them, activities that address conservation must be linked to those concerned with

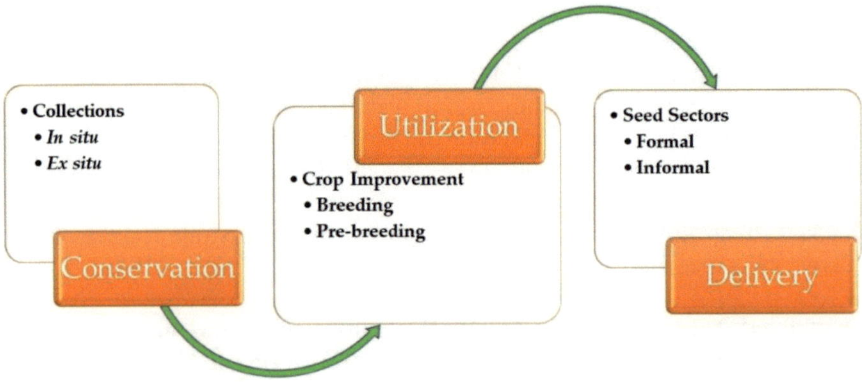

Figure 4.
Continuum of crop diversity, showing the linkages between conservation, sustainable use and seed systems.

plant breeding which in turn must feed into seed delivery systems. In many countries and regions, there is a lack of these linkages between these three modules of the PGRFA management continuum [131] (**Figure 4**).

This continuum approach is also relevant for the efforts to leverage farmers' varieties/landraces to enhance on-farm crop diversity and will require the concerted actions of extension workers, researchers, breeders, seed enterprises and farmers. Similarly, greater cooperation at different stages in the production chain, from the development and testing of new varieties, through value-adding activities, to the opening up of new markets is essential.

5.2 The enabling environment

In order to have long-term impact on the ground, clear and non-conflictual policies are needed, together with effective delivery systems. The policies must be evidence-based and offer relevant interventions that can rapidly be deployed on the ground. Often policies can be at variance with one another, with a resulting negative impact on crop diversity, livelihoods and/or diets. For example, subsidies for promoting staple crops may have a negative impact on the cultivation of minor, but highly nutritious and resilient crops and varieties [16]. Addressing this, FAO developed *Guidelines for Developing a National Strategy for Plant Genetic Resources for Food and Agriculture* [132]. These guidelines support countries in developing national strategies for PGRFA, which include identifying a national vision, goals and objectives, and the corresponding plan of action, including responsibilities, resources, and timeframes for activities. They take into account each country's needs, capacities and constraints.

Efforts must continue to target the development of appropriate national strategies and policies to promote the diversification of cropping systems, including the on-farm conservation and use of underutilized species, enable R&D and the uptake of their outputs. The Second GPA [5] highlights the importance of conservation and sustainable use of crop diversity in terms of policy and capacity development. National policies should aim to strengthen capacities in crop improvement in order to produce varieties that are specifically adapted to local environments. These policies may include appropriate for the protection of new varieties – as applicable, varietal release and seed certification – or other appropriate quality assurance regimes. These would promote and strengthen their use and ensure that they are included in national agricultural development strategies.

Building national programmes and institutional capacities is critically important as a means to promote public awareness on the importance of the diversity of PGRFA [5, 131]. The support to policy-makers as well as training and capacity building for scientists, breeders, extension specialists, seed producers, farmers, indigenous peoples and local communities on themes that enable the promotion of the development and commercialization of all crop varieties, primarily farmers' varieties, landraces and underutilized species, is recognized as a fundamental necessity [3]. Relevant topics for such training and capacity building activities include activities that promote the increased on-farm management of crop diversity such as the identification of all suitable materials and the development and implementation of sustainable management practices, postharvest processing and marketing methods and the documentation of relevant local and traditional knowledge. Additional activities include those that promote establishing, running and advising local small-scale seed enterprises.

The Second GPA [5] provides guidance on the human and institutional capabilities that should be strengthened for the conservation and sustainable use of PGRFA, including farmers' varieties/landraces. These are summarized below:

- Priority Activity 13 focuses on developing national programmes, recognizing that efforts to coordinate national planning, priority setting and fundraising are needed. Emphasis is placed on enhancing collaboration between the public and private sectors, national and international cooperation, strengthening links between PGRFA conservation and use, developing information systems and publicly accessible databases, identifying gaps in the conservation and use of PGRFA, increasing public awareness and implementing national policies and legislation and international treaties and conventions.

- Promoting and strengthening networks for PGRFA, as described in Priority Activity 14, are crucial for improved coordination, communication and organizational skills. Resources and capacity should be available for activities such as planning, communications, travel, meetings, network publications such as newsletters and meeting reports, and network strengthening, including the preparation of successful proposals for submission to donors.

- Information systems for PGRFA facilitates evidence-based decision making for their effective conservation and use. Priority Activity 15 provides guidance for national and regional programmes, including for strengthening and harmonizing documentation, characterization and evaluation of germplasm.

- In order to monitor and safeguard genetic diversity and minimize genetic erosion of crop diversity, capacities must to be strengthened for gathering and interpreting information in conducting inventories and surveys (Priority activity 16). Training on monitoring should be provided to breeders, farmers and indigenous and local communities. It is important to develop training materials, including self-teaching tools, in local languages as needed.

- As described in Priority activity 17, the long-term availability of adequate human resources capacity in all areas of PGRFA conservation and use, including management, legal and policy aspects, must be developed and strengthened. This includes support for enabling national and regional organizations and programmes to update curricula, provide advanced education and strengthen research and technical capacities in all relevant areas.

- Communicating effectively about the many benefits of crop diversity to food security and sustainable livelihoods is critical to the success of any intervention. Priority Activity 18 highlights the importance of national public awareness programmes and the development of international links and collaborative mechanisms such as networks, involving different sectors, agencies and stakeholders. The aim is to increase the value of crop diversity by bringing this information to the attention of policy-makers and the general public.

6. Conclusions

Five years after the world committed through the SDG to end hunger, food insecurity and all forms of malnutrition, we are not on track to achieve these objectives by 2030. The sense of urgency is even more pressing due to the looming 2030 deadline of the SDGs, which underscores the need to 'think outside of the box'. Options for addressing food insecurity and malnutrition should include increasing the diversity of crops and varieties cultivated. This chapter highlighted the danger of the continued overreliance on a few crops and their varieties. It prescribed the means for incorporating a wider diversity of farmers' varieties/landraces into crop production systems. These local crop genetic resources tend to be adapted to low input production systems, which is prevalent in many food insecure countries of the world. The underlying premise is that improving agricultural production while using the diverse plant genetic resources available can benefit directly the livelihoods of smallholder farmers and farming communities. The ensuing result is a positive impact on food security and nutrition, environmental resilience and effective management of crop diversity.

The Priority Activities of the Second GPA provide guidance for the enhanced integration of farmers' varieties/landraces into cropping systems. These include recommendations for promoting on-farm crop diversity directly and the conservation of these critical resources in genebanks. The Second GPA also addresses continued genetic improvement of germplasm and suitable seed delivery systems, especially those that are community-based and are tailored to low input production systems. Advances in molecular genetics, phenotyping and computing capacities enhance the prospects of generating compelling R&D outputs. In the same vein, policies and strategic partnerships – at local, national, regional and global levels – that facilitate the participation of a multiplicity of stakeholders are also critically important.

Notes/thanks/other declarations

Author details

Bonnie Furman*, Arshiya Noorani and Chikelu Mba
Food and Agriculture Organization of the United Nations, Rome, Italy

*Address all correspondence to: bonnie.furman@fao.org

References

[1] FAO, IFAD, UNICEF, WFP and WHO. The State of Food Security and Nutrition in the World. Transforming food systems for affordable healthy diets. Rome: Food and Agriculture Organization of the United Nations; 2020. 289 p.

[2] United Nations. Transforming our world: the 2030 Agenda for Sustainable Development. Division for Sustainable Development Goals. New York: United Nations; 2015. 40 p.

[3] FAO. Voluntary Guidelines for the Conservation and Sustainable Use of Farmers' Varieties/Landraces. Rome: Food and Agriculture Organization of the United Nations; 2019. 135 p. DOI: 10.4060/CA5601EN

[4] CBD. Secretariat of the Convention on Biological Diversity. Global Biodiversity Outlook 5. Montreal: CBD; 2020. 208 p.

[5] FAO. Second Global Plan of Action for Plant Genetic Resources for Food and Agriculture. Rome: Food and Agriculture Organization of the United Nations; 2011. 91 p.

[6] FAO. The Second Report on the State of the World's Plant Genetic Resources for Food and Agriculture. Rome: Food and Agriculture Organization of the United Nations; 2010. 369 p.

[7] Rao VR, Hodgkin T. Genetic diversity and conservation and utilization of plant genetic resources. Plant cell, tissue and organ culture. 2002 Jan 1;68(1):1–19. DOI: 10.1023/A: 1013359015812

[8] Jarvis DI, Hodgkin T, Brown AH, Tuxill JD, Noriega IL, Smale M, Sthapit B. Crop Genetic Diversity in the Field and on the Farm: Principles and applications in research practices. New Haven: Yale University Press; 2016. 395 p.

[9] Li X, Siddique KH, editors. Future Smart Food. Rediscovering hidden treasures of neglected and underutilized species for Zero Hunger in Asia. Bangkok: Food and Agriculture Organization of the United Nations; 2018. 178 p.

[10] Ceccarelli S. Specific adaptation and breeding for marginal conditions. In: Rognli OA, Solberg E, Schjelderup I, editors. Breeding Fodder Crops for Marginal Conditions. Developments in Plant Breeding, vol 2. Dordrecht: Springer; 1994. p. 101–127. DOI: 10.1007/978-94-011-0966-6_11

[11] Knüpffer H, Terentyeva I, Hammer K, Kovaleva O. Ecogeographical diversity-a Vavilovian approach In: Von-Bothmer R, Van-Hintum T, Knüpffer H and Sato K, editors. Diversity in Barley (*Hordeum vulgare*). Amsterdam: Elsevier; 2003. p. 53–76.

[12] Chalak L, Mzid R, Rizk W, Hmedeh H, Kabalan R, Breidy J, Hamadeh B, Machlab H, Rizk H, Elhajj S. Performance of 50 Lebanese barley landraces (*Hordeum vulgare L.* subsp. *vulgare*) in two locations under rainfed conditions. Annals of Agricultural Sciences. 2015 Dec 1;60(2): 325–334. DOI: 10.1016/j.aoas.2015.11.005

[13] RBG Kew. The State of the World's Plants Report – 2016. Kew: Royal Botanic Gardens; 2016. 80 p.

[14] Cheek M, Nic Lughadha E, Kirk P, Lindon H, Carretero J, Looney B, Douglas B, Haelewaters D, Gaya E, Llewellyn T, Ainsworth AM. New scientific discoveries: Plants and fungi. Plants, People, Planet. 2020 Sep;2(5): 371–388. DOI: 10.1002/ppp3.10148

[15] Food Plant International. Edible Plants of the World [Internet]. 2020.

Available from https://fms.cmsvr.com/fmi/webd/Food_Plants_World [Accessed 2020-12-01]

[16] FAO. Neglected and underutilized crops species. Information document COAG/2018/INF/7. FAO Committee on Agriculture Twenty-sixth Session; 1–5 October 2018; Rome. Rome: Food and Agriculture of the United Nations; 2018. 5 p. Available from: http://www.fao.org/3/mx479en/mx479en.pdf

[17] Mansfeld's World Database of Agriculture and Horticultural Crops [Internet]. 2020. Available from: https://www.re3data.org/repository/r3d100010097 [Accessed 2020-12-07]

[18] FAO. FAO Statistics [Internet]. 2018. Available from: http://www.fao.org/faostat/en/#home [202020–12-01]

[19] Brown AH, Hodgkin T. Indicators of genetic diversity, genetic erosion, and genetic vulnerability for plant genetic resources. In: Ahuja MR, Jain SM, editors. Genetic diversity and erosion in plants. Switzerland: Springer; 2015. p. 25–53.

[20] FAO. Save and grow: A policymaker's guide to sustainable intensification of smallholder crop production. Rome: Food and Agriculture Organization of the United Nations; 2011. 101 p.

[21] Bruford, M.W., Davies, N., Dulloo, M.E., Faith, D.P. and Walters, M. Monitoring changes in genetic diversity. In: Walters M, Scholes RJ, editors. The GEO handbook on biodiversity observation networks. Switzerland: Springer Nature; 2017. P. 107–128.

[22] Dulloo ME, Thormann I, Drucker AG. 39 What DoWe Have To Lose? Monitoring Crop Genetic Diversity. In: Maxted N, Dulloo ME, Ford-Lloyd BV, editors. Enhancing Crop Genepool Use: Capturing Wild Relative and Landrace Diversity for Crop Improvement. Oxfordshire: CABI; 2016. p. 421–435

[23] Ahmed I, Ullah A, ur Rahman MH, Ahmad B, Wajid SA, Ahmad A, Ahmed S. 2019. Climate change impacts and adaptation strategies for agronomic crops. In: Hussain S. Editor. Climate Change and Agriculture. IntechOpen; 2019. p. 1–14. DOI: /10.5772/intechopen.82697

[24] IPCC. Climate Change 2014: Synthesis Report. Contribution of Working Groups I, II and III to the Fifth Assessment Report of the Intergovernmental Panel on Climate Change [Core Writing Team, R.K. Pachauri RK, Meyer LA, editors]. Geneva: IPPC; 2014. 151 p.

[25] IPPC. IPCC Special Report on the Ocean and the Cryosphere in a Changing Climate. [Pörtner HO, Roberts DC, Masson-Delmotte V, Zhai P, Tignor M, Poloczanska E, Mintenbeck K, Nicolai M, Okem A, Petzold J, Rama B, Weyer N, editors. Geneva: IPPC; 2019. 755 p.

[26] Tripathi A, Tripathi DK, Chauhan DK, Kumar N, Singh GS. Paradigms of climate change impacts on some major food sources of the world: a review on current knowledge and future prospects. Agriculture, ecosystems & environment. 2016 Jan 15;216:356–373. DOI: 10.1016/J.AGEE.2015.09.034

[27] Iizumi T, Ramankutty N. Changes in yield variability of major crops for 1981–2010 explained by climate change. Environmental Research Letters. 2016 Feb 26;11(3):034003. DOI: 10.1088/1748-9326/11/3/034003

[28] Brouder SM, Volenec JJ. Future climate change and plant macronutrient use efficiency. In: Hossain MA, Kamiya T, Burritt D, Tran LS, Fujiwara T, editors. Academic Press; 2017. p. 357–379. DOI: 10.1016/B978-0-12-811308-0.00020-X

[29] Sugiura T, Ogawa H, Fukuda N, Moriguchi T. Changes in the taste and textural attributes of apples in response to climate change. Scientific reports. 2013 Aug 15;3:2418. DOI: 10.1038/srep02418

[30] Yan Y, Wang YC, Feng CC, Wan PH, Chang KT. Potential distributional changes of invasive crop pest species associated with global climate change. Applied geography. 2017 May 1;82:83–92. DOI: 10.1016/j.apgeog.2017.03.011

[31] Quiroz R, Ramírez DA, Kroschel J, Andrade-Piedra J, Barreda C, Condori B, Mares V, Monneveux P, Perez W. Impact of climate change on the potato crop and biodiversity in its center of origin. Open Agriculture. 2018 Aug 1;3 (1):273–283. DOI: 10.1515/opag-2018-0029

[32] Chakraborty S, Murray GM, Magarey PA, Yonow T, O'Brien RG, Croft BJ, Barbetti MJ, Sivasithamparam K, Old KM, Dudzinski MJ, Sutherst RW. Potential impact of climate change on plant diseases of economic significance to Australia. Australasian Plant Pathology. 1998 Mar 1;27(1):15–35. DOI: 10.1071/AP98001

[33] Pautasso M, Döring TF, Garbelotto M, Pellis L, Jeger MJ. Impacts of climate change on plant diseases - opinions and trends. European Journal of Plant Pathology. 2012 May 1;133(1): 295–313. DOI: 10.1007/s10658-012-9936-1

[34] Juroszek P, von Tiedemann A. Climate change and potential future risks through wheat diseases: a review. European Journal of Plant Pathology. 2013 May 1;136(1):21–33. DOI: 10.1007/s10658-012-0144-9

[35] Zhao C, Liu B, Piao S, Wang X, Lobell DB, Huang Y, Huang M, Yao Y, Bassu S, Ciais P, Durand JL. Temperature increase reduces global yields of major crops in four independent estimates. Proceedings of the National Academy of Sciences. 2017 Aug 29;114(35):9326–9331. DOI: 10.1073/pnas.1701762114

[36] Meldrum G, Mijatović D, Rojas W, Flores J, Pinto M, Mamani G, Condori E, Hilaquita D, Gruberg H, Padulosi S. Climate change and crop diversity: farmers' perceptions and adaptation on the Bolivian Altiplano. Environment, Development and Sustainability. 2018 Apr 1;20(2):703–730. DOI: 10.1007/s10668-016-9906-4

[37] van Etten J, de Sousa K, Aguilar A, Barrios M, Coto A, Dell'Acqua M, Fadda C, Gebrehawaryat Y, van de Gevel J, Gupta A, Kiros AY. Crop variety management for climate adaptation supported by citizen science. Proceedings of the National Academy of Sciences. 2019 Mar 5;116(10):4194–4199. DOI: 10.1073/pnas.1813720116

[38] Morales-Castilla I, de Cortázar-Atauri IG, Cook BI, Lacombe T, Parker A, van Leeuwen C, Nicholas KA, Wolkovich EM. Diversity buffers winegrowing regions from climate change losses. Proceedings of the National Academy of Sciences. 2020 Feb 11;117(6):2864–2869. DOI: /10.1073/pnas.1906731117

[39] Corrado G, Rao R. Towards the genomic basis of local adaptation in landraces. Diversity. 2017 Dec;9(4):51. DOI: 10.3390/d9040051

[40] Fenzi M, Jarvis DI, Reyes LM, Moreno LL, Tuxill J. Longitudinal analysis of maize diversity in Yucatan, Mexico: influence of agro-ecological factors on landraces conservation and modern variety introduction. Plant Genetic Resources. 2017 Feb;15(1):51–63. DOI: 10.1017/S1479262115000374

[41] Ficiciyan A, Loos J, Sievers-Glotzbach S, Tscharntke T. More than

yield: Ecosystem services of traditional versus modern crop varieties revisited. Sustainability. 2018 Aug;10(8):2834. DOI: 10.3390/su10082834

[42] Coto A, de Sousa K, Fadda C, Gebrehawaryat Y, van de Gevel J, Gotor E, Gupta A, Madriz B, Mathur P, Mengistu DK, Paliwal A. Seeds for Needs: crop diversity for resilience; Technical Report – May 2019. Rome: Bioversity International; 2019. 4 p. DOI: 10.13140/RG.2.2.24932.22408

[43] FAO. The future of food and agriculture: alternative pathways to 2020. Rome: Food and Agriculture Organization of the United Nations; 2018. 224 p.

[44] Davis DR. Declining fruit and vegetable nutrient composition: What is the evidence?. HortScience. 2009 Feb 1; 44(1):15–19. DOI: 10.21273/HORTSCI.44.1.15

[45] Foley JA, Ramankutty N, Brauman KA, Cassidy ES, Gerber JS, Johnston M, Mueller ND, O'Connell C, Ray DK, West PC, Balzer C. Solutions for a cultivated planet. Nature. 2011 Oct;478(7369):337–342. DOI: 10.1038/nature10452

[46] Garnett T, Appleby MC, Balmford A, Bateman IJ, Benton TG, Bloomer P, Burlingame B, Dawkins M, Dolan L, Fraser D, Herrero M. Sustainable intensification in agriculture: premises and policies. Science. 2013 Jul 5;341(6141):33–34. DOI: 10.1126/science.1234485

[47] Jones AD. Critical review of the emerging research evidence on agricultural biodiversity, diet diversity, and nutritional status in low-and middle-income countries. Nutrition reviews. 2017 Oct 1;75(10):769–782. DOI: 10.1093/nutrit/nux040

[48] Padulosi S, Heywood V, Hunter D, Jarvis A. Underutilized species and climate change: current status and outlook. In: Yadav SS, Redden RJ, Hatfield JL, Lotze-Campen H, Hall AJ, editors. Crop adaptation to climate change. Chichester: John Wiley & Sons; 2011. p. 507–521.

[49] Padulosi S, Hodgkin T, Williams JT, Haq N. Underutilized crops: trends, challenges and opportunities in the 21st century. In Engels JMM, Ramanatha Rao V, Brown AHD, Jackson MT, editors. Managing plant genetic diversity. Proceedings of an international conference, Kuala Lumpur, Malaysia, 12–16 June 2000. Wallingford: CAB International; 2002. p. 323–338.

[50] Padulosi S, Thompson J, Rudebjer P. Fighting poverty, hunger and malnutrition with neglected and underutilized species: needs, challenges and the way forward. Rome: Bioversity International; 2013. 56 p.

[51] Noorani A, Bazile D, Diulgheroff S, Kahane R, Nono-Womdim R. Promoting neglected and underutilized species through policies and legal frameworks. In: de Ron M, coordinator. Proceedings of the EUCARPIA International Symposium on Protein Crops, V Meeting AEL [V Jornadas de la AEL], May 2015; Pontevedra, Spain. Spain: Spanish Association for Legumes (AEL); 2015, p. 107–111. DOI: 10.5281/zenodo.1254012

[52] Oniang'o R, Grum M, Obel-Lawson E., editors. Developing African leafy vegetables for improved nutrition: Regional workshop, 6–9 December 2005. Nairobi: Rural Outreach Program; 2008. 160 p.

[53] Ulian T, Diazgranados M, Pironon S, Padulosi S, Liu U, Davies L, Howes MJ, Borrell JS, Ondo I, Pérez-Escobar OA, Sharrock S. Unlocking plant resources to support food security and promote sustainable agriculture. Plants, People,

Planet. 2020 Sep;2(5):421–445. DOI: 10.1002/ppp3.10145

[54] Kahane R, Hodgkin T, Jaenicke H, Hoogendoorn C, Hermann M, Hughes JD, Padulosi S, Looney N. Agrobiodiversity for food security, health and income. Agronomy for sustainable development. 2013 Oct 1;33(4):671–693. DOI: 10.1007/s13593-013-0147-8

[55] WHO. Guidelines: Vitamin A supplementation in infants and children 6–59 months of age. Geneva: World Health Organization; 2011. 24 p.

[56] Imdad A, Mayo-Wilson E, Herzer K, Bhutta ZA. Vitamin A supplementation for preventing morbidity and mortality in children from six months to five years of age. Cochrane Database of Systematic Reviews. 2017(3). 108 p. DOI: 10.1002/14651858.CD008524.pub3

[57] Davey MW, Van den Bergh I, Markham R, Swennen R, Keulemans J. Genetic variability in Musa fruit provitamin A carotenoids, lutein and mineral micronutrient contents. Food Chemistry. 2009 Aug 1;115(3):806–813. DOI: 10.1016/j.foodchem.2008.12.088

[58] Englberger L, Lyons G, Foley W, Daniells J, Aalbersberg B, Dolodolotawake U, Watoto C, Iramu E, Taki B, Wehi F, Warito P. Carotenoid and riboflavin content of banana cultivars from Makira, Solomon Islands. Journal of food composition and analysis. 2010 Sep 1;23(6):624–632. DOI: 10.1016/j.jfca.2010.03.002

[59] Englberger L, Marks GC, Fitzgerald MH. Insights on food and nutrition in the Federated States of Micronesia: a review of the literature. Public Health Nutrition. 2003 Feb;6(1): 5–17. DOI: 10.1079/PHN2002364

[60] Burlingame B, Dernini S., editors. Sustainable diets and biodiversity directions and solutions for policy, research and action. Rome: Food and Agriculture Organization of the United Nations; 2012. p. 309.

[61] Abukutsa-Onyango M. The diversity of cultivated African leafy vegetables in three communities in Western Kenya. In: Oniang'o R, Grum M, Obel-Lawson E., editors. Developing African leafy vegetables for improved nutrition: Regional workshop, 6–9 December 2005. Nairobi: Rural Outreach Program; 2005. p. 85–91.

[62] Guarino L, editor. Traditional African Vegetables. Promoting the conservation and use of underutilized and neglected crops. 16. Proceedings of the IPGRI International Workshop on Genetic Resources of Traditional Vegetables in Africa: Conservation and Use, 29–31 August 1995, ICRAF-HQ, Nairobi, Kenya. Gatersleben: Institute of Plant Genetics and Crop Plant Research/ Roe: International Plant Genetic Resources Institute; 1997.

[63] Maundu PM. The status of traditional vegetable utilization in Kenya. In Guarino L., editor. Traditional African vegetables; Proceedings of the IPGRI international workshop on genetic resources of traditional vegetables in Africa: conservation and use. IPGRI International Workshop on Genetic Resources of Traditional Vegetables in Africa: Conservation and Use; ICRAF-HQ, Nairobi 29–31 Aug, 1995 Kenya. Rome: IPGRI; 1997. p. 66–71.

[64] Fanzo J, Remans R, Pronyk PM, Negin J, Wariero J, Mutuo P, Masira J, Diru W, Lelerai E, Kim D, Nemser B. A 3-year cohort study to assess the impact of an integrated food-and livelihood-based model on undernutrition in rural western Kenya. In: Thompson B, Amoroso L, editors. Combating micronutrient deficiencies: food-based approaches. Oxfordshire: CABI; 2011. p. 76–91.

[65] Neugart S, Baldermann S, Ngwene B, Wesonga J, Schreiner M.

Indigenous leafy vegetables of Eastern Africa—A source of extraordinary secondary plant metabolites. Food Research International. 2017 Oct 1;100: 411–422. DOI: 10.1016/j. foodres.2017.02.014

[66] Gido EO, Ayuya OI, Owuor G, Bokelmann W. Consumption intensity of leafy African indigenous vegetables: towards enhancing nutritional security in rural and urban dwellers in Kenya. Agricultural and Food Economics. 2017 Dec 1;5(1):14. DOI: 10.1186/ s40100-017-0082-0

[67] Mercer KL, Perales HR. Evolutionary response of landraces to climate change in centers of crop diversity. Evolutionary applications. 2010 Sep;3(5–6):480–493. DOI: 10.1111/ j.1752-4571.2010.00137.x

[68] FAO. WIEWS - World Information and Early Warning System on Plant Genetic Resources for Food and Agriculture [Internet]. 2020. Available from: http://www.fao.org/wiews/en/

[69] FAO. Genebank Standards for Plant Genetic Resources for Food and Agriculture. Rev. ed. edition. Rome: Food and Agriculture Organization of the United Nations; 2014. 166 p.

[70] Khoury C, Laliberté B, Guarino L. Trends in *ex situ* conservation of plant genetic resources: a review of global crop and regional conservation strategies. Genetic Resources and Crop Evolution. 2010 Apr 1;57(4):625–639. DOI: 10.1007/s10722-010-9534-z

[71] Ceccarelli S, Guimarães EP, Weltzien E, editors. Plant breeding and farmer participation. Rome: Food and Agriculture Organization of the United Nations; 2009. 685 p.

[72] Westengen OT, Winge T, editors. Farmers and Plant Breeding: Current Approaches and Perspectives. London: Routledge; 2019 Oct 2. 329 p.

[73] FAO. Global Farmer Field School Platform [Internet]. 2020. Available from: http://www.fao.org/farmer-field-schools/overview/en/ [Accessed 2020-12-01]

[74] Vernooy R, Clancy E, Diulgheroff S, Furman B, González Santos R, Kajtna B, Marino M, Mushita A, Shrestha P, Song Y, Egon Sosinski E. Joining forces to strengthen community seedbanks worldwide. Rome: Bioversity International; 2108. 8 p.

[75] Vernooy R, Shrestha P, Sthapit B, editors. Community seed banks: Origins, evolution and prospects. New York: Routledge; 2015. 270 p.

[76] Manalo C. SEARICE Highlights Importance of CSBs in International Event. In Southeast Asia Regional Initiatives for Community Empowerment [Internet]. 2019. Available from: https://www.searice.org .ph/searice-highlights-importance-of-cs [Accessed 2020-12-07]

[77] Tin HQ, Cuc NH, Be TT, Ignacio N, Berg T. Impacts of seed clubs in ensuring local seed systems in the Mekong Delta, Vietnam. Journal of Sustainable Agriculture. 2011 Oct 1;35 (8):840–854. DOI: 10.1080/ 10440046.2011.611746.

[78] FAO. Quality Declared Seed System. Rome: Food and Agriculture Organization of the United Nations; 2010. 243 p.

[79] Visser B. The impact of national seed laws on the functioning of small-scale seed systems: A country-case study. The Hague: Oxfam Novib. 2017. 38 p.

[80] FAO. Quality Declared Planting Material: Protocols and standards for vegetatively propagated crops. Rome: Food and Agriculture Organization of the United Nations; 2010. 126 p.

[81] FAO, AfricaSeeds. Seeds Toolkit Module 1: Development of small-scale

seed enterprises. Rome: Food and Agriculture Organization of the United Nations; 2018. 109 p.

[82] FAO, AfricaSeeds. Seeds Toolkit Module 2: Seed processing: principles, equipment and practice. Rome: Food and Agriculture Organization of the United Nations; 2018. 79 p.

[83] FAO, AfricaSeeds. Seeds Toolkit Module 3: Seed quality assurance. Rome: Food and Agriculture Organization of the United Nations; 2018. 111 p.

[84] FAO, AfricaSeeds. Seeds Toolkit Module 4: Seed Sector Regulatory Framework. Rome: Food and Agriculture Organization of the United Nations; 2018. 64 p.

[85] FAO. Seeds Toolkit Module 5: Seed marketing. Rome: Food and Agriculture Organization of the United Nations; 2018. 97 p.

[86] FAO. Seeds Toolkit Module 6: Seed storage. Rome: Food and Agriculture Organization of the United Nations; 2018. 102 p.

[87] FAO. Voluntary Guide for National Seed Policy Formulation. Rome: Food and Agriculture Organization of the United Nations; 2015. 60 p.

[88] Alipour H, Bihamta MR, Mohammadi V, Peyghambari SA, Bai G, Zhang G. Genotyping-by-sequencing (GBS) revealed molecular genetic diversity of Iranian wheat landraces and cultivars. Frontiers in Plant Science. 2017; 8:1293. DOI: 10.3389/fpls.2017.01293

[89] Bioversity International. Descriptors [Internet]. 2020. Available from: https://www.bioversityinternational.org/e-library/publications/descriptors/[Accessed 2020-12-01]

[90] UPOV. Test Guidelines [Internet]. 2020. Available from: https://www.upov.int/test_guidelines/en/ [Accessed 2020-12-01]

[91] USDA ARS. GRIN-Global - U.S. National Plant Germplasm System [Internet]. 2020. Available from: https://npgsweb.ars-grin.gov/gringlobal/descriptors [Accessed 2020-12-01]

[92] Alercia A, Diulgheroff S, Mackay M. FAO/bioversity multi-crop passport descriptors V. 2.1 [MCPD V. 2.1]-December 2015. Rome: Bioversity International; 2015. 11 p.

[93] GRIN Global. The GRIN-Global Project [Internet]. 2020. Available from: https://www.grin-global.org/ [Accessed 2020-12-01]

[94] Genesys. Genesys PGR [Internet]. 2020. Available from: https://www.genesys-pgr.org/ [Accessed 2020-12-01]

[95] Cattivelli L, Rizza F, Badeck FW, Mazzucotelli E, Mastrangelo AM, Francia E, Marè C, Tondelli A, Stanca AM. Drought tolerance improvement in crop plants: an integrated view from breeding to genomics. Field crops research. 2008 Jan 2;105(1–2):1–4. DOI: 10.1016/j.fcr.2007.07.004

[96] Challinor AJ, Koehler AK, Ramirez-Villegas J, Whitfield S, Das B. Current warming will reduce yields unless maize breeding and seed systems adapt immediately. Nature Climate Change. 2016 Oct;6(10):954–958. DOI: 10.1038/nclimate3061

[97] Wang W, Mauleon R, Hu Z, Chebotarov D, Tai S, Wu Z, Li M, Zheng T, Fuentes RR, Zhang F, Mansueto L. Genomic variation in 3,010 diverse accessions of Asian cultivated rice. Nature. 2018 May;557(7703):43–49. DOI: 10.1038/s41586-018-0063-9

[98] Atlin GN, Cairns JE, Das B. Rapid breeding and varietal replacement are critical to adaptation of cropping

systems in the developing world to climate change. Global food security. 2017 Mar 1;12:31–7. DOI: 10.1016/j.gfs.2017.01.008

[99] FAO. Climate-smart agriculture sourcebook. Rome: Food and Agriculture Organization of the United Nations; 2017.

[100] Saini P, Saini P, Kaur JJ, Francies RM, Gani M, Rajendra AA, Negi N, Jagtap A, Kadam A, Singh C, Chauhan SS. Molecular approaches for harvesting natural diversity for crop improvement. In: Salgotra RK, Zargar SM, editors Rediscovery of Genetic and Genomic Resources for Future Food Security. Singapore: Springer; 2020. p. 67–169. DOI: 10.1007/978-981-15-0156-2_3

[101] FAO. E-learning course on Pre-breeding [Internet]. 2011. Available from: https://elearning.fao.org/course/view.php?id=487 [Accessed 2020-12-01]

[102] Mba C, Guimaraes EP, Ghosh K. Re-orienting crop improvement for the changing climatic conditions of the 21st century. Agriculture & Food Security. 2012;1(1):7. DOI: 10.1186/2048-7010-1-7

[103] Mba C. Induced mutations unleash the potentials of plant genetic resources for food and agriculture. Agronomy. 2013;3(1):200–31. DOI: 10.3390/agronomy3010200

[104] Yang W, Feng H, Zhang X, Zhang J, Doonan JH, Batchelor WD, Xiong L, Yan J. Crop phenomics and high-throughput phenotyping: past decades, current challenges, and future perspectives. Molecular Plant. 2020 Feb 3;13(2):187–214. DOI: 10.1016/j.molp.2020.01.008

[105] Araus J., Kefauver SC, Zaman-Allah M, Olsen MS, Cairns JE. Translating high-throughput phenotyping into genetic gain. Trends in Plant Science. 2018; 23(5):451–466. DOI: 10.1016/j.tplants.2018.02.001

[106] Kim J, Kim KS, Kim Y, Chung YS. A short review: Comparisons of high-throughput phenotyping methods for detecting drought tolerance. Scientia Agricola. 2021;78(4). DOI: 10.1016/j.molp.2020.01.008

[107] Normanly J, editor. High-throughput phenotyping in plants: methods and protocols. Berlin: Humana Press; 2012. 362 p.

[108] Tattaris M, Reynolds MP, Chapman SC. A direct comparison of remote sensing approaches for high-throughput phenotyping in plant breeding. Frontiers in Plant Science. 2016 Aug 3;7:1131. DOI: 10.3389/fpls.2016.01131

[109] Chawade A, van Ham J, Blomquist H, Bagge O, Alexandersson E, Ortiz R. High-throughput field-phenotyping tools for plant breeding and precision agriculture. Agronomy. 2019 May;9(5):258. DOI: 10.3390/agronomy9050258

[110] White JW, Andrade-Sanchez P, Gore MA, Bronson KF, Coffelt TA, Conley MM, Feldmann KA, French AN, Heun JT, Hunsaker DJ, Jenks MA. Field-based phenomics for plant genetics research. Field Crops Research. 2012 Jul 11;133:101–112. DOI: 10.1016/j.fcr.2012.04.003

[111] Singh D, Wang X, Kumar U, Gao L, Noor M, Imtiaz M, Singh RP, Poland J. High-throughput phenotyping enabled genetic dissection of crop lodging in wheat. Frontiers in plant science. 2019 Apr 3;10:394. DOI: 10.3389/fpls.2019.00394

[112] Desta ZA, de Koning DJ, Ortiz R. Molecular mapping and identification of quantitative trait loci for domestication traits in the field cress (*Lepidium*

campestre L.) genome. Heredity. 2020 Apr;124(4):579–591. DOI: 10.1038/s41437-020-0296-x

[113] Seo JH, Kang BK, Dhungana SK, Oh JH, Choi MS, Park JH, Shin SO, Kim HS, Baek IY, Sung JS, Jung CS. QTL Mapping and Candidate Gene Analysis for Pod Shattering Tolerance in Soybean (*Glycine max*). Plants. 2020 Sep;9(9):1163. DOI: 10.3390/plants9091163

[114] Rivers J, Warthmann N, Pogson BJ, Borevitz JO. Genomic breeding for food, environment and livelihoods. Food Security. 2015 Apr 1;7(2):375–382. DOI: 10.1007/s12571-015-0431-3

[115] Gogorcena Y, Sanchez G, Moreno-Vázquez S, Pérez S, Ksouri N. Genomic-based breeding for climate-smart peach varieties. In: Cole C, editor. Genomic Designing of Climate-Smart Fruit Crops. Switzerland: Springer, Cham; 2020. p. 271–331. DOI: 10.1007/978-3-319-97946-5_8

[116] Kearsey MJ, Farquhar AG. QTL analysis in plants; where are we now?. Heredity. 1998 Feb;80(2):137–142. DOI: 10.1046/j.1365-2540.1998.00500.x

[117] Camargo AV, Mackay I, Mott R, Han J, Doonan JH, Askew K, Corke F, Williams K, Bentley AR. Functional mapping of quantitative trait loci (QTLs) associated with plant performance in a wheat MAGIC mapping population. Frontiers in plant science. 2018 Jul 9;9:887. DOI: 10.3389/fpls.2018.00887

[118] Kersey PJ. Plant genome sequences: past, present, future. Current opinion in plant biology. 2019 Apr 1;48:1–8. DOI: 10.1016/j.pbi.2018.11.001

[119] Jamnadass R, Mumm RH, Hale I, Hendre P, Muchugi A, Dawson IK, Powell W, Graudal L, Yana-Shapiro H, Simons AJ, Van Deynze A. Enhancing African orphan crops with genomics. Nature Genetics. 2020 Apr;52(4):356–60. DOI: 10.1038/s41588-020-0601-x

[120] AOCC. African Orphan Crops Consortium On-going Projects [Internet]. 2010. Available from: http://africanorphancrops.org/ongoing-projects/.

[121] Diulgheroff S. A global overview of assessing and monitoring genetic erosion of crop wild relatives and local varieties using WIEWS and other elements of the FAO Global System on PGR. In: Ford-Lloyd B, Dias SR, Bettencourt E, editors. Genetic erosion and pollution assessment methodologies. Rome: Bioversity International; 2006. p. 5–14.

[122] Noorani A., Bazile D, Diulgheroff S., Kahane R., Nono-Womdim R. Promoting neglected and underutilized species through policies and legal frameworks. Poster presentation at: de Ron M, coordinator. Proceedings of the EUCRPIA International Symposium on Protein Crops, V Meeting AEL [V Jornadas de la AEL], May 2015; Pontevedra, Spain. Spain: Spanish Association for Legumes (AEL); 2015b.

[123] FAO. Interpretation of the international undertaking on plant genetic resources. FAO Conference Twenty-fifth Session; 11–30 November 1989; Rome. Rome: Food and Agriculture of the United Nations; 1989. 10 p. Available from: http://www.fao.org/3/z4968en/z4968en.pdf

[124] FAO. The Global Plan of Action for the Conservation and Sustainable Utilization of Plant Genetic Resources for Food and Agriculture and the Leipzig Declaration. Rome: Food and Agriculture Organization of the United Nations; 1996. 63 p.

[125] FAO. International Treaty on Plant Genetic Resources for Food and Agriculture. Rome: Food and

Agriculture Organization of the United Nations; 2009. 68 p.

[126] Crop Trust. The Global Crop Diversity Trust [Internet]. 2020. Available from: https://www.croptrust.org/ [Accessed 2020-12-01]

[127] FAO. Cordoba Declaration on Promising Crops for the XXI Century. International Seminar on Traditional and New Crops to Meet the Challenges of the XXI Century. Cordoba, Spain, 10–13 December 2012. Rome: Food and Agriculture Organization of the United Nations; 2012b. 7 p.

[128] FAO. Conference Outcome Document: Framework for Action. Second International Conference on Nutrition Rome, 19–21 November 2014. Rome: Food and Agriculture Organization of the United Nations; 2014a. 8 p.

[129] FAO, WHO. United Nations Decade on Nutrition: Towards country-specific SMART commitments for action on nutrition. Rome: Food and Agriculture Organization of the United Nations; 2016. 4 p.

[130] FAO. FAO and the SDGs Indicators: Measuring up to the 2030 Agenda for Sustainable Development. Rome: Food and Agriculture Organization of the United Nations; 2017b. 40 p.

[131] Mba C, Guimaraes EP, Guei GR, Hershey C, Paganini M, Pick B, Ghosh K. Mainstreaming the continuum approach to the management of plant genetic resources for food and agriculture through national strategy. Plant Genetic Resources. 2012a Apr 1;10 (1):24–37. DOI:10.1017/S1479262111000943

[132] FAO. Guidelines for Developing a National Strategy for Plant Genetic Resources for Food and Agriculture: Translating the Second Global Plan of Action for Plant Genetic Resources for Food and Agriculture into National Action. Rome: Food and Agriculture Organization of the United Nations; 2015a. 55 p.

Chapter 2

Wild Progenitor and Landraces Led Genetic Gain in the Modern-Day Maize (*Zea mays* L.)

Devender Sharma, Rajesh K. Khulbe, Ramesh S. Pal, Jeevan Bettanaika and Lakshmi Kant

Abstract

Maize (*Zea mays ssp. mays*) originated from Mexico and Central America and grew worldwide for food, feed and industrial products components. It possesses ten chromosomes with a genome size of 2.3 gigabases. Teosinte (*Z. mays* ssp. *parviglumis*) is the probable progenitor of the modern-day maize. The maize domestication favored standing gain of function and regulatory variations acquired the convergent phenotypes. The genomic loci *teosinte branched 1* (*tb1*) and *teosinte glume architecture 1* (*tga1*) played a central role in transforming teosinte to modern-day maize. Under domestication and crop improvement, only 2% (~1200) genes were undergone selection, out of ~60000 genes. Around ~98% of the genes have not experienced selection; there is enormous variation present in the diverse inbred lines that can be potentially utilized to identify QTLs and crop improvement through plant breeding. The genomic resources of wild relatives and landraces harbor the unexplored genes/alleles for biotic/abiotic tolerance, productivity and nutritional quality. The human-made evolution led to the transformation of wild relatives/landraces to the modern-day maize. This chapter summarized the maize's wild relatives/landraces and the genetic gain over time in biotic/abiotic, productivity, and nutritional quality traits.

Keywords: maize, teosinte, landrace, domestication, *Zea mays*

1. Introduction

Maize (*Zea mays ssp. mays*) is a member of the Maydae tribe of the Poaceae family, originated in Mexico and Central America. Maize with somatic chromosome 2n = 20, 2.3 gigabase of genome size and more than 60000 genes [1]. After Columbus entered the New World and introduced maize to Europe, it gradually spread worldwide [2]. Its production exceeds wheat (*Triticum aestivum*) and rice (*Oryza sativa*) nowadays (http://www.fao.org). It has emerged as a crop of global importance due to its use as human food, livestock feed and various industrial products. A significant portion of maize production is utilized as animal consumption as it serves as important source of calories and protein in the developing countries [3]. A debate emerged at the beginning of the last century concerning the origin of maize. Maize ought to be derived from the cross between a close relative of maize and *Tripsacum* or the oldest wild maize is the progenitor of maize [4]. Subsequent archeological and genetical evidences indicated that teosinte

(*Z. mays* ssp. *parviglumis*) is the only ancestor of maize and is widely accepted [5, 6]. The discrepancy between teosinte and modern day maize were found around the Balsas River in southwestern Mexico around 9000 years ago [7].

The ancestor teosinte originated from Mexico, the selection by Native Americans for improved plant types and seed types become corn. According to modern breeding, different generations of selection turned teosinte into landraces and ultimately to the modern-day maize. The modern day maize differs from teosinte in the key traits; for example, teosinte is characterized by multi-branched, tiny reproductive parts and two rows of seed. In comparison, the modern-day maize possesses 20–22 kernel rows. The reproductive separation is not complete in primitive strains. The modern-day maize reproductive separation is complete, i.e., ear and tassels, taller plant height, erect stature, more light interception and more photosynthesis [4]. Domestication leads to the evolution of wild progenitor species to early domesticated landraces and ultimately to modern cultivars. The loss of genetic diversity often accompanies domestication. Typically landraces are heterogeneous (non-uniform) and, therefore, a good source of genetic diversity. Landraces are usually less diverse than wild relatives but more diverse than modern-day cultivars [8].

Typically, domesticated plant species' wild relatives do not have all the desirable characteristics for normal agricultural production and use. Only a small portion of the genome's selection and the key genes/QTLs/transcription factors involved under domestication have been identified and cloned [1]. In transforming teosinte into modern maize, genetic loci such as *teosinte branched 1* (*tb1*) and *teosinte glume architecture 1* (*tga1*) has played a pivotal role [9, 10]. The loci involved in transforming plant architecture and morphology were shown to have a pleiotropic effect on other traits. This lead to the development of convergent phenotype of the modern-day maize. Maize later spread from the centre of origin to various parts of the globe, including America, Europe, Africa and Asia. Genetic resources, especially wild relatives and landraces, harbor novel alleles/genes to impart resistance/tolerance to various biotic/abiotic stresses and boost productivity and nutrition quality. Teosintes and *Tripsacum* are native to Mexico and Central America among wild families, while in Southeast Asia, *Coix, Chionachne, Sclerachne, Trilobachne* and *Polytoca* originated. Landrace accessions with unexplored alleles/genes function as important donors with substantial characteristics.

This chapter summarizes the early crop domestication process from thousands of years ago to modern-day plant breeders' success in plant improvement. Understanding the domestication of crops and plant breeding provides a background for the importance, significance and usage of wild relatives of crops maintained either in situ or in gene banks. The importance of landrace accessions and wild relatives of maize in supplying useful genes for different essential traits has been addressed.

2. Evolution of maize

Maize (*Zea mays ssp. mays*, Taino: mahiz, Spanish: maíz), also known as corn (North American English), is a cereal grain that was first domesticated by indigenous people around 10,000 years ago in southern Mexico. The ancient farmers from Mexico took the initiative to domesticate the maize by only looking at their kernels. They observed that all plants are not the same; some kernels look better, taste better or easier to grind. They saved the kernels based on the beneficial characteristics and used them to plant in the next season for their harvest. This forms artificial selection or selective breeding. Over time, with more rows of kernels, maize cobs grew bigger, gradually taking the form of modern maize. The identity of maize's progenitor or wild ancestor remained a mystery for a while compared to other crops. Although there are apparent wild relatives of other grains such as wheat and rice, there is no wild plant that looks like maize, with smooth, starchy kernels arranged along the cob. To explain the origin of maize, various researchers explained several theories/hypotheses: tripartite hypothesis, catastrophic theory of sexual transmutation, *Tripsacum-Zea diploperennis* hypothesis, and teosinte hypothesis were discussed and discussed in depth by various scientists.

The tripartite hypothesis stated that the progenitor of maize was the extinct popcorn. The crosses between corn and related genera *Tripsacum* lead to teosinte formation, with further crosses giving rise to the diversity of maize we observe today [11]. Among the theories, the teosinte hypothesis is the most accepted one. Teosinte does not look much like maize, particularly when its kernels are compared to maize kernels. But at the DNA level, the two are surprisingly alike. Both possess the same number of chromosomes and similar gene arrangements. The hybrids between teosinte and maize are fertile and can reproduce naturally. Beadle [4, 12, 13] was one of the first scientists to establish the close relationship between maize and teosinte. He proposed that ancient people cultivated teosinte for food. During the cultivation of teosinte, mutations arose and being selected by the people. A set of five major mutations transformed teosinte into maize. Beadle [4] has studied the advanced generations of teosinte × maize derived hybrids. In the F_2 population of 5000 plants, the frequency of parental types was 1 in 500 plants. He concluded that 5 major loci/genes are responsible for maize domestication based on simple Mendelian genetics. Later five major quantitative trait loci (QTLs) and QTLs with minor effect for the key traits differ for maize and teosinte [9]. Wright et al. [14] reported 2–4% of genes had been selected during evolution/domestication by investigating around 774 genes. Out of a total of 3900–42000 protein-coding genes, only 800–1700 (2–4%) protein-coding genes underwent selection during the process of domestication. With the advent of next-generation sequencing (NGS) techniques, Hoffard et al. [15] identified 484 domesticated loci, of which 107 loci were further selected during improvement. The evidence mentioned above suggests that only a small portion of the genome was selected during maize domestication and improvement.

Genetic studies have provided firm evidence that maize was domesticated from Balsas teosinte (*Zea mays* subspecies *parviglumis*). This wild relative is endemic to the mid-to lowland regions of southwestern Mexico. Thus, genetic data point to the primary diffusion of domesticated maize from the highlands rather than from the region of initial domestication. The gene flow between maize and its wild relatives meaningfully impacts geographic origins [16].

3. Genes selected under domestication

Selection during evolution, whether natural or artificial, acts through the phenotype. For multifaceted phenotypes such as plant and inflorescence architecture, the underlying genetic architecture comprises a complex network of interacting genes rather than single genes that act independently to determine the trait. As such, selection acts on entire gene networks [17]. A set of genes/loci were selected during domestication knowingly or unknowingly by farmers then breeders. The earlier selection mainly focused on plant morphology, ear size, seed type and single stalk etc., for transforming its wild progenitor to the modern-day maize. Only a few genes, i.e., 2% genes (1200 genes) of the 60000 genes of maize, have been selected during the process of domestication. Those genes that have experienced artificial selection have greatly reduced genetic diversity in modern germplasm. Therefore cannot contribute to the variation for agronomically important traits. Artificial selection has impacted maize diversity during its domestication from teosinte (*Zea mays ssp. parvglumis*) to landraces and plant breeding from landraces to modern inbred lines. Artificial selection has impacted protein, oil, starch and amino acid content.

Maize domestication started around 10000 years ago. Early farmers selected and planted seeds from plants with beneficial traits while eliminating the undesirable ones. As a result of good alleles, i.e., alleles of genes controlling the favored traits, the frequency has been increased within the population. At the same time, bad/deleterious alleles frequency decreased. Such selection is made possible due to the tremendous availability of natural genetic variability in the teosintes. Over time, with current agricultural practices, certain combinations of genes have been selected. This includes major and minor gene mutations distinguishing from wild ancestors. That's why only few genes are responsible for the transformation. Beadle and Doebley revealed that only five genes might be responsible for the dramatic morphological changes. The "one gene-one trait" model for such genes is still questionable. Although a small number of genes has striking effects, on-ear and plant morphology resulted in the maize evolution. However, the vast majority of genes have an only a modest effect. Thousand of genes were likely necessary to contribute to the transformation like, increase in the size of the ear, adapting maize to the modern agricultural practices and an increase in the maize kernel's nutrient status.

3.1 Traits modified under domestication

3.1.1 Glume

Teosinte was characterized by hard glumes, the seed, which were passed through the digestive tract of the ruminants and not get digested. This ultimately serves as propagating material for the next generation under natural conditions. The locus *tga1, teosinte glume architecture* leads to a decrease in the size of glumes (**Figure 1**). It encodes the squamosa-promoter binding protein (SBP) transcription factor. This QTL has been mapped on chromosome 4 [18] and cloned [19]. Interestingly, in the *tga1* promoter region, *tga1* is regulated by *tb1 via* direct binding of *tb1* to two GGNCCC motifs [20].

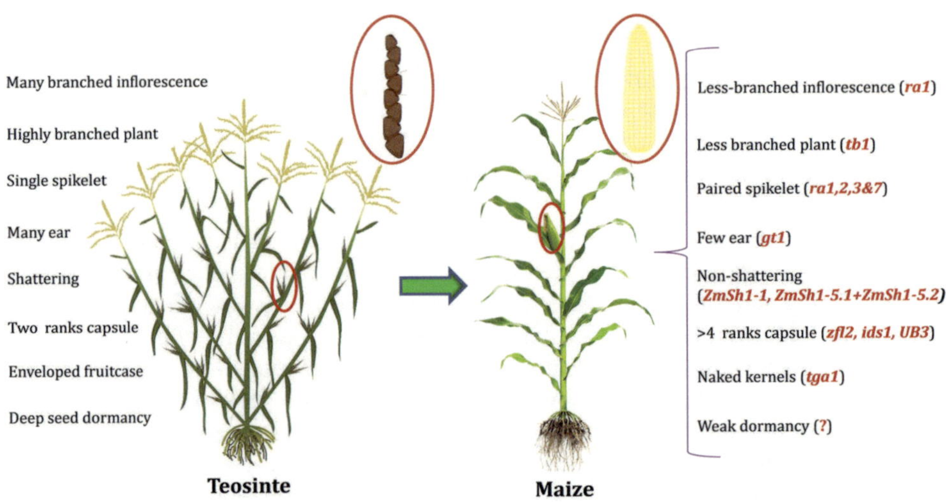

Figure 1.
Morphological changes during maize domestication and the underlying key genes.

3.1.2 Plant architecture

The loci teosinte branched 1 (*tb1*) responsible for the transformation from many tillers, many inflorescence to the single stalk with single inflorescence in the modern day maize. Concentrating the energy resources in a single ear and stalk made it possible to increase the ear size. The *tb1* encodes the TCP family transcription factor, i.e., *TEOSINTE BRANCHED1/CYCLOIDEA/PROLIFERATING CELL NUCLEAR ANTIGEN FACTOR*, which is mapped for apical dominance and inflorescence development [10]. The *tb1* was mapped by transposon tagging, which leads to the inhibition of axillary bud formation and transformation to female inflorescence [21, 22]. The causal variant of this phenotype was transposable element (TE) in the upstream of *tb1* suggests the role of long range chromatin insertions in the maize domestication [23]. The ortholog *tb1* locus also has been reported in the other crops suggesting its conserved nature in other plant species. In rice *OsTB1* with negative regulation [24], *BRC1* in *Arabidopsis thaliana* [25], *HvTB1* for higher tiller number in barley [26]; *TaTB1* along with *FLOWERING LOCUS T1* regulates inflorescence architecture in wheat [27]. These conserved loci of *tb1* explained the evidence for its role during domestication.

3.1.3 Others

QTLs at genes responsible for shattering vs. solid cobs, single vs. paired spikelets and distichous (two rank ear) vs. polystichous (> four ranks ear). The gene *Zea floricaula leafy2* (*zfl2*) primarily regulates the teosinte ear's two ranks [9, 28]. Sweet maize and popcorn retain tillering growth habit during maize diversification. However, the underlying molecular genetic mechanism remains unknown. The retention of maize tillering is controlled by a major quantitative trait locus (QTL), *tin1*, which encodes a C2H2-zinc-finger transcription factor that acts independently of *tb1*. *tin1* is involved in multiple pathways, directly represses two tiller-related genes, *gt1* and *Laba1/An-2*, and interacts with three TOPLESS proteins to regulate the tiller buds' outgrowth. Therefore maize *tin1*, derived from a standing variation in wild progenitor teosinte population, determines tillering retention during maize diversification [29].

Characterization of variations in the Four-row Wax landrace of China reported kernel row number (KRN) related genes and KRN QTL regions revealed potential causal mutations in *fea2, td1, kn1*, and *te1* [30].

3.1.4 Starch

In the initial phase of domestication, the focus was mainly on the plant shape and ear morphology. Many additional traits were acts as the targets from recent years. Grain yield, ear size (increased from 2 cm to 30 cm), quality and starch. Starch is the major byproduct of maize, constitutes ~73% of kernels' total weight. The three loci, i.e., *Su1, bt2* and *ae2,* are targets of selection during maize domestication and improvement. T*assel*, *seed2* and *dwarf8* as the targets of selection based on a screen of genes on chromosome1 [31, 32].

3.2 QTLs beyond the domesticated genes

Various QTLs were reported beyond the domesticated genes/loci affecting the morphological traits (**Table 1**). Around 314 QTLs were identified for 22 morphological traits involved in domestication and improvement [34]. Out of 314 QTLs, only 14 QTLs explained phenotypic variation >10 percent, affecting the

SN.	Trait	Gene/QTLs	Chromosome	Phenotype	Reference
1.	Plant architecture	*teosinte branched1 (tb1), grassy tillers 1 (gt1)*	1	Number of basal branches or tillers, Limited number of large ears	[28, 33]
2.	Glume hardness	*teosinte glume architecture1 (tga1)*	4	Inhibits secondary sexual traits in the female flower, preventing glumes from hardening	[9, 28, 34]
3.	Paired and single spikelets of maize	*ramosa1, ramosa2, ramosa3, ramosa7*	7 3	High number of kernels in each row of the ear of modern maize parents	[35, 36]
4.	Distichous and polystichous ear	*Zea floricaula leafy2(zfl2), Zea floricaula leafy1(zfl1)*	2 10	Multiple ear ranks along the inflorescence meristem	[9, 28, 34]
5.	Disarticulating rachides and non-disarticulating rachises	*ZmSh1–1, ZmSh1–5.1 + ZmSh1–5.2, Zga1*	1 5	Shattering, ear size	[33, 37–39]

Table 1.
QTLs/gene their chromosome location and phenotype.

morphological traits. Further research leads to the cloning of some of these QTLs, enabling identifying differences in the morphological traits between maize and teosinte. Plant architecture grassy tillers1 (*gt1*) encodes the Homeodomain leucine zipper transcription factor [40]. Plant architecture is impacted by the enhanced expression of BTB/POZ ankyrin repeat protein and Homeodomain leucine zipper transcription factor encoded by *tru1 and grassy tillers1* (*gt1*), respectively [33, 41]. The increased expression of transcription factors encoded by *gt1* (*grassy tillers1)* and *tru1* (*tassels replace upper ears1*) encodes Homeodomain leucine zipper transcription factor BTB/POZ ankyrin repeat protein [33, 40, 41]. *Tassels replace the upper ears1* (*tru1*) confers a sexual conversion of the terminal lateral inflorescence in teosinte to ear (pistillate) in maize from tassel (staminate). Other genes for seed filling *ZmSWEET4c* [42], *UB3*, *ids1/Ts6* for kernel row number [43], shattering *ZmSh1–1, ZmSh1–5.1* + *ZmSh1–5.2* [37] and for inflorescence architecture *ra1* [44], were cloned, key domesticated genes of maize. Most of these domesticated genes were the transcription factors that were unregulated during domestication. A maize-teosinte-derived BC_2S_3 population, the QTLs *UPA1* (*Upright Plant Architecture1*) and *UPA2*, which confer on upright plant architecture, were identified. The teosinte allele at *UPA2*, which reduces leaf angle, was lost during maize domestication [45]. More compact plants and improved yields under high planting densities could be developed by incorporating this allele into modern maize hybrids [45, 46].

4. Pleiotropic gene interactions during smaize domestication and improvement

Pleiotropy generally describes the effect of an allele of a gene for producing an unrelated phenotype. It affects the path of evolution as it facilitates if a directional selection of a phenotype affects other beneficial phenotypes' fitness or restricts if the allele has the deleterious effect on another phenotype. Generally, the developmental traits reveal pleiotropy by explaining the association among flowering time in male and female flowers [47], ear and tassel developmental traits [48], leaf length and flower length [49]. Apart from these, QTLs responsible for tassel and ear development are also responsible for flowering time [50].

Understanding pleiotropy and the association between phenotypes will help to explain the selection outcome constraints. For example, the maize allele at *zfl2* is responsible for spiral ear phyllotaxy that increases the kernel number and is involved in other traits like early flowering [51]. So in a well-adapted environment stabilizing selection for flowering might limit directional selection for kernel number. Therefore such pleiotropy may limit domestication alleles when selection disallows variation.

Various researchers have reported that domestication alleles were pleiotropic [11, 12, 52, 53]. The *teosinte branched1 (tb1)* is pleiotropic across many traits; apical dominance, growth of leaves on the lateral branches, length of lateral branches, ear and root architecture [54]. *tb1*, as a transcription factor, binds to many locations in the genome. It directly regulates *gt1* by binding to its promoter. Still, it directly affects the cell cycle by suppressing many cell cycle genes (*proliferating cell nuclear antigen2 (pcna2)* and *minichromosome maintenance2/prolifera (mcm2/prl*) [20]. *zga1* is a MADS-box transcription factor associated with ear size and has a pleiotropic effect on flowering time [33]. *tga1*, a glume architecture allele shown to have pleiotropic effect on lateral branch lengthy, ear phyllotaxy and ear disarticulation [55].

5. Effect of maize domestication on genetic diversity (Domestication Bottleneck)

Both domestication and artificial selection during crop improvement led to selecting only desirable/beneficial traits, resulting in the reduced genetic diversity of the unselected genes. During the process of domestication, nearly all crop species experience "Domestication Syndrome" or "Domestication Bottleneck" [22, 56]. These effects happen in two stages; i) initial bottleneck effect, when a subset of crop wild species population brought under cultivation and ii) subsequent reduction in the genetic diversity through selective breeding for the desirable traits during crop improvement is improvement bottleneck. Among crop species, maize experiences a relatively mild genetic bottleneck, as domesticated maize retains around ~81% of the genetic diversity of teosinte [15]. Approximately 2–4% of genes were the target during the initial domestication and crop improvement stage [14, 15]. It is established that genetic diversity generally declines with the domestication of teosinte to the landraces. Subsequently, modern plant breeding reduces the genetic diversity of modern-day maize inbred lines relative to the landraces (**Figure 2**). Therefore such genes strongly influenced by domestication or improvement are enriched in modern improved varieties in the subset of genes that show low nucleotide diversity [14]. Yamasaki et al. [57] proposed a model containing three types of genes: 'neutral genes that demonstrate diversity reduction by general bottleneck effects, domestication genes in which diversity by selection between the teosintes and landraces is significantly reduced, and improvement genes in which diversity by selection between landraces and inbreds is significantly reduced (**Figure 2**).'

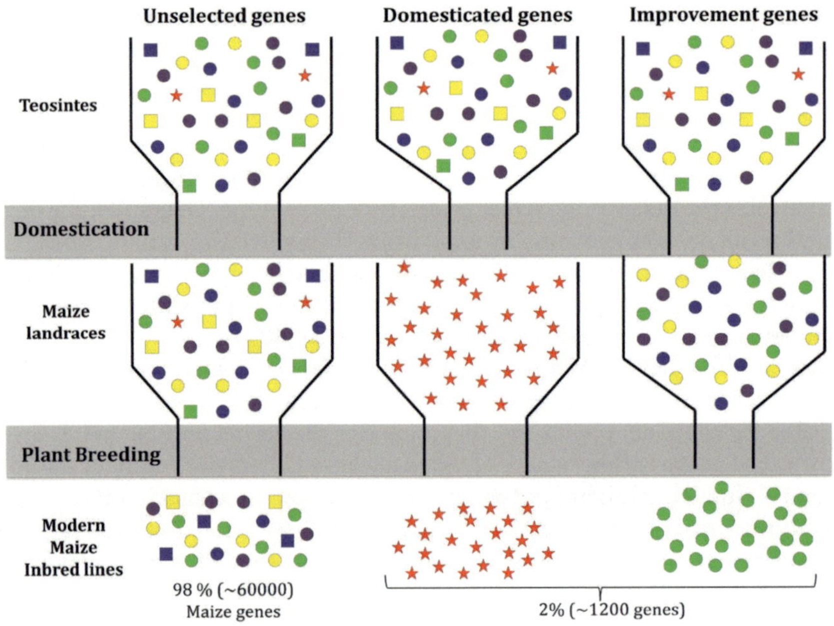

Figure 2.
Domestication and plant breeding effects on genetic diversity of maize genes [redrawn from Yamasaki et al. [57]]: Shapes with different color represents different genes and shaded area depicts the bottleneck effect.

The genes that experienced artificial selection during domestication and crop improvement have significantly reduced genetic diversity in the modern germplasm and cannot contribute to the agro morphological traits. Therefore the selected genes are difficult to identify in the genetic screens and may not be useful in traditional breeding programmes. If we need to utilize the selected genes fully, the new variation must be reintroduced from teosintes. Additionally, for the 98% of genes, which do not experience selection during domestication and crop improvement, there are huge genetic variations in the diverse inbred lines that could be utilized by identifying the genetic loci/QTLs and improvement through plant breeding.

6. Wild relatives of maize

Wild relatives of crops are the species of wild plants that are genetically linked to cultivated crops. Unattended by humans, they continue to grow in the wild, developing traits that farmers and breeders can cross with domesticated crops to produce new varieties, such as drought tolerance or pest resistance. The *Zea* genus of grass consists of seven genera with different chromosome numbers divided into two groups: viz. old-world and new world groups. *Chionachne, Coix, Polytoca, Sclerachne* and *Trilobachne* originated in Southeast Asia and belonged to the old-world group. The new world group consists of *Zea* and *Tripsacum* and originated in Mexico and Central America. *Zea mays ssp. mays* is the only species of economic importance and other species referred to as teosintes.

7. Landraces of maize

A landrace is defined as 'dynamic population(s) of a cultivated plant that has a historical origin, distinct identity, and lacks formal crop improvement and often being genetically diverse, locally adapted, and associated with traditional farming systems' [58]. Compared to other crops, maize has tremendous genetic diversity, which offers potential for crop improvement for biotic/abiotic stresses, nutritional quality and grain yield. In landraces, the diversity/genetic variations lie within-population rather than among populations. Worldwide, the landraces have been characterized both morphologically and molecularly. The genetic variability present in the available landraces has been utilized to improve agro morphological traits, biotic/abiotic stresses and specialty traits. In crop centers of origin and diversity, often biotic and abiotic conditions vary across the landscape, creating the possibility of local adaptation of crops. Local landraces perform better than non-local ones under local conditions. Some of the examples of their utilization are given below:

7.1 Agromorphological traits

The conservation of landraces is fundamental to safeguarding crop diversity, food security, and sustainable production. 'Jala' is a particular maize landrace from the region in and around the Jala Valley of Mexico that produces the largest ear and tallest plant of all maize landraces in the world. Changing socio-economic and environmental conditions in the Jala Valley could lead to the genetic erosion of the ancestral 'Jala' landrace, leading to global consequences [59]. In southwest China, Four-row Wax landrace, with four rows of kernels on the cob.

7.2 Tolerance to abiotic stresses

Maize landrace accessions constitute an invaluable gene pool of unexplored alleles that can be harnessed to mitigate the challenges of the narrowing genetic base, declined genetic gains, and reduced resilience to abiotic stress in modern varieties developed from repeated recycling of few superior breeding lines. Some landraces of Mexico origin that imparts abiotic stress tolerance are Bolita, Breve de Padilla, Conica, Conica Nortena, Chalqueno × Ancho de Tehuacan cross (alkalinity tolerant), La Posta Sequia, Nal Tel, Oloton (acid soil tolerant) and Tuxpeno (drought tolerant) [60]. At CIMMYT, the production of inbred lines, drought-tolerant population-1 (DTP-1) and drought-tolerant population-2 (DTP-2) is exploited for imparting drought tolerance. Some of the inbred lines derived from 'La Posta Sequia' were reported to have drought and heat tolerance [61]. Maize landraces L25, L14, L1, and L3, are reported as the most valuable source of drought tolerance [62]. A higher transcript accumulation in shoot tissues of *ZmATG* genes reported in landrace 'Argentino Amarelo' under the osmotic stress conditions compared to landrace 'Taquarão' [63]. Nelimor et al. [64] identified extra-early maize landraces that express tolerance to drought and heat stress. Root system architecture plays a crucial role in water and nutrient acquisition in maize. *ZmCKX5* (cytokinin oxidase/dehydrogenase) was resequenced in maize landraces and revealed its importance in developing the maize root system [65].

7.3 Resistance to biotic stresses

Maize crops encounter a lot of diseases due to their wide distribution. Among fungal diseases, Turcicum leaf blight (TLB) and Maydis leaf blight (MLB) results in the decline In maize production throughout the world. A subpopulation Tuxpeno Crema derived from the landrace Tuxpeno known to possess resistance to the foliar diseases [66]. Palomero Toluqueno, a landrace of popcorn reported to have resistance to the maize weevil [67], few Carrebian landraces possess resistance to larger grain borer [68]. Two Kenyan maize landraces (Jowi and Nyamula) and one Latin

American landrace (Cuba 91) shown a lower number of eggs and egg batches deposition of *C. partellus* due to production of herbivore-induced plant volatiles (HIPVs) [69, 70]. The fall armyworm *Spodoptera frugiperda* J. E. Smith (Lepidoptera: Noctuidae) is one of the most damaging maize production pests in tropical areas. The maize landraces 'Chimbo' and 'Elotillo' had the lowest leaf damage, calculated by the area under the severity progress curve [71]. The maize landrace 'Pérola' from Brazil showed resistance to fall armyworm in the winter and summer seasons [72].

7.4 Enhancement of specialty traits

A northeastern Indian landrace, 'Murlimakkai,' was utilized to develop Baby Corn composite VL Baby Corn [60]. Several landraces, viz., Azul, Bolita, Tlacoya, Pepitilla and Oaxaqueno, were very popular and utilized for tortilla quality. Mexican popcorn landrace 'Palomero,' utilized to understand the landrace structure and improvement in the popping quality. Landraces had significantly higher values than checks for oil content, oleic acid, MUFA and tocopherol contents. Genetic analyses suggest that the kernel quality traits could be successfully manipulated using the investigated plant material [73].

7.5 Unlocking the genetic variability present in the landraces

Using landraces for broadening the genetic base of elite maize germplasm is hampered by heterogeneity and high genetic load. Production of DH line libraries can help to overcome these problems. Landraces of maize (*Zea mays* L.) represent a vast reservoir of genetic diversity untapped by breeders. Genetic heterogeneity and a high genetic load hamper their use in hybrid breeding. Production of doubled haploid line libraries (DHL) by the in vivo haploid induction method promises to overcome these problems. Böhm et al. [74] developed doubled haploid lines from European flint landraces and reported considerable breeding progress. This reveals that there is tremendous potential of landraces for broadening the narrow genetic base of elite germplasm. DH technology's use demonstrated broadening the flint heterotic pool's narrow genetic base [75]. Altogether, the DH technology also provides new opportunities for characterizing and utilizing the genetic diversity present in gene bank accessions of maize [76].

8. Conclusions

The domestication and crop improvement processes lead to converting teosinte into landraces and subsequently to the modern-day maize inbred. During domestication, based on genetic evidence, it is clear that selection was mainly focused on five genes. This leads to the change in the architecture and morphology of teosinte into maize. Maize has evolved distinct genetic solution towards domestication: domestication of maize has involved distinct genetic and regulatory networks have been used to acquire convergent phenotypes. During domestication and artificial selection, only a small part of the genome underwent selection, which ultimately led to the modern-day maize. So, wild relatives and landraces encompassing the unselected genes possess enormous potential as the donor for beneficial genes/alleles. The derived inbred lines from such material could not be directly utilized in the breeding programme. They must be utilized as a donor for the specific traits, i.e., tolerance to biotic/abiotic stresses and nutritional quality traits. The utilization of wild relatives and landraces in the breeding programmes is not that easy; utilization of bridging species and embryo rescue provides the solution to this problem.

The pre-breeding programme helps to utilize wild relatives and landraces to enrich the ongoing crop improvement programmes. The availability of the genome sequence of 'B 73' and 'Palmoreo' landrace and strong pre-breeding programme can potentially enhance unexplored germplasm in the maize breeding programme.

Author details

Devender Sharma*, Rajesh K. Khulbe, Ramesh S. Pal, Jeevan Bettanaika and Lakshmi Kant
ICAR—Vivekananda Parvatiya Krishi Anusandhan Sansthan, Almora, Uttarakhand, India

*Address all correspondence to: devender.kumar1@icar.gov.in

References

[1] Liu J, Fernie AR, Yan J. The Past, Present, and Future of Maize Improvement: Domestication, Genomics, and Functional Genomic Routes toward Crop Enhancement. Plant Commun. 2020;1(1):100010. doi:10.1016/j.xplc.2019.100010

[2] Nunn N and Nancy Q. The Columbian exchange: a history of disease, food, and ideas. Journal of Economic Perspectives. 2010; 24:163-188.

[3] Shiferaw B, Prasanna BM, Hellin J, Banziger M. Crops that feed the world. 6. Past successes and future challenges to the role played by maize in global food security. Food Secience. 2011; 3:307-327.

[4] Beadle GW. The mystery of maize. Chicago. Field Museum of Natural History. 1972;43:2-11.

[5] Piperno DR and Flannery KV. The earliest archaeological maize (*Zea mays* L.) from highland Mexico: new accelerator mass spectrometry dates and their implications. Proc. Proceedings of the National Academy of Sciences. Proceedings of the National Academy of Sciences USA. 2001; 98:2101-2103.

[6] Vallebueno-Estrada M, Rodrıguez-Arevalo I, Rougon-Cardoso A, Martınez Gonza'lez J, Garcıa Cook, A, Montiel R, Vielle- Calzada, JP. The earliest maize from San Marcos Tehuaca´ n is a partial domesticate with genomic evidence of inbreeding. 2016; Proceedings of the National Academy of Sciences USA. 2016; 113:14151-14156.

[7] Matsuoka Y, Vigouroux Y, Goodman MM, Sanchez GJ, Buckler E, Doebley J. A single domestication for maize shown by multilocus microsatellite genotyping. Proceedings of the National Academy of Sciences USA. 2002; 99:6080-6084.

[8] Tanksley SD, McCouch SR. Seed banks and molecular maps: unlocking genetic potential from the wild. Science. 1997;277:1063-1066.

[9] Doebley JF and Stec A. Inheritance of the morphological differences between maize and teosinte: compari- son of results for two F_2 populations. Genetics. 1993;134:559-570.

[10] Doebley J, Stec A and Gustu C. Teosinte branched1 and the origin of maize: evidence for epistasis and the evolution of dominance. Genetics. 1995;141:333-346.

[11] Mangelsdorf PC, Reeves RG. The origin of Indian corn and its relatives. Texas AES Bull. 1939;574:1-315.

[12] Beadle GW. Teosinte and the origin of maize. J Hered. 1939; 30:24-247.

[13] Beadle GW. Teosinte and the origin of maize. In: Walden DB (ed) Maize breeding and genetics. 1978; Wiley, New York, pp 113-128.

[14] Wright SI, Bi IV, Schroeder SG, Yamasaki M, Doebley JF, McMullen MD and Gaut BS. The effects of artificial selection on the maize genome. Science. 2005; 308:1310-1314.

[15] Hufford MB, Xu X, van Heerwaarden J, Pyhajarvi T, Chia JM, Cartwright RA, Elshire RJ, Glaubitz JC, Guill KE, Kaeppler SM, et al. Comparative population genomics of maize domestication and improvement. Nature Genetic. 2012; 44:808-811.

[16] Van Heerwaarden J, Doebley J, Briggs WH, et al. Genetic signals of origin, spread, and introgression in a large sample of maize landraces. Proceedings of National Academy of Science. U S A. 2011; 108: https://doi.org/10.1073/pnas.1013011108.

[17] Studer A, Zhao Q, Ross-Ibarra J and Doebley J. Identification of a functional transposon insertion in the maize domestication gene tb1. Nature Genetics. 2011;43:1160-1163.

[18] Dorweiler J, Stec A, Kermicle J and Doebley J. Teosinte glume architecture 1: a genetic locus controlling a key step in maize evolution. Science. 1993;262:233-235.

[19] Wang H, Nussbaum-Wagler T, Li B, Zhao Q, Vigouroux Y, Faller M, Bomblies K, Lukens L and Doebley JF. The origin of the naked grains of maize. Nature. 2005;436:714-719.

[20] Studer AJ, Wang H, Doebley JF. Selection during maize domestication targeted a gene network controlling plant and inflorescence architecture. Genetics. 2017;207:755-765.

[21] Doebley J. The genetics of maize evolution. Annual Review of Genetics. 2004; 38:37-59.

[22] Doebley J, Gaut B and Smith B. The molecular genetics of crop domestication. Cell. 2006;127:1309-1321.

[23] Li E, Liu H, Huang L, Zhang X, Dong X, Song W, Zhao H and Lai J. Long-range interactions between proximal and distal regulatory regions in maize. Nature Communications. 2019;10:2633.

[24] Takeda T, Suwa Y, Suzuki M, Kitano H, Ueguchi-Tanaka M, Ashikari M, Matsuoka M and Ueguchi C. The OsTB1 gene negatively regulates lateral branching in rice. Plant Journal. 2003;33:513-520.

[25] Aguilar-Martı'nez JA, Poza-Carrio´ n C and Cubas P. Arabidopsis BRANCHED1 acts as an integrator of branching signals within axillary buds. Plant Cell. 2007;19:458-472.

[26] Ramsay L, Comadran J, Druka A, Marshall DF, Thomas WT, Macaulay M, MacKenzie K, Simpson C, Fuller J, Bonar N, et al. INTERMEDIUM-C, a modifier of lateral spikelet fertility in barley, is an ortholog of the maize domestication gene TEOSINTE BRANCHED 1. Nature Genetics. 2011; 43:169-172.

[27] Dixon LE, Greenwood JR, Bencivenga S, Zhang P, Cockram J, Mellers G, Ramm K, Cavanagh C, Swain SM and Boden SA. TEOSINTE BRANCHED1 regulates inflorescence architecture and development in bread wheat (*Triticum aestivum*). Plant Cell. 2018; 30:563-581.

[28] Doebley JF and Stec A. Genetic analysis of the morphological differences between maize and teosinte. Genetics. 1991;129:285-295.

[29] Zhang X, Lin Z, Wang J, et al. The tin1 gene retains the function of promoting tillering in maize. Nature Communications. 2019;10:1-13. https://doi.org/10.1038/s41467-019-13425-6.

[30] Liu H, Wang X, Wei B, et al. Characterization of genome-wide variation in four-row wax, a waxy maize landrace with a reduced kernel row phenotype. Front Plant Science. 2016;7: https://doi.org/10.3389/fpls.2016.00667

[31] Tenaillon MI, Sawkins MC, Long AD, Gaut RL, Doebley JF and Gaut BS. Patterns of DNA sequence polymorphism along chromosome 1 of maize (*Zea mays* L. ssp. mays). Proceedings of National Academy of Science. USA. 2001;98:9161-9166.

[32] Tenaillon MI, U'Ren, J, Tenaillon O and Gaut BS. Selection versus demography: A multilocus investigation of the domestication process in maize. Molecular Biology Evolution. 2004; 2:1214-1225.

[33] Wills DM, Whipple CJ, Takuno S, Kursel LE, Shannon LM, Ross-Ibarra J, Doebley JF. From many, one: genetic control ofprolificacy during maize domestication. PLoS Genetics. 2013;9: e1003604.

[34] Briggs WH, McMullen MD, Gaut BS, Doebley J. Linkage mapping of domestication loci in a large maize-teosinte backcross resource. Genetics. 2007;177:1915-1928.

[35] Vollbrecht E, Springer PS, Goh L, Buckler ES IV, Martienssen R. Architecture offloral branch systems in maize and related grasses. Nature. 2005;436:1119-1126.

[36] McSteen P. Branching out: the ramosa pathway and the evolution of grass inflorescence morphology. Plant Cell. 2006;18:518-522.

[37] Lin Z, Li X, Shannon LM, Yeh CT, Wang ML, Bai G, Peng Z, Li J, Trick HN, Clemente TE, et al. Parallel domestication of the Shattering1 genes in cereals. Nature Genetics. 2012;44:720-724.

[38] Weber AL, Briggs WH, Rucker J, Baltazar BM, de Jes?us S?anchez-Gonzalez J, Feng P, Buckler ES, Doebley J. The genetic architecture ofcomplex traits in teosinte (Zea mays ssp. parviglumis): new evidence from association mapping. Genetics. 2008;180:1221-1232.

[39] Wills DM, Fang Z, YorkAM, Holland JB, Doebley JF. Defining the role of the MADS-box gene, Zea Agamous-like1, a target of selection during maize domestication. Journal of Heredity. 2018; 109:333-338.

[40] Dong Z, Li W, Unger-Wallace E, Yang J, Vollbrecht E, ChuckG. Ideal crop plant architecture is mediated by tasselsreplaceupperears1,a btb/pozankyrin repeat gene directly targeted by teosintebranched1. Proceedings of the National Academy of Sciences, USA. 2017;114: E8656-E8664.

[41] Whipple CJ, Kebrom TH, Weber AL, Yang F, Hall D, Meeley R, Schmidt R, Doebley J, Brutnell TP and Jackson DP. Grassy tillers1 promotes apical dominance in maize and responds to shade signals in the grasses. Proceedings of the National Academy Sciences, U S A. 2011;108:E506-E512.

[42] Sosso D, Luo D, Li QB, Sasse J, Yang J, Gendrot G, Suzuki M, Koch KE, McCarty DR, Chourey PS, et al. Seed filling in domesticated maize and rice depends on SWEET-mediated hexose transport. 2015;47:1489-1493.

[43] Wang J, Lin Z, Zhang X, Liu H, Zhou L, Zhong S, Li Y, Zhu C and Lin Z. krn1, a major quantitative trait locus for kernel row number in maize. New Phytologist. 2019;223:1634-1646.

[44] Sigmon B and Vollbrecht E. Evidence of selection at the ramosa1 locus during maize domestication. Molecular Ecology. 2010;19:1296-1311.

[45] Tian J, Wang C, Xia J, Wu L, Xu G, Wu W, Li D, Qin W, Han X, Chen Q, et al. Teosinte ligule allele narrows plant architecture and enhances high-density maize yields. Science. 2019; 365:658-664.

[46] Kong D, Wang B and Wang H. UPA2 and ZmRAVL1: promising targets of genetic improvement of maize plant architecture. Journal of Integrative Plant Biology. 2020;62:394-397.

[47] Buckler ES, Holland JB, BradburyPJ, AcharyaCB, Brown PJ, Browne C, Ersoz E, Flint-Garcia S, Garcia A, Glaubitz JC, et al. The genetic architecture of maize flowering time. Science. 2009;325: 714-718.

[48] Brown PJ, Upadyayula N, Mahone GS, Tian F, Bradbury PJ, Myles S, Holland JB, Flint-Garcia S,

McMullen MD, Buckler ES, et al. Distinct genetic architectures for male and female inflorescence traits ofmaize. PLoS Genetics. 2011;7: e1002383.

[49] Tian F, Bradbury PJ, Brown PJ, Hung H, Sun Q, Flint-Garcia S, Rocheford TR, McMullenMD, Holland JB, Buckler ES. Genome-wide association study of leaf architecture in the maize nested association mapping population. Nature Genetics. 2011;43:159-162.

[50] Xu G, Wang X, Huang C, Xu D, Li D, Tian J, Chen Q, Wang C, Liang Y, Wu Y etal. 2017. Complex genetic architecture underlies maize tassel domestication. New Phytologist214: 852-864.

[51] Bomblies K and Doebley JF. Pleiotropic effects of the duplicate maize FLORICAULA/LEAFY genes zfl1 and zfl2 on traits under selection during maize domestication. Genetics. 2006;172:519-531.

[52] Collins GN and Kempton JH. A teosinte-maize hybrid. Journal of Agriculural Research. 1920;19:1-38.

[53] Langham DG. The inheritance of intergeneric differences in Zea-Euchlaena hybrids. Genetics. 1940;25:88-108.

[54] Gaudin AC, McClymont SA, Soliman SS, Raizada MN. The effect ofaltered dosage of a mutant allele of teosinte branched1 (*tb1-ref*) on the root system of modern maize. BMC Genetics. 2014;15: 23.

[55] Wang Q, Dooner HK. Remarkable variation in maize genome structure inferred from haplotype diversity at the bz locus. Proceedings of the National Academy of Sciences, USA. 2006;103: 17644-17649.

[56] Olsen KM and Wendel JF. A bountiful harvest: genomic insights into crop domestication phenotypes. Annual Review of Plant Biology. 2013;64:47-70.

[57] Yamasaki M, Tenaillon MI, Bi IV, et al. A large-scale screen for artificial selection in maize identifies candidate agronomic loci for domestication and crop improvement. Plant Cell. 2005; 17:2859-2872. https://doi.org/10.1105/tpc.105.037242.

[58] Camacho Villa TC, Maxted N, Scholten M, Ford-Lloyd B. Defining and identifying crop landraces. Plant Genetics Resource. 2005; 3:373-384.

[59] Ocampo-Giraldo V, Camacho-Villa C, Costich DE, et al. Dynamic conservation of genetic resources: Rematriation of the maize landrace Jala. Food Security. 2020;12: https://doi.org/10.1007/s12571-020-01054-7

[60] Prasanna BM. Diversity in global maize germ- plasm: characterization and utilization. Journal of Bioscience. 2012;37:843-855.

[61] Cairns JE, Crossa J, Zaidi PH et al. Identifi cation of drought, heat, and combined drought and heat tolerant donors in maize. Crop Science. 2013;53:1335-1346.

[62] Andjelkovic V, Kravic N, Babic V, et al. Estimation of drought tolerance among maize landraces from mini-core collection. Genetika. 2014; 46:775-788.

[63] Tejeda LHC, Viana VE, Maltzahn LE, et al. Abiotic stress and self-destruction: ZmATG8 and ZmATG12 gene transcription and osmotic stress responses in maize. Biotechnology Research Innovation. 2020; https://doi.org/10.1016/j.biori.2019.12.001

[64] Nelimor C, Badu-Apraku B, Tetteh AY, et al. Assessing the potential of extra-early maturing landraces for improving tolerance to drought, heat, and both combined stresses in

maize. Agronomy. 2020;10: https://doi.org/10.3390/agronomy10030318

[65] Wang H, Sun H, Xia H, et al. Natural variation and domestication selection of ZmCKX5 with root morphological traits at the seedling stage in maize. Plants. 2021;10: https://doi.org/10.3390/plants10010001

[66] Rodriguez MG, Miguel-Chavez RS, Larque-Saavedra A. Physiological aspects in Tuxpeno maize with improved drought tolerance. Maydica. 1998; 43:137-141.

[67] Arnason JT, Baum B, Gale J, et al. Variation in resistance of Mexican landraces of maize to maize weevil *Sitophilus zeamais* , in relation to taxonomic and biochemical parameters. Euphytica. 1994;74:227-236.

[68] Kumar H. Resistance in maize to the larger grain borer, Prostephanus truncates (Horn) (Coleoptera: Bostrichidae). Journal of Stored Products Research. 2002;38:267-280.

[69] Mutyambai DM, Midega CAO, Bruce TJA, et al. Behaviour and biology of Chilo partellus on maize landraces. Entomologia Experimentalis et Applicata. 2014;153: https://doi.org/10.1111/eea.12237

[70] Magara HJO, Mutyambai DM, Charles MAO, et al. Responses of stemborer Chilo partellus to volatiles emitted by maize landraces exposed to signal grass (Brachiaria brizantha). Journal of Plant Interaction. 2020;15: https://doi.org/10.1080/17429145.2020.1827056

[71] dos Santos LFC, Ruiz-Sánchez E, Andueza-Noh RH, et al. Leaf damage by *Spodoptera frugiperda* J. E. Smith (Lepidoptera: Noctuidae) and its relation to leaf morphological traits in maize landraces and commercial cultivars. Journal of Plant Diseases and Protection. 2020; 127:https://doi.org/10.1007/s41348-019-00276-y

[72] Costa EN, Fernandes MG, Medeiros PH, Evangelista BMD. Resistance of maize landraces from Brazil to fall armyworm (Lepidoptera: Noctuidae) in the winter and summer seasons. Bragantia. 2020;79. https://doi.org/10.1590/1678-4499.20200034

[73] Kahriman F, Aktaş F, Songur U, et al. Screening Turkish maize landraces for kernel oil content and oil quality traits. Plant Genetics Resource Characterisation Utilisation. 2020; https://doi.org/10.1017/S1479262120000258

[74] Böhm J, Schipprack W, Utz HF, Melchinger AE. Tapping the genetic diversity of landraces in allogamous crops with doubled haploid lines: a case study from European flint maize. Theoretical and Applied Genetics. 2017;130:861-873. https://doi.org/10.1007/s00122-017-2856-x

[75] Brauner PC, Schipprack W, Utz HF, et al. Testcross performance of doubled haploid lines from European flint maize landraces is promising for broadening the genetic base of elite germplasm. Theoretical and Applied Genetics. 2019; https://doi.org/10.1007/s00122-019-03325-0

[76] Strigens A, Schipprack W, Reif JC, Melchinger AE. Unlocking the Genetic Diversity of Maize Landraces with Doubled Haploids Opens New Avenues for Breeding. PLoS One. 2013; 8: https://doi.org/10.1371/journal.pone.0057234

Chapter 3

Genetic Diversity of *Coffea arabica* L.: A Genomic Approach

José Luis Spinoso-Castillo, Tarsicio Corona-Torres, Esteban Escamilla-Prado, Victorino Morales-Ramos, Víctor Heber Aguilar-Rincón and Gabino García-de los Santos

Abstract

Coffea arabica L. produces a high-quality beverage, with pleasant aroma and flavor, but diseases, pests and abiotic stresses often affect its yield. Therefore, improving important agronomic traits of this commercial specie remains a target for most coffee improvement programs. With advances in genomic and sequencing technology, it is feasible to understand the coffee genome and the molecular inheritance underlying coffee traits, thereby helping improve the efficiency of breeding programs. Thanks to the rapid development of genomic resources and the publication of the *C. canephora* reference genome, third-generation markers based on single-nucleotide polymorphisms (SNPs) have gradually been identified and assayed in *Coffea*, particularly in *C. arabica*. However, high-throughput genotyping assays are still needed in order to rapidly characterize the coffee genetic diversity and to evaluate the introgression of different cultivars in a cost-effective way. The DArTseq™ platform, developed by Diversity Arrays Technology, is one of these approaches that has experienced an increasing interest worldwide since it is able to generate thousands of high quality SNPs in a timely and cost-effective manner. These validated SNP markers will be useful to molecular genetics and for innovative approaches in coffee breeding.

Keywords: *Coffea* spp., high throughput genotyping, molecular markers, plant breeding, DArTseq

1. Introduction

Coffee is an important crop and the second most traded commodity in the world (after petroleum) providing a living to more than 125 million people. Commercial coffee production is controlled by only two species belonging to the *Coffea* genus: *Coffea arabica* L. (Arabica coffee) and *Coffea canephora* Pierre ex A. Froehner (Robusta coffee), which supplied 60 and 40% of the world coffee production in 2018/19, respectively [1]. Although *C. canephora* does not have the cup quality of the more popular *C. arabica*, it continues to be widely grown, especially in regions where farming is low intensive because of its tolerance to diseases and pests as well as abiotic stresses [2].

C. arabica produces a high-quality beverage, with pleasant aroma and flavor, but a range of biotic and abiotic stresses often affect its yield [3]. Therefore, improving important agronomic traits of both commercial species remains a target for most coffee breeding programs. Advances in genomic and sequencing technology, make possible to understand the coffee genome and the molecular inheritance underlying coffee traits, thereby helping improve the efficiency of coffee breeding [3].

The development of new genomic tools can help us explore, more deeply and more precisely, the genomic diversity at intra and inter-specific levels [4]. Two examples of high-throughput platforms include next-generation sequencing (NGS) [5] and the development of DNA microarrays [6]. Compared to a whole-genome sequencing methodology, an SNP array approach provides time-effective, low-cost and more straightforward genotyping technology for germplasm screening [7, 8].

Thanks to the rapid development of genomic resources and the publication of the reference genome [9], third-generation markers based on single-nucleotide polymorphisms (SNPs) have gradually been identified and assayed in *Coffea*, particularly in *C. arabica* [10, 11].

2. Genetic diversity of *Coffea arabica* L.

The *Coffea* genus belongs to the Rubiaceae family that includes around 124 species, most of them are diploids (2n = 2x = 22). The only allotetraploid is *C. arabica* L. with 2n = 4x = 44 [12], which was originated from the natural cross between *Coffea eugenioides* S. Moore and *C. canephora* Pierre ex A. Froehner [13], *C. arabica* is the only self-fertile among the other cultivated species. This specie is genetically less diverse when compared to the diploid species [14, 15], a situation that has been associated with its susceptibility to the common coffee diseases [16].

C. arabica is mainly native to the highlands of southwestern Ethiopia, South Sudan (Boma plateau), and north Kenya (Mount Marsabi). *C. arabica* cultivars grown all around the world are derived from either 'Typica' or 'Bourbon' genetic base [17]. Studies report wide agronomic diversity of Arabica coffee accessions collected in these regions of Ethiopia regarding leaf size, height, biotic and abiotic stresses tolerance and yield [18, 19]. In addition, studies using molecular markers indicated the presence of higher genetic variability of Ethiopian (ET) accessions compared with cultivars, demonstrating the potential of these accessions for breeding purposes [10, 20–23]. These accessions also showed a great variability of metabolite profiles contents of coffee beans for cup quality improvement [10, 24].

The assessment of population structure and genetic relationships of these ET accessions, among themselves and in relation to traditional cultivars is fundamental for efficient use of genetic diversity of these genotypes in Arabica coffee breeding programs [25]. However, selection of genetically diverse parental lines based on morphological and agronomic traits is often difficult because of a high degree of morphological similarities [26].

During the past 30 years, molecular markers have been increasingly used in germplasm diversity assessment of various crops [27, 28]. The molecular information allows gaining insight into the genetic structure of individual genotypes, and eventually helps in accurate selection of superior genotypes for maximizing selection gains [29].

3. *C. arabica* diversity assessment by molecular markers

Several works on the assessment of Arabica genetic diversity have been carried out with different results. Generally, among different types of material (cultivars, accessions, hybrids, and spontaneous genotypes) practically all studies show a very low genetic variation by using different marker systems [3]. Arabica's genetic diversity has been evaluated by a range of molecular markers, such as Random Amplified Polymorphic DNA (RAPD) [30, 31], Inter Simple Sequence Repeat (ISSR) [32], Simple Sequence Repeat (SSR) [23, 29, 33, 34], SSR and Amplified Fragment Length Polymorphism (AFLP) [35, 36].

In a recent study presented in the World Coffee Research annual report a genetic diversity assessment of 800 Arabica's accessions from the collection at CATIE, Costa Rica, shows the least genetic diversity of *C. arabica* compared to other major crops [37]. This study also found that coffee cultivars contain almost 45% of the genetic diversity found in the 800 above-mentioned accessions indicating the limitation of variability for breeding programs [3]. Therefore, it is crucial to assess the population structure and its genetic diversity in *Coffea* genus.

Of course, all *C. arabica* germplasm available in ex situ collections may represent only a fraction of the total genetic diversity of the remaining wild and semi-wild forest coffees in S.W. Ethiopia [38]. However, Arabica's breeders do have already an idea of the potential and limits of ET germplasm, in particular in regard to host resistances to diseases and pests. For example, none of the modern Arabica cultivars with host resistances to CLR derive from these ET germplasm [39]. Also cultivars resistant to CBD outside Ethiopia do not have ET germplasm as progenitors [40], while nematode resistance found in ET accessions provide only limited protection to the severe nematode problems in Central America [41].

In contrast, ET germplasm may be a good source for sensory quality traits in cup. The cup quality profile of the new Arabica's F1-hybrids developed for Central America is said to derive largely from one of the two progenitors, being a selected ET accession of the FAO-1964 pool [42]. Silvarolla et al. found three coffee plants in offspring of ET germplasm, which were nearly caffeine-free [43]. Male sterility has been detected in a few ET accessions, a character useful for F1-hybrid seed production [44].

4. Next generation sequencing techniques in *C. arabica*

NGS incorporate technologies which, at low cost and in short time, produce millions of short DNA sequence. The most commonly used platforms for high-throughput, useful genomic research, especially in non-model plant species include second generation sequencing techniques (SGseqTs): Illumina/Solexa, 454/Roche, ABI/SOLiD, and Helicos (read mostly in the range of 25 and 700 bp in length) [45]. Results obtained from such research point to the fact that NGS techniques (NGSTs) should not be restricted to the genomes of model organisms only as non-model plants have provided useful resources for genomic studies [45].

In contrast to classical molecular markers, SNPs are the most abundant markers, particularly in the non-coding regions of the genome [46]. NGS used jointly with different complexity reduction methods, Genotyping by sequencing (GBS) and DArTseq™ (Sequencing-based diversity array technology) methods, enable a large-scale discovery of SNPs in a wide variety of non-model organisms [47–49]. These techniques provide measures of genetic divergence and diversity within the major genetic clusters that comprise crop germplasm [50].

The genotyping profiles of SNPs can be compared across laboratories and sequencing platforms. These benefits have resulted in the increasing use of SNPs as high-quality markers for genotype identification in a wide range of crops [51], as recently demonstrated in cacao (*Theobroma cacao*; [52]), pummelo (*Citrus maxima*; [53]), tea (*Camellia sinensis*; [54]), longan (*Dimocarpus longan*; [55]), and litchi (*Litchi chinensis*; [56]).

Although significant, the number of reports concerning genomic resources in *Coffea*, even for a specie of commercial importance, such as *C. arabica*, is still low. Already, genotyping profiles of SNPs were identified and tested in *C. arabica* by Moncada et al. [57], Sousa et al. [29], Sant'Ana et al. [10] and Merot-L'anthoene et al. [4]. High-throughput genotyping assays are still needed in order to rapidly characterize the coffee genetic diversity and to evaluate the introgression of different cultivars in a cost-effective way. Measures must be taken to construct high-density genetic maps in *Coffea* [57, 58]. However, the use of SNP markers to generate denser maps is still low.

5. DArTseq™: an effective tool for genome diversity in *C. arabica*

The DArTseq™ technology, developed by DArT company (https://www.diversityarrays.com), is one of those methods that have received increasing interest worldwide since it can generate thousands of high-quality SNPs in a timely and cost-effective manner [59, 60]. The DArTseq™ method, a variation of GBS, implements complexity reduction methods that effectively targets low-copy sequences of the genome [61]. Besides, this process is optimized for each organism and type of study, by using combinations of restriction enzymes (REs) and selecting the most effective in reducing genome complexity [59].

The DArTseq™ technology has been utilized in diploid but more often in polyploid plant species, such as rice (*Oryza sativa*; [62]), barley (*Hordeum vulgare*; [63]) and maize (*Zea mays*; [64]), because SNP detection is facilitated by high fidelity REs, rather than relying on the annealing of primers to genomic targets in the presence of homologous annealing sequences [65].

In coffee, we have reported a genetic diversity study in 87 accessions of *Coffea* spp. These accessions were selected from the National Coffee Germplasm Bank located at 19°10′ 27" N and 96° 57′ 50" W and 1345 masl, in Huatusco, Veracruz, Mexico. Accessions were previously characterized by DArTseq™ method and SNP markers in Spinoso-Castillo et al. [66].

As a result, 16,995 SNP markers, derived from 34,000 unique sequences, were obtained by DArTseq™ from 87 accessions of different *Coffea* spp. After removing the markers with more than 10% of the missing data and MAF <5%, there were 1,739 polymorphic SNP markers for the analysis. After imputation and elimination of markers based on MAF, a heat map of the 87 accessions was obtained by using the genomic relations matrix *G* (**Figure 1**).

For the heat map, the genomic relations matrix ***G*** can be easily calculated using the following expression:

$$\mathbf{G} = \frac{\mathbf{ZZ'}}{\mathbf{p}}, \tag{1}$$

where ***Z*** is the matrix of markers of dimension n = 87 rows (individuals) and p = 1,739 columns (markers), which is obtained by centering and standardizing the columns of the matrix of markers. The model-based Bayesian cluster analysis in STRUCTURE visualized the population structure under examination (**Figure 2**). Five distinct sub-populations were found across cultivars.

The sub-populations were denoted as Pop1, Pop2, Pop3, Pop4 and Pop5. The first group clustered *C. liberica* (84) and *C. canephora* (85, 86 and 87) species, the second group clustered mostly *C. arabica* accessions of the central collection, which evidenced the greater dissimilarity of these accessions with *C. liberica* and *C. canephora* species; the third group clustered CIRAD's F1-hybrids (74–79). Fourth and fifth clusters compiled different *C. arabica* accessions among them. Also, it was shown by

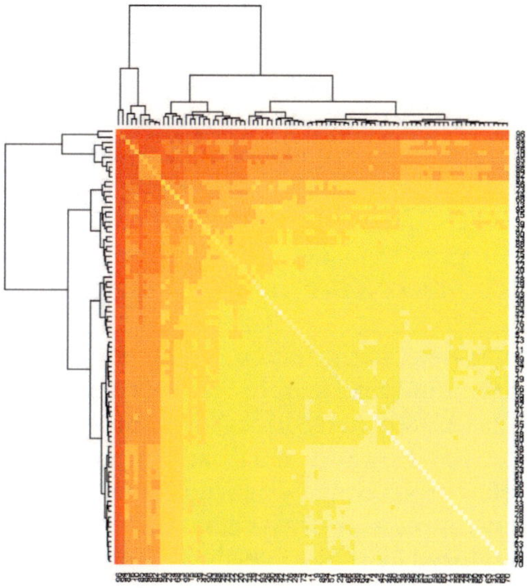

Figure 1.
Heat map for the 87 accessions of Coffea *spp. from the National Bank of Coffee Germplasm in Mexico using DArTseq Technology. Red small squares indicate an individual's genetic relatedness to itself, dark orange color represents high kinship relations while lighter colors (yellow) represent weaker relations.*

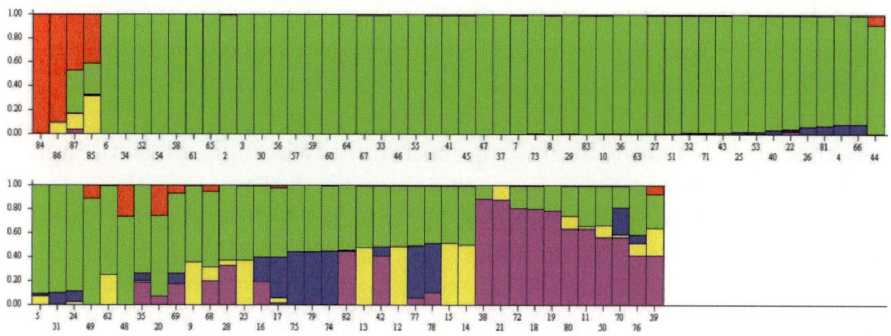

Figure 2.
Bar graphic of the STRUCTURE software used to study the diversity of the 87 coffee accessions using SNP marker data. The 87 genotypes are represented below the graphic, and were divided into five (K = 5) groups.

Steiger et al. [67], using AFLP markers, that *C. canephora* and *C. arabica* were more genetically similar, revealing inter-species diversity even though *C. arabica* resulted from a recent hybridization between *C. canephora* and *C. eugenioides* [13].

The results obtained from this *Coffea* spp. central collection are similar to those reported in the study of Sant'Ana et al. [10] who found in the population structure analyses the presence of two to three groups (K = 2 and K = 3), corresponding to the east and west sides of the Great Rift Valley and an additional group formed by wild *C. arabica* accessions collected in the western forests. Sousa et al. [29] analyzed the population structure of coffee genotypes of interest for breeding studies, they used 11,187 SNP markers from which two groups (K = 2) were obtained.

6. Advantages and disadvantages of NGS techniques in *C. arabica* genomics

High quality reference genome assemblies accelerate plant breeding by selecting desirable genes with improved agronomic traits, including high yield, tolerance to various abiotic and biotic stresses, and resistance to pathogens [68]. However, draft genomes are suffering from unknown sequences and ambiguous assembly due to homologous sequences, while high-quality genomes are required for comparative genomics and functional annotation to crop improvement [68, 69].

These NGSTs are classified as second and third generation. The success of these NGSTs is mainly due to advancement in nanofluidics and automated single molecule imaging [69]. SGseqTs refer to those methods which require a PCR step for signal intensification prior to sequencing and third generation sequencing techniques (TGSeqTs) are those which can perform single molecule sequencing (SMS) [70].

As an advantage, in SGseqT the variation is different in their sequencing chemistry, cost, accuracy, speed and read length; SGseqTs produce thousands to billions of nucleotide long reads (25–800 nucleotides) as compared to first generation sequencing method [69, 70]. However, as a disadvantage, the accuracy of SGseqTs differs due to dependence on several multiplication steps during library preparation, each manipulation causes various artifacts in DNA measurements; additionally, the small reads produced by these procedures are not suitable for de novo genome assembly [69, 70].

Therefore, novel technologies are being designed in such a way that involve a minimum or no manipulation of the natural DNA molecule; TGSTs are able to analyze natural DNA/RNA molecules without any manipulation and without amplification [70] TGSTs have average read length longer to 10 kb, the availability of long reads constitutes a great advantage.

The first SMS technology, was developed by Quake and commercialized in 2009 by Helicos BioSciences; it worked similar to Illumina sequencers, but without any bridge amplification [70, 71]. However, it was slow, expensive and produced relatively short reads, around 35 bp long; therefore, two single-molecule approaches were technologically advanced to overcame these disadvantages [72].

The first approach, Single Molecule Real-Time (SMRT) sequencing was developed by Craighead, Korlach, Turner and Webb and was further refined and commercialized by Pacific Biosciences (PacBio) since 2011 [73]. The second approach, Nanopore sequencing, was first hypothesized in the 1990s and further developed and commercialized by Oxford Nanopore Technologies (ONT) since 2005; the advantages of SMRT sequencing over NGS have come at the price of higher per base sequencing costs [70].

Finally, DArTseq™ technique is based on genomic complexity reduction. This technique benefitted from the development in NGSTs and now DArTseq™ markers are replaced by NGS-DArT markers. Sansaloni et al. [60] found that the combined use of DArTseq™ with NGS make available more quantity of markers than conventional DArT method. DArTseq™ markers in combination with other molecular techniques have been used to create deeper genetic maps in *C. arabica* to perform association studies [4, 74, 75].

7. A future in genomic resources of *C. arabica*

Arabica's cultivars and landraces are generally propagated by seed. The mating system is primarily based on self-fertilization. Thereby, autogamy leads to high levels of inbreeding. Besides, an effective clonal propagation system is being adopted but limited for F1 Arabica hybrids. It is evident that molecular analyses of genetic diversity are needed to support this scenario [74, 75].

The development of a new coffee variety takes about 25 years. An efficient selection can be addressed when sequencing approaches are adopted in the variety development process [66, 76]. In the 1990s, Marker-Assisted Selection (MAS) was proposed, which enabled selecting individuals with specific alleles. However, MAS has shown to be inefficient in polygenic and/or low heritability traits [77]. Due to its potential and importance, genome-wide selection (GS) was developed by Meuwissen et al. [78].

With the development of NGSTs, GS has become a reality for several economically important species. However, the procedure requires precaution for polyploid species, which have subgenomes with duplicate regions or with high similarity, such as *C. arabica* [77]. Despite the economic importance of *C. arabica*, GS works in Arabica coffee are scarce. Coffee trees have been selected based on biometric analyses using phenotypic data of yield and resistance to biotic and abiotic stresses. However, due to the complexity and number of genes that control most of the agronomic traits of this *Coffea* spp., GS studies are promising for they allow estimating the effects of all loci that explain the genetic variation and the genomic estimated breeding value (GEBV) [74, 75, 77].

Genome sequencing initiatives of Arabica accessions have been launched by several research groups (https://coffeegenome.ucdavis.edu/, among others) but an open-access genome assembly, with a reliable sorting of homologous sequences, is not yet available [77, 79]. Decoding the allotetraploid genome of *C. arabica* is therefore required to have accurate GS studies in this species.

8. Conclusions

DArTseq™ technology identifies thousands of high quality SNP polymorphic markers in a timely and cost-effective manner. Our study confirmed that the genotyping method by DArTseq™ can be successfully used in studies of genetic diversity specially in coffee. In addition, trait-associated-SNPs identified by GWAS may be helpful to develop strategies aiming to improve the biochemical quality of coffee or another important trait. These SNPs markers may be useful for marker-assisted selection (MAS) in Arabica coffee breeding programs and genomic selection.

Acknowledgements

This work was supported by the FONDO SECTORIAL SAGARPA-CONACYT [2016-2101-277838].

Conflict of interest

The authors have no conflicting interests, and all authors have approved the manuscript and agree with its submission to IntechOpen.

Author details

José Luis Spinoso-Castillo[1], Tarsicio Corona-Torres[1*], Esteban Escamilla-Prado[3], Victorino Morales-Ramos[2], Víctor Heber Aguilar-Rincón[1] and Gabino García-de los Santos[1]

1 Postgraduate College, Campus Montecillo, Texcoco, State of Mexico, Mexico

2 Postgraduate College, Campus Córdoba, Amatlán de los Reyes, Veracruz, Mexico

3 Autonomous University of Chapingo, Eastern Regional University Center, Veracruz, Mexico

*Address all correspondence to: tcoronat@gmail.com

References

[1] ICO. International coffee organization. 2020. Available from: https://www.ico.org/prices/po-production.pdf. [Accessed: 2020-11-05]

[2] Oxfam. The Coffee Market: A Background Study; Oxfam: London, UK; 2001

[3] Tran HT, Lee LS, Furtado A, Smyth H, Henry RJ. Advances in genomics for the improvement of quality in coffee. J Sci Food Agric. 2016;96:3300-3312

[4] Merot-L'anthoene V, Tournebize R, Darracq O, Rattina V, Lepelley M, Bellanger L, et al. Development and evaluation of a genome-wide Coffee 8.5 K SNP array and its application for high density genetic mapping and for investigating the origin of *Coffea arabica* L. Plant Biotechnol J. 2019;1-13

[5] Davey JW, Hohenlohe PA, Etter PD, Boone JQ, Catchen JM, Blaxter ML. Genome-wide genetic marker discovery and genotyping using next-generation sequencing. Nat. Rev. Genet. 2011;12:499-510

[6] Gupta PK, Rustgi S, Mir RR. Array-based high-throughput DNA markers for crop improvement. Heredity. 2008;101:5-18

[7] You Q, Yang X, Peng Z, Xu L, Wang J. Development and applications of a high throughput genotyping tool for polyploidy crops: single nucleotide polymorphism (SNP) array. Front. Plant Sci. 2018;9:104

[8] Yu H, Xie W, Li J, Zhou F, Zhang, Q. A whole-genome SNP array (RICE6K) for genomic breeding in rice. Plant Biotechnol. J. 2014;12:28-37

[9] Denoeud F, Carretero-Paulet L, Dereeper A, Droc G, Guyot R, Pietrella M, Zheng C, et al. The coffee genome provides insight into the convergent evolution of Caffeine biosynthesis. Science. 2014;345:1181-1184

[10] Sant'Ana GC, Pereira LFP, Pot D, Ivamoto ST, Domingues DS, Ferreira RV, Pagiatto NF, et al. Genome-wide association study reveals candidate genes influencing lipids and diterpenes contents in *Coffea arabica* L. Sci. Rep. 2018;8:465.

[11] Tran HTM, Ramaraj T, Furtado A, Lee LS, Henry RJ. Use of a draft genome of coffee (*Coffea arabica*) to identify SNPs associated with caffeine content. Plant Biotechnol. J. 2018;16:1756-1766

[12] Davis AP, Tosh J, Ruch N, Fay M. Growing coffee: *Psilanthus* (Rubiaceae) subsumed on the basis of molecular and morphological data; implications for the size, morphology, distribution and evolutionary history of *Coffea*. Bot J Linn Soc. 2011;167:357-377

[13] Lashermes P, Combes MC, Robert J, et al. Molecular characterization and origin of the *Coffea arabica* L. genome. Mol Gene Genet. 1999;261:259-266

[14] Pruvot-Woehl S, Krishnan S, Solano W, Schilling T, Toniutti L, Bertrand B, Montagnon C. Authentication of *Coffea arabica* varieties through DNA fingerprinting and its significance for the coffee sector. Journal of AOAC International. 2020;103(2):325-334

[15] Baruah A, Naik V, Hendre PS, Rajkumar P, Aggarwal RK. Isolation and characterization of nine microsatellite markers from *Coffea arabica* L. showing wide cross-species amplifications. Mol Ecol Notes. 2003;3:647-650

[16] Prakash NS, Combes MC, Somanna N, Lashermes P. AFLP analysis

of introgression in coffee cultivars (*Coffea arabica* L.) derived from a natural interspecific hybrid. Euphytica. 2002;124:265-271

[17] Benti T, Gebre E, Tesfaye K, Berecha G, Lashermes P, Kyallo M, Yao NK. Genetic diversity among commercial arabica coffee (*Coffea arabica* L.) varieties in Ethiopia using simple sequence repeat markers. Journal of Crop Improvement. 2020

[18] Bertrand B, Etienne H, Cilas C, Charrier A, Baradat P. *Coffea arabica* hybrid performance for yield, fertility and bean weight. Euphytica. 2005;141(3):255-262

[19] Pot D, Scholz MBS, Lannes SD, Del Grossi L, Pereira LFP, Vieira LG, Sera T. Phenotypic analysis of *Coffea arabica* accessions from Ethiopia: contribution to the understanding of *Coffea arabica* diversity. In: 22nd International Conference on Coffee Science, Campinas. Anais. Campinas; 2008. p 165

[20] Silvestrini M, et al. Genetic diversity and structure of Ethiopian, Yemen and Brazilian *Coffea arabica* L. accessions using microsatellites markers. Genet. Resour. Crop Ev. 2007;54:1367-1379

[21] López-Gartner G, Cortina H, Couch MC, Susan R, Moncada MDP. Analysis of genetic structure in a sample of coffee (*Coffea arabica* L.) using fluorescent SSR markers. Tree Genet Genomes. 2009;5:435-446

[22] Teressa A, Crouzillat D, Petiard V, Brouhan P. Genetic diversity of Arabica coffee (*Coffea arabica* L.) collections. EJAST. 2010;1:63-79

[23] Aerts R, Berecha G, Gijbels P, Hundera K, Glabeke S, Vandepitte K, Muys B, Roldán RI, Honnay O. Genetic Variation and Risks of Introgression in the Wild *Coffea* Arabica Gene Pool in South-Western Ethiopian Montane Rainforests. Evolutionary Applications. 2013;6(2):243-252

[24] Scholz MBS, Kitzberger CSG, Pagiatto NF, Pereira LFP, Davrieux F, Pot D, et al. Chemical composition in wild Ethiopian Arabica coffee accessions. Euphytica. 2016;209(2):429-438

[25] da Silva, BSR, Sant'Ana GC, Chaves CL, et al. Population structure and genetic relationships between Ethiopian and Brazilian Coffea arabica genotypes revealed by SSR markers. Genetica. 2019;147:205-216

[26] Jingade P, Huded AK, Kosaraju B, Mishra MK. Diversity Genotyping of Indian Coffee (*Coffea Arabica* L.) Germplasm Accessions by Using SRAP Markers. Journal of Crop Improvement. 2019;33:327-345

[27] Kesawat MS, Das Kumar B. Molecular Markers: It's Application in Crop Improvement. Journal of Crop Science and Biotechnology. 2009;12:169181

[28] Garrido-Cardenas JA, Mesa-Valle C, Manzano-Agugliaro F. Trends in Plant Research Using Molecular Markers. Planta. 2018;247:543

[29] Sousa TV, Caixeta ET, Alkimim ER, de Oliveira ACB, Pereira AA, Zambolim L, Sakiyama NS. Molecular Markers Useful to Discriminate Coffea Arabica Cultivars with High Genetic Similarity. Euphytica. 2017;213

[30] Lashermes P, Trouslot P, Anthony F, Combes MC, Charrie A. Genetic Diversity for RAPD Markers between Cultivated and Wild Accessions of Coffea Arabica. Euphytica. 1996;87:59-64

[31] Anthony F, Bertrand B, Quiros O, Lashermes P, Berthaud J, Charrier A. Genetic Diversity of Wild Coffee (*Coffea arabica* L.) Using Molecular Markers. Euphytica. 2001;118:53-65

[32] Aga E, Bekele E, Bryngelsson T. Inter-Simple Sequences Repeat Variation in Forest Coffee Trees (*Coffea arabica* L.) Populations from Ethiopia. Genetica. 2005;124:213-221

[33] Tornincasa P, Dreos R, Nardi D, Barbara E, Asquini J, Devasia M, Mishra K, et al. Genetic Diversity of Commercial Coffee (*C. Arabica* L.) From America, India and Africa Assessed by Simple Sequence Repeats (SSR). Proceedings of 21st International Association for Coffee Science (ASIC), Montpellier, France. 2007; pp. 778-785

[34] Geleta M, Herrera I, Monzón A, Bryngelsson T. Genetic Diversity of Arabica Coffee (*Coffea arabica* L.) In Nicaragua as Estimated by Simple Sequence Repeat Markers. The Scientific World Journal. 2012:1-11

[35] Dessalegn Y, Herselman L, Labuschagne M. Comparison of SSR and AFLP Analysis for Genetic Diversity Assessment of Ethiopian Arabica Coffee Genotypes. South African Journal of Plant and Soil. 2009;26(2):119-125

[36] Maluf MP, Silvestrini M, Ruggiero LM, de Guerreiro-Filho CO, Colombo CA. Genetic Diversity of Cultivated *Coffea* Arabica Inbred Lines Assessed by RAPD, AFLP and SSR Marker Systems. Scientia Agricola. 2005;62:366-373

[37] World Coffee Research (WCR), Assessment of genetic diversity in *Coffea arabica*. World Coffee Research 2014 Annual Report, Texas, USA.;2014

[38] Labouisse JP, Bellachew B, Kotecha S, Bertrand B. Current status of coffee (*Coffea arabica*) genetic resources in Ethiopia: implications for conservation. Genet Res Crop Evol. 2008;55:1079-1093

[39] Silva MC, Várzea V, Guerra-Guimaraés L, Gil Azinheira H, Fernandez D, Petitot AS, Bertrand B, Lashermes Ph, Nicole M. Coffee resistance to the main diseases: leaf rust and coffee berry disease. Braz J Plant Physiol. 2006;18:119-147

[40] Van der Vossen HAM. The cup quality of disease-resistant cultivars of arabica coffee (*Coffea arabica*). Expl Agric. 2009;45:323-332

[41] Anzueto F, Bertrand B, Sarah JL, Eskes AB, Decazy B. Resistance to Meloidogyne incognita in Ethiopian *Coffea arabica* origins: detection and study of resistance transmission. Euphytica. 2001;118:1-8

[42] Bertrand B, Vaast Ph, Alpizar E, Etienne H, Davrieux F, Charmetant P. Comparison of bean biochemical composition and beverage quality of arabica hybrids involving Sudanese-Ethiopian origins with traditional varieties at various elevations in Central America. Tree Physiol. 2006;26:1239-1248

[43] Silvarolla MB, Mazzafera P, Fazuoli LC. A naturally decaffeinated arabica coffee. Nature. 2004;429:826

[44] Georget F, Alpizar E, Courtel P, Hidalgo JM, Dechamp E, Poncon C, Etienne H, Bertrand *B. Arabica* F1 hybrid seed production based on genetic male sterility. Proceedings if the 25th ASIC international conference on coffee science, Armenia, Colombia. 2015

[45] Unamba CIN, Nag A, Sharma RK. Next Generation Sequencing Technologies: The Doorway to the Unexplored Genomics of Non-Model Plants. Front. Plant Sci. 2015;6:1074

[46] Mondini L, Noorani A, Pagnotta MA. Assessing Plant Genetic Diversity by Molecular Tools. Diversity. 2009;1(1):19-35.

[47] Baird NA, Etter PD, Atwood TS, Currey MC, Shiver AL, Lewis ZA, Selker EU, Cresko WA, Johnson EA.

Rapid SNP Discovery and Genetic Mapping Using Sequenced RAD Markers. PLoS ONE. 2008;3(10):e3376

[48] Elshire RJ, Glaubitz JC, Sun Q, Poland JA, Kawamoto K, Buckler ES, MitchellSE. ARobust, SimpleGenotyping-by-Sequencing (GBS) Approach for High Diversity Species. PLoS ONE. 2011;6(5):e19379

[49] Cruz VMV, Kilian A, Dierig DA. Development of DArT Marker Platforms and Genetic Diversity Assessment of the U.S. Collection of the New Oilseed Crop Lesquerella and Related Species. PLoS ONE. 2013;8(5):e64062

[50] Kumar S, Banks TW, Cloutier S. SNP Discovery through Next-Generation Sequencing and Its Applications. Int J Plant Genomics. 2012;2012:15

[51] Zhou L, Vega FE, Tan H, Ramírez Lluch AE, Meinhardt LW, Fang W, Mischke S, Irish B, Zhang D. Developing Single Nucleotide Polymorphism (SNP) Markers for the Identification of Coffee Germplasm. Tropical Plant Biol. 2016;9:82-95

[52] Ji K, Zhang D, Motilal L, et al. Genetic diversity and parentage in farmer varieties of cacao (*Theobroma cacao* L.) from Honduras and Nicaragua as revealed by single nucleotide polymorphism (SNP) markers. Genet Resour Crop Evol. 2013;60:441-453

[53] Wu GA, Prochnik S, Jenkins J, et al. Sequencing of diverse mandarin, pummelo and orange genomes reveals complex history of admixture during citrus domestication. Nat Biotechnol. 2014;32:656-662

[54] Fang W, Meinhardt LW, Tan H, et al. Varietal identification of tea (*Camellia sinensis*) using nanofluidic array of single nucleotide polymorphism (SNP) markers. Hortic Res. 2014;1:14035

[55] Wang B, Tan H, Fang W, et al. Developing single nucleotide polymorphism (SNP) markers from transcriptome sequences for identification of longan (*Dimocarpus longan*) germplasm. Hortic Res. 2015;2:14065

[56] Liu W, Xiao ZD, Bao XL, et al. Identifying litchi (*Litchi chinensis* Sonn.) cultivars and their genetic relationships using single nucleotide polymorphism (SNP) markers. PLoS ONE. 2015;10(8):e0135390

[57] Moncada MD, Tovar E, Montoya JC, Gonzalez A, Spindel J, McCouch S. A genetic linkage map of coffee (*Coffea arabica* L.) and QTL for yield, plant height, and bean size. Tree Genet. Genomes. 2016;12:5

[58] Pearl HM, Nagai C, Moore PH, Steiger DL, Osgood RV, Ming R. Construction of a genetic map for Arabica coffee. Theor. Appl. Genet. 2004;108: 829-835

[59] Nguyen NH, Premachandra HKA, Kilian A, Knibb W. Genomic prediction using DArT-Seq technology for yellowtail kingfish *Seriola lalandi*. BMC Genomics. 2018;19(107):1-9

[60] Sansaloni C, Petroli C, Jaccoud D, Carling J, Detering F, Grattapaglia D, Kilian A. Diversity arrays technology (DArT) and next-generation sequencing combined: genome-wide, high throughput, highly informative genotyping for molecular breeding of Eucalyptus. BMC Proc. 2011;5(S7):54

[61] Melville J, Haines ML, Boysen K, Hodkinson L, Kilian A, Smith Date KL, Potvin DA, Parris KM. Identifying hybridization and admixture using SNPs: application of the DArTseq platform in phylogeographic research on vertebrates. R. Soc. Open Sci. 2017;4(7):161061

[62] Courtois B, Audebert A, Dardou A, Roques S, Ghneim-Herrera T, Droc G, Frouin J, Rouan L, Gozé E, Kilian A, Ahmadi N, Dingkuhn M. Genome-wide association mapping of root traits in a japonica rice panel. PLoS One. 2013;8(11):e78037

[63] Dracatos PM, Khatkar MS, Singh D, Park RF. Genetic mapping of a new race specific resistance allele effective to *Puccinia hordei* at the Rph9/Rph12 locus on chromosome 5HL in barley. BMC Plant Biol. 2014:14(1598):1-12

[64] Dos Santos JPR, Pires LPM, De Castro Vasconcellos RC, Pereira GS, Von Pinho RG, Balestre M. Genomic selection to resistance to *Stenocarpella maydis* in maize lines using DArTseq markers. BMC Genet. 2016;17(86):1-10

[65] Akbari M, Wenzl P, Caig V, Carling J, Xia L, Yang S, Uszynski G, Mohler V, Lehmensiek A, Kuchel H, Hayden MJ, Howes N, Sharp P, Vaughan P, Rathmell B, Huttner E, Kilian A. Diversity Arrays Technology (DArT) for highthroughput profiling of the hexaploid wheat genome. Theor. Appl. Genet. 2006;113(8):1409-1420

[66] Spinoso-Castillo JL, Escamilla-Prado E, Aguilar-Rincón VH, Morales Ramos V, García de los Santos G, Pérez-Rodríguez P, Corona-Torres T. Genetic diversity of coffee (*Coffea* spp.) in Mexico evaluated by using DArTseq and SNP markers. Genet Resour Crop Evol. 2020. DOI: doi.org/10.1007/s10722-020-00940-5

[67] Steiger DL, Nagai C, Moore PH, Morden CW, Osgood RV, Ming R. AFLP analysis of genetic diversity within and among Coffea arabica cultivars. Theor Appl Genet. 2002;105(2-3):209-215

[68] Li C, Lin F, An D, Wang W, Huang R. Genome Sequencing and Assembly by Long Reads in Plants. Genes. 2018;9(1):6

[69] Yadav P, Vaidya E, Rani R, et al. Recent Perspective of Next Generation Sequencing: Applications in Molecular Plant Biology and Crop Improvement. Proc. Natl. Acad. Sci., India, Sect. B Biol. Sci. 2018;88:435-449

[70] Giani AM, Gallo GR, Gianfranceschi L, Formenti G. Long walk to genomics: History and current approaches to genome sequencing and assembly. Computational and Structural Biotechnology Journal. 2020;18:9-19

[71] Harris TD, Buzby PR, Babcock H, Beer E, Bowers J, et al. Single-molecule DNA sequencing of a viral genome. Science. 2008;320:106-109

[72] Salazar AN, Nobrega FL, Anyansi C, Aparicio-Maldonado C, Costa AR, et al. An educational guide for nanopore sequencing in the classroom. PLOS Computational Biology. 2020;16(1):e1007314

[73] Deamer D, Akeson M, Branton D. Three decades of nanopore sequencing. Nat Biotechnol. 2016;34:518-524

[74] Scalabrin S, Toniutti L, Di Gaspero G, et al. A single polyploidization event at the origin of the tetraploid genome of Coffea arabica is responsible for the extremely low genetic variation in wild and cultivated germplasm. Sci Rep. 2020;10:4642

[75] Alkimim ER, Caixeta ET, Sousa TV, et al. Selective efficiency of genome-wide selection in *Coffea canephora* breeding. Tree Genetics & Genomes. 2020;16:41

[76] Gimase JM, Thagana WM, Omondi CO, et al. Genome-Wide Association Study identify the genetic loci conferring resistance to Coffee Berry Disease (*Colletotrichum kahawae*) in *Coffea arabica* var. Rume Sudan. Euphytica. 2020;216:86

[77] Sousa TV, Caixeta ET, Alkimim ER, Oliveira ACB, Pereira AA, Sakiyama NS, Zambolim L, Resende MDV. Early Selection Enabled by the Implementation of Genomic Selection in *Coffea arabica* Breeding. Front. Plant Sci. 2019;9:1934

[78] Meuwissen TH, Hayes BJ, Goddard ME. Prediction of total genetic value using genome-wide dense marker maps. Genetics. 2001;157:1819-1829

[79] Huang L, Wang X, Dong Y, et al. Resequencing 93 accessions of coffee unveils independent and parallel selection during *Coffea* species divergence. Plant Mol Biol. 2020; 103:51-61 (2020)

Chapter 4

Vegetable Landraces: The "Gene Banks" for Traditional Farmers and Future Breeding Programs

Radu-Liviu Sumalan, Sorin-Ion Ciulca, Renata-Maria Sumalan and Sorina Popescu

Abstract

Crop diversity of vegetable species is threatened by the current homogenization of agricultural production systems due to specialization of plant breeders and increasing globalization in the seed sector. With the onset of modern agriculture, most traditional vegetable cultivars were replaced by highly productive and often genetically uniform commercial breeds and hybrids. This led to the loss of landraces, especially in countries with a super-intensive agriculture. The agricultural biodiversity erosion represents a huge risk for food safety and security. Vegetable landraces are associated with the cultural heritage of their place of origin being adapted to local agro-ecological areas and are more resilient to environmental stress than commercial cultivars. The chapter aim to highlight the importance of keeping and using vegetable landraces as valuable sources of genes for traditional farmers, but also for future breeding processes. We analyze the historical role of landraces, genetic diversity, high physiological adaptability to specific local conditions in association with traditional farming systems, as well as the breeding perspectives and evaluation of genetic diversity based on molecular markers.

Keywords: old local populations, biodiversity, food security, stress tolerance, quality, tomatoes, onion, breeding, molecular markers

1. Introduction

In 1996 World Food Summit stated that "food security is ensured when the entire population has at all times, physical and economic access to sufficient food resources, safe and of high nutritional value, to meet food needs and preferences providing an active and healthy life".

Food security has long been associated with the abundance of cereal products, roots and tubers, vegetables and fruits from the main agricultural crops, which could provide affordable sources of nutritional energy. But this image has changed as the concept of nutritional security has become the essential element of food safety, and nutritional diversity has become the basic component to ensuring the human population health. Healthy diets, qualitatively superior, determine the consumption of a variety of foods in optimal quantities [1].

The vegetables are an affordable and relatively inexpensive source of fiber, vitamins and minerals. In general, they have the highest nutritional value when are eaten fresh. Unfortunately, a large part of primary (unprocessed) horticultural products have a relatively short life before they begin to degrade. The extent to which the nutritional value of vegetables deteriorates during harvesting, processing and storage depends both on the type of product (species, organ, ripening level) and on the used technologies [2].

Also, the vegetables are recognized as essential for food and nutritional security of humanity. Producing them offers multiple economic opportunities, reducing poverty and unemployment in rural areas especially, and is also an essential component of plant biodiversity maintaining strategies. The systematic production of vegetables for local markets not only provides income for small farmers, but also contributes to strengthening their resilience to external risks. Diversification of vegetable crops, short cycles of growth and development, the use of local, environmentally friendly inputs and the efficient use of fertilizers, pesticides and irrigation can reduce farmers' vulnerability to climate changes. For economic resilience, farmers may choose either to integrate vegetables into existing large crop systems or to focus exclusively on specialized vegetable production.

Vegetable production has increased more than twice in the last 25 years and the economic value generated by their cultivation has exceeded the commercial value of cereals [3].

2. A brief analysis of the production, consumption and trade of vegetables

The global market of vegetables is still predominantly local because only about 5% of vegetables grown worldwide are marketed internationally. However, this percentage continues to increase quite a lot from one year to other. Easy access to a booming global market is essential for export vegetable producing countries, such as Mexico, Spain or The Netherlands. For example, over the past two decades, Mexico has strengthened its leading position of vegetable exports in the North American market and EU domestic trade has continued to grow, particularly on the basis of products from the two European countries mentioned above.

Declared revenues on the global vegetable market were around 1.249.8 billion US$ in 2018, and their market share increased at an average annual rate of +4.1% between 2007 and 2018. Overall vegetable consumption reached the maximum value in 2018 and is expected to increase continuously between 2020 and 2025 [4].

The quantities of vegetables exported worldwide in 2018 (**Figure 1**), reached a level of about 47 million tonnes, the total volume of exports increasing at an average annual rate of 1.7% between 2007 and 2018. In terms of value, vegetable exports amounted to 42.3 billions US$. The world's most important exporters were; The Netherlands (6.1 million tonnes), Mexico (5.8 million tonnes), Spain (5.1 million tonnes), China (4.3 million tonnes), France (3.5 million tonnes), Germany (2.7 million tonnes) and the United States (2.4 million tonnes) accounting for about 64% of total vegetable exports in 2018.

Vegetables import levels have also had an upward trend over the past decade. Statistical data show that in 2018 the greatest importers was the US with 7.4 million tonnes, followed by Germany (3.8 million tonnes), the Netherlands (3.1 million tonnes) Russia and the United Kingdom (2.2 million tonnes). An interesting trend has been the emergence in recent years of new countries with high requirements on imports of vegetables such as India, China or the United Arab Emirates. Russia has also seen an increase in trade, despite the imposition of economic sanctions on imports since 2014. The main countries providing vegetables to Russia are Belarus, Morocco, China, Armenia and Azerbaijan [4].

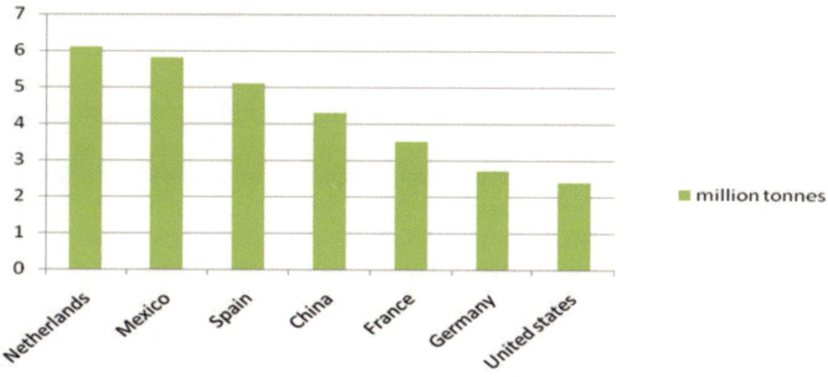

Figure 1.
The main global exporters of vegetables, and the volume of their exports for 2018. Processed by; World - Vegetable - Market Analysis, Forecast, Size, Trends and Insights (researchandmarkets.com).

2.1 Fresh or conserved vegetables?

It is estimated that 70% of vegetables grown around the world are sold fresh and whole as primary (unprocessed) horticultural products. Processing of vegetables by preserving, freezing and drying is the main purpose of storage technologies, the possibility of long-distance transport, long lasting storage and the reduction of damage losses. However, the global consumption of preserved vegetables has decreased over the past decade, which attests to consumers' preferences for fresh vegetables against the background of reduced time from harvest to market (concept from field to fork). Has increased however the demand for frozen vegetables over the past decade by an average of about 1% annually [5].

2.2 The vegetables and chain food waste

Due to the relatively high level of perishability, primary horticultural products are exposed to loss in a significant percentage. With 1 in 8 people on Earth starving (about 759 million people), the loss of vegetables and fruits is a component with major social effects. According to the FAO, about 14% of globally produced foods are lost between harvest and retail trade, with significant quantities also being wasted at the retail and consumption level. The value is higher in the case of fruit and vegetables where losses range from 20 to 40 % [6]. Analysis of the data presented shows that significant losses of fresh vegetables and fruits occur in the production process (Europe, North America, Oceania and Latin America), in processing (Africa, South Asia and South-East) and to the final consumer (Europe, North America and Industrialized Asia).

Recent studies haves shown that in European Union around 7.2 million tonnes of fruits and vegetables are discarded annually, which is the equivalent of 14.2 kg/person/year. Of this quantity, avoidable waste (edible parts) accounted for almost half, and the inevitable waste (shells, seeds, stalk, etc.) was the difference [7–9]. These wastes, if are not properly treated, pose major environmental hazards because their decomposition eliminates an important quantity of various greenhouse gases [10].

Therefore, the reducing of food waste is the main way to close the gap between food supply and demand [11]. On the basis of this argument, one of the specific targets of the UN Sustainable Development Goals is to halve food losses along the production and supply chain by 2030 (Objective 12.3) [12]. The European Commission is committed to respect the objective 12.3. and considers food waste as a priority area in its Circular Economy Action Plan [13]. Moreover, to underline the importance of reducing food loss, the UN declared 29 September as "International Day of Food Lost and Waste".

2.3 Ecological and organic vegetables, increasingly sought after in rich societies

The global market share of organic foods is growing from year to year. The share of trade in organic and ecologic fruit and vegetables (out of the total trade in fresh fruit and vegetables) has increased by around 10% in some european countries with high standards of living such as; Switzerland, Sweden, Austria and Denmark. In the United States, this rate is around 9%, but there has been recorded intense growth rates in the last years. Although, income per capita appears to be a determining factor in the consumption of these products, this is not the only one. The consumer education level, supermarket policies on the category of organic vegetables, the price and availability of conventional or traditional products, cultural factors, etc. can be important vectors that influence the consumption of organic and ecologic vegetables products [5].

2.4 Seed vegetables market

Vegetable quality assurance is achieved by a succession of attributes related to biological material and cultivation technologies, harvesting, conditioning, processing, storage and marketing. Seed quality is the basic appropriation that characterizes the biological material. The demand of growers for quality seeds is increasing. The world market for vegetable seeds accounts for about 11% of the total plant seed market. The estimated value of the vegetable seed market in 2017 was 8.02 billion US$, reaching 12.6 billion US$ by 2021, with a cumulative annual rate of 8.1 [3].

3. Vegetable genetic resources and biodiversity preservation

In general, plant genetic resources are defined as that part of biodiversity used to generate productivity and quality in agriculture. In addition to commercial genotypes (varieties and hybrids), the genetic resources of a cultivated species include breeding lines, genetic forms obtained by various technologies by deliberate breeding (natural or induced mutant lines, substitution and addition lines, inter-specific hybrids, etc.), wild descendants, related species and local races, also referred to as 'farmers, local or primitive varieties' [14].

Plant Genetic Resources (PGR's) represents an important component of the conservation of plant biodiversity and the food security of the human population [15]. PGRs are actually the expression of natural variability in plants, variability that has sustained the human species for millennia. The multitude of plant species, with all existing genotypes, are especially important for ensuring food security, but also because they represent energy sources, medicines, animal feed, fiber, ecosystem services, etc. All these aspects are essential in the context of the global challenges currently facing life on Earth, in particular due to climate change and resource shortages. In the light of this, the efficient conservation and sustainable use of the PGR's is extremely important and has never been more necessary [16].

Thus, according to The Second Report on the State of the World's Plant Genetic Resources [17], approximately 7.4 million genotypes, sources of germplasm, belong to over 16,500 species of plants are currently stored in 1750 gene banks and collections around the world.

Vegetable genetic resources (VGR's) are the foundation on which vegetable cultivation techniques and food chains integrated with them have been developed, and the genetic diversity present in small farms and germplasm collections is essential in efforts to eradicate hunger and poverty. They are the main gene reservoir for the production of new vegetables cultivars and the main supplier of genetic diversity [18]. Therefore, plant genetic resources offer a huge diversity and variability, widely used in genetic studies and plant breeding programs, with undeniable benefits for global food production [19, 20].

Vegetable genetic resources (VGR's) are used both by traditional farmers to obtain safe and quality production and by researchers as the initial biological material for obtaining new cultivars. The genetic resources are also a reservoir of biodiversity that acts as an element of balancing sudden economic and environmental changes. Recent studies have shown that the main factor in the erosion of PGR's and biodiversity loss is the replacement in cultivation of local genotypes (old varieties, local populations) with modern cultivars [21].

Unfortunately, VGR's natural pools are strongly affected by the modern society activities – urbanization, habitat degradation through intensive exploitation, deforestation and arson, increased pressure from diseases and pests, to name just some of these activities.

Modern industrial agriculture based on improved hybrids and cultivars limited and marginalized the use of landraces, causing a serious loss of genetic variability. The high genetic erosion of vegetable landraces was highlighted by Hammer and Laghetti [22], who found that from 1950 till 1986 in Southern Italy only 27.2% of the landraces were still grown. Also, Dias [23] reported that, during the last 50-60 years the genetic diversity of vegetables has been severely eroded all over the world, so that the vegetable genetic resources are disappearing yearly on a global scale with a rate of 1.5-2.0%. This genetic erosion represents an alarm signal for the breeding activities in order to streamline the vegetable production under stressful environments [24].

As genetic erosion continues "in situ" and on farms due to the reasons already mentioned and climate change as well as by replacing old local varieties with improved, super-productive genotypes, it is necessary to intensify the efforts of collection, characterization and conservation with a major focus on the wild relatives of cultivated plants and on the breeds of vegetables poorly represented by the major and minor groups of this class. The conservation of the diversity of local and underutilized plant crops should also be given greater attention [25].

4. Landraces – definition and their importance in traditional farms and breeding programs

Widely used in the literature, the term "landrace" encompasses different concepts, variable in time and space, depending on trends prevalent in the use and conservation of genetic resources. After a period of beginning when the issue of preserving and maintaining biodiversity was prevalent, today the commercial message is clear and promotes the higher nutritional and sensorial qualities of local vegetable landraces [26]. Due to their complex nature and huge diversity landraces are extremely difficult to be characterized by an all-encompassing definition (**Figure 2**).

However, over time, different authors have tried to define landraces on the basis of the characterization of their main attributes. Kiessling [27] in 1912 defined landraces as a mixture of shapes (phenotypes) with a certain degree of external uniformity, specific composition and a high adaptability to the natural, technical and economic conditions of the region of origin [28].

An interesting definition has been proposed by Prospéri et al. [29] in 1994 which attest that a landrace represents a set of genotypes belonging to the same species,

Figure 2.
Vegetable landraces diversity.

that a grower in a given region, uses specific cultivation methods and carries out mass selection, more or less targeted, over several generations.

Zeven [28] said that a "landrace" is a variety with high tolerance to biotic and abiotic stressors, manifested by medium but stable productive yield, under low technological inputs conditions. Landraces have also been defined as dynamic populations of a cultivated plant of distinct historical origin and identity, with genetic variability and high adaptability to specific local conditions (soil, climate, biotic stressors) adapted to cultivation technologies specific to local farmers [30].

Vegetable landraces are considered local old varieties with distinctive characteristics resulting from archaic selection and adaptation over time to pedo-climatic conditions specific to a localized geographical region, which usually exhibit greater genetic diversity than the types subjected to the usual breeding techniques. According to the definition developed by Dwivedi et al. [24] landraces represent heterogeneous, local adaptations of some cultivated species and therefore provide genetic resources adapted to the current challenges posed by biotic and abiotic stress factors.

The analysis of these definitions attests to the existence of some common elements in the characterization of landraces in cultivated plants such as; local character, historical origin, adaptability to soil, climate and stress factors, genetic variability, harvest stability, reduced inputs, traditional farms. Landraces through their long selection process by farmers during the pre-intensive agricultural period provide a great opportunity to find appropriate combinations of genes and phenotypes tolerant to complex situations [31].

In conclusion, landraces are dynamic populations usually associated with traditional farming systems. As such, their evolution was based on both natural and farmers' selection in low-input cultivation systems [32]. During long period of cultivation, farmers greatly contributed to the diversification of vegetable crops by selecting populations with moderate yield and well adapted to the specific agro-climatic conditions of different regions. The diversity of landraces is usually lower than at their wild ancestors, but considerably higher than at modern cultivars produced by plant breeding [33]. The vegetable landraces are valuable genetic resources to identify genes for increasing yield and adaptation to abiotic stress under the current and future climate changes [34].

Compared with modern varieties, the vegetable landraces have a low presence on the market, due to their lower yields, disease sensitivity, and poorer postharvest shelf life [35]. In the last period, amid an increasing interest of the consumers for traditional and healthy products of the local growers, the landraces are reconsidered both as a source of food and as a source of useful genes [36, 37].

5. Breeding perspectives of vegetable landraces

The breeding of plants is as old as their cultivation. The first vegetable growers exploited the favourable variability of landraces of the main attributes such as productivity and high tolerance to environmental stress factors. Much later, probably after a few millennia, mankind developed new methods of breeding and multiplication, including hybridization techniques, and the peak was reached through the use of molecular tools, all of which led to the creation of modern vegetable genotypes with high yielding performance characters [38].

Therefore, an important source of genes that is increasingly used in breeding programs are landraces, old varieties adapted to the conditions of a specific pedoclimatic area [39]. Due to the stronger genetic proximity to modern varieties than their wild relatives, landraces show huge potential to improve modern genotypes by increasing stress tolerance and as sources of healthy and nutritive food [20, 40–43].

Featuring by a good stress tolerance and high adaptability to different conditions, despite the lack of pathogen tolerance genes, vegetable landraces are still a reservoir of genetic diversity, in particular for certain attributes of interest, such as; tolerance to abiotic stress and high fruit quality [44]. For these reasons, studies carried out on some heterogeneous tomato populations have shown that they have been, are and will continue to represent very important genetic resources used in breeding processes [28]. The genetic profiles of landraces are clearly different from those of modern genotypes [45]. It has been observed that numerous morpho-anatomical, physiological and biochemical traits record significant levels of phenotypic and genotypic diversity [46]. However, information on the variation within vegetable landraces is still limited.

The antioxidant content of the edible organs of wild vegetable species is significantly different from landraces. These compositions have been associated with the features of the organs, the geographical origin and altitude at which they are found. For example, in high-rise areas of northwestern Argentina, local tomato populations with the highest concentration of antioxidants have been identified [47].

Recovering and rendering these qualities in adapted landraces to the original communities will contribute to the sustainable maintenance of these varieties [48, 49]. For example, tomato landraces are characterized by excellent fruit quality, high content in metabolites [50], antioxidants [20, 47] and volatile organic compounds [51]. Landraces and old varieties have a typical flavour that consumers appreciate and demand, although the availability of their seeds is increasingly low [52].

The vegetable landraces are particularly important because they exhibit high heterogeneity (for improvement), are adapted to biotic and abiotic stress conditions, have excellent taste qualities, thus justifying a higher recovery price than commercial varieties [53].

One strategy to highlight the genetic treasure represented by the landraces is to identify the size of genetic variability for primary and secondary metabolites and to establish existing links between biochemical composition of edible products, genetic basis and consumer preferences [54]. Studies from last decade [20, 55, 56] showed that in Romania it still exists many vegetable landraces that need to be preserved and evaluated for further use in breeding programs.

5.1 Case studies: Romanian landraces of tomatoes (*Solanum lycopersicum*) and onion (*Allium cepa*) gene source for breeding programs

5.1.1 Tomatoes landraces

In order to obtain appropriate tomato yield under environmental stress conditions, the plants must show tolerance during the developmental stages from seed germination to flowering and fruit maturity [57]. Characterized by a good adaptability and stress tolerance amid a lack of diseases resistance genes, the landraces still represent an important reservoir of genetic diversity especially for traits associated with abiotic stress resistance and fruit quality [58].

The genetic structure of tomato landraces is quite different from those of modern tomato cultivars [32, 42, 59, 60], while the morphological variation of tomato landraces is higher compared to cultivars [61]. The heterogeneous structure of landraces was highlighted by Terzopoulos and Bebeli [32] who found a wide intra-population phenotypic diversity at 34 Greek tomato landraces for 33 morphological traits except for stem pubescence and foliage density, or plant growth type, respectively. Also, Manzano et al. [62] found a wide phenotypic diversity among 39 Spanish tomato landraces both in terms of morphological traits and postharvest quality of fruits, under organic greenhouse conditions. Analyzing the diversity between 75 landraces and 25 tomato varieties from Southern Italy, Corrado et al [58] revealed that the genetic structures of the landraces were mainly related with the fruit traits

The intra- and inter-populations variability may occur even in case of landraces from a small area, for morphological, agronomical and quality traits [63]. Based of farmer's activities, different selections of the same landrace can be made. These populations will evolve in different environmental conditions thus contributing to phenotypic diversity of tomato landraces [64, 65]. The diversity/variability between tomato landraces could be attributed both to genetic background and environmental conditions where these genotypes were evolved [66]. The analysis of landraces genetic variability will be useful for a better understanding of fruit shape and size and can help to identify valuable alleles for improving productivity, adaptation and quality [67–69].

Even, during the last decades the tomato landraces were replaced by new cultivars, in different regions of Romania these landraces are still cultivated for local consumption and market. They have especially distinctive morphological and quality traits of the fruits, considering that the fruits quality is highly appreciated by local consumers.

Within the project S-Stress 82 tomato landraces from two regions of Romania were evaluated using ISSR markers in order to establish the degree of similarity between them. The literature data show that this category of markers could be successfully used for evaluation of tomato variability.

The genetic variability was evaluated based on amplification with 8 ISSR markers namely: UBC 808 – $(AG)_8C$, UBC810 - $(GA)_8T$, UBC811- $(GA)_8C$, UBC840- $(GA)_8YT$, UBC841- $(GA)_8YC$, UBC843- $(CT)_8RA$, UBC884- $HBH(AG)_7$, UBC886- $VDV(CT)_7$, where Y = C or T, R = G or A, H = non G, B = non A, D = non C and V = non T.

In the case of primers such as UBC843, molecular fingerprints revealed major differences between the analyzed populations, while other markers, such as UBC 840, generated very similar fingerprints (**Figure 3**).

The results indicated the existence of a wide diversity, both between landraces from the two regions and from the same region, arguing the wide genetic basis of these landraces (**Figure 4**). Based on these results, combined with the analysis of

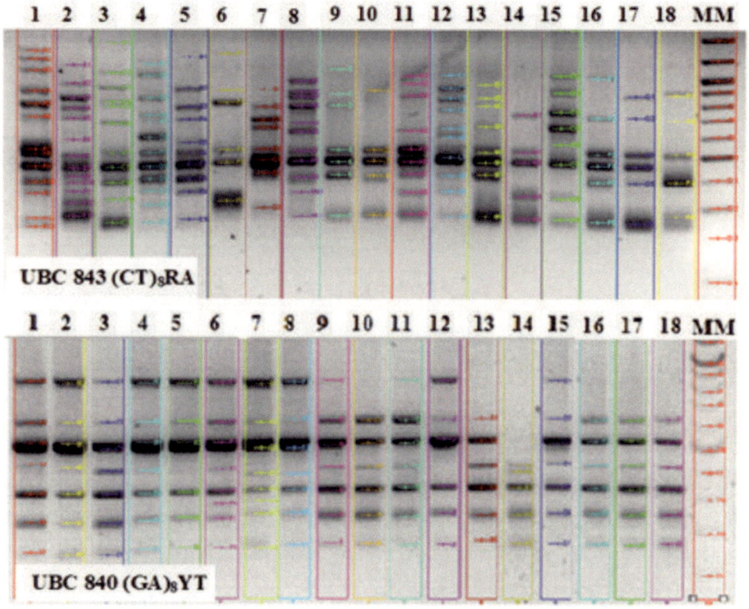

Figure 3.
Analysis of amplification products for UBC843 and UBC840 primers.

fruit traits, divergent landraces were crossed together and finally five commercial hybrids were homologated.

Given that the fruit traits were the main selection criteria used by the farmers during the evolution of tomato landraces, the maintenance of some landraces in a specific ecological region was mainly due to social factor, thus influencing the diversity of tomato landraces from different regions [70].

The landraces with wild specifics characteristics like; high number of branches and fruits per plants, lower values of fruit weight and small pericarp thickness, exhibit a better disease resistance [71]. In this regard, the modern cultivars for fresh market are characterized by large and round fruits with suitable firmness and shelf-life, amid uniformity of size, shape and colour of the fruits [72]. After a comparative study of tomato landraces and advances lines, Carrillo-Rodriguez et al. [73] suggests that it is possible to select tomato landraces with healthy plants and similar performance to that of advanced breeding lines.

Amid the increasing of consumer's interest in fruit quality, landraces with fruits appreciated for flavour and aroma should be considered both for production and for breeding activities. Crossings among varieties and landraces or among landraces can provide a useful variability for different plant and fruit traits [46, 65, 74]. Studying the Mexican tomato landraces Martinez-Vazquez et al. [75] found crosses derived between landraces and commercial lines with values of important traits like firmness, yield and fruit size, close to a commercial hybrid. As such, tomato landraces are a valuable source to obtain breeding lines with high general combining ability, possessing important alleles for yield traits, suitable to be used in breeding programs.

Considering that the landraces are genetically closer to modern cultivars than to their wild relatives, they represent an important source of genes for improvement of adaptation to abiotic stress [43]. In this regard, Massaretto et al. [76] highlighted the potential of tomato landraces from Southeast of Spain to improve the fruit quality and also to maintain the yield stability under salt stress conditions. Studying tomato landraces from Romanian areas with medium and high levels of soil salinity,

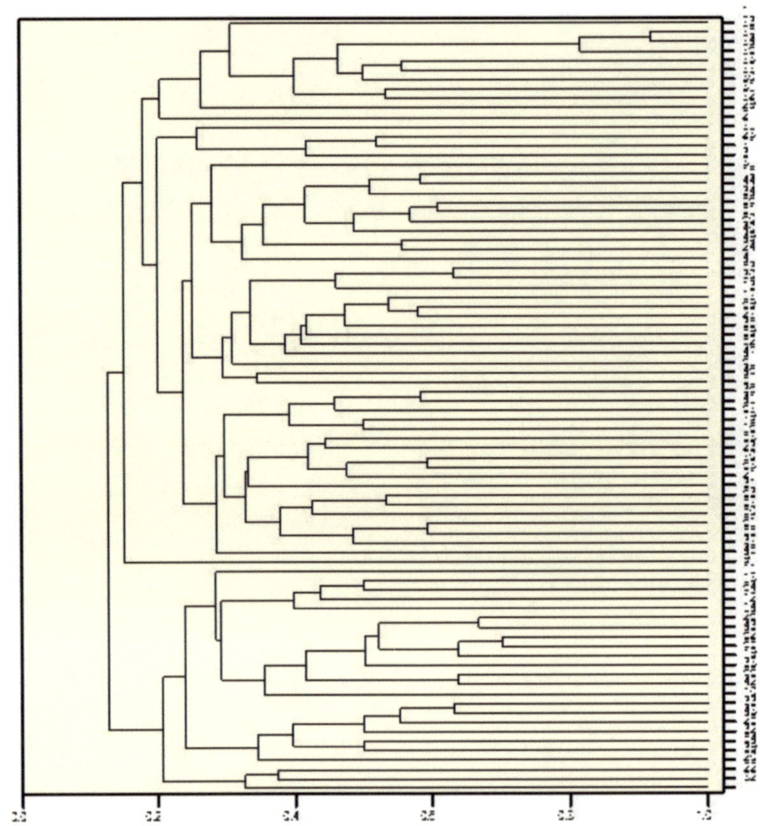

Figure 4.
UPGMA clustering of 82 tomato landraces using ISSR markers (Landraces 1 to 70 from S-W Romania; landraces 71 to 82 from N-E of Romania).

Sumalan et al. [20] found that landraces with tolerance to soil salinity have a high ability to accumulate large amounts of antioxidants in the ripe fruits, increasing their nutraceutical value. Taking into account that the growing conditions have a high influence on plant morphology, chemical composition of the fruits and agronomic performances, Figas et al [77] suggest that long–shell life landraces from Mediterranean basin could be a useful material for improvement of tomato adaptation to greenhouse cultivation, or to predicted climate change conditions, especially drought [78].

Breeding of tomato focused on yield led to a loss of genetic diversity and a decrease of nutritional value and disease resistance [79]. Under a low diversity and a narrow genetic base of disease resistance, the cultivation of tomato becomes vulnerable and dependent to widespread use of pesticides [80]. Given that the preservation of tomato landraces is influenced by both natural and human selection, these populations can be considered a suitable breeding material for the identification of genes with supposed adaptive value [81].

5.1.2 Onion landraces

Due to the replacement of landraces and old varieties with modern varieties and in particular F1 hybrids the genetic basis of onion has been considerably reduced, so that many genes with adaptive value contained in the landraces and old varieties are in danger of being lost [82].

The success of onion breeding programs, among others depends mainly on the availability of genetic variability for different traits of interest. The use of wild Allium species for genetic improvement of cultivated varieties is a very long-term process that can take up to 20 years [83]. As such the onion landraces are a more suitable material for breeding of adaptive traits like bulbing and flowering, controlled by multiple genes [84–87].

For an effective use of onion landraces it is necessary to characterize and evaluate these germplasm at both molecular and at morphological level. In this regard et al. [82] found a 69% diversity between 85 Spanish onion landraces based of pungency, day length requirements, and skin colour, without being established a relation among the diversity at molecular and at morphological or physico-chemical level. Similar results have been reported by other studies: Hanci and Gökçe [88] for Turkish onions; Mitrová et al. [89] for Czech onions; González-Pérez et al. [90] for Galician onions. The landraces possessing high genetic diversity have an important selection potential for the development of new onion cultivars with favorable yield, adaptive and quality traits.

Likewise, the molecular diversity of Indian onions studied by Khar et al [91] was not related with colour, growing season and geographical origin. The exchange among farmers from different regions could be an explanation for the lack of relation between clustering of landraces and their geographical origin.

Following the molecular evaluation of 43 onion landraces from two regions of Romania using ISSR markers within the S-Stress project, a high level of diversity (around 80 %) was found, associated with a clear separation of the landraces in two clusters, related with their geographical origin (**Figure 5**). Amid a lack o biological material exchange between two regions, it is assumed that the landraces have had a distinct evolution under the influence of local ecological conditions. As such, these onion landraces are important sources of genetic diversity, containing valuable genes for different yield and adaptive traits under salt stress conditions.

High levels of heterozigosity associated with low allele number reported by several studies [90, 92–94] represents a consequence of out-crossing and continuous

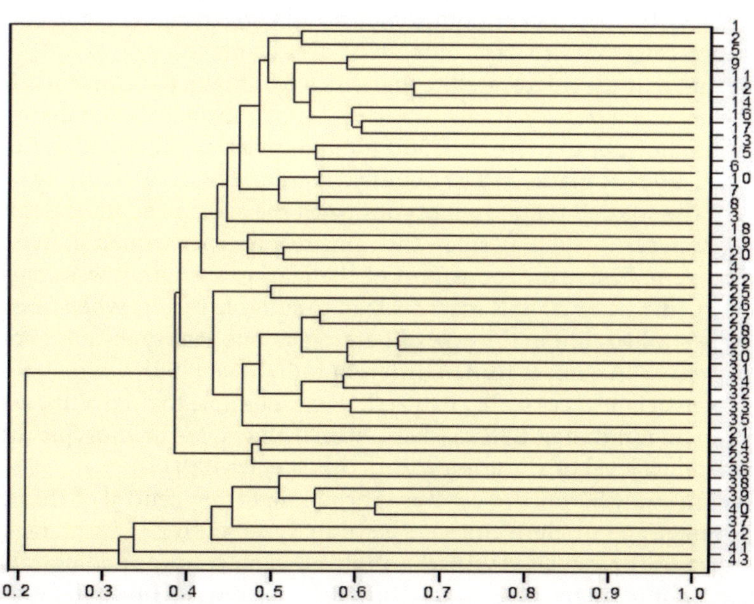

Figure 5.
UPGMA clustering of 43 onion landraces using ISSR markers (Landraces 1 to 35 from S-W Romania; landraces 36 to 43 from N-E of Romania).

gene flow in small geographical regions where the onion landraces have evolved. In order to capitalize the genetic variation of onion landraces in breeding programs, it is necessary to ensure a certain degree of out-crossing on the selected genotypes [95, 96]. The breeding potential of onion landraces was also revealed by Porta et al [97], who found transgressive segregation for different bulb traits in selfing (S1) lines, compared to original population. The high variance within and among S1 lines for all traits, confirm the heterogeneous structure of landraces and efficiency of their use as a selection material.

6. The evaluation of the tomatoes landraces genetic diversity based on molecular markers

A representative of the horticultural plants studied in our research were tomatoes landraces, due to their importance as food in Romania and because it is one of the first crop assessed by molecular markers for variability evaluation. The genetic study of local landraces is based on the evaluation of their genetic variability to determine the degree of similarity. Next, it is necessary to correlate the molecular fingerprints with the phenotypic traits in order to identify genotypes of interest for plant breeding.

Over the time, the variability was evaluated with morphological markers followed by biochemical ones, developed on the basis of isoenzymes. The biochemical markers had a major disadvantage because they are affected by the phenological development stage, being possible to detect a percentage of only 0.1% of the variability. For this reason, the DNA markers have gained increasing importance and have been used on a very large scale today. They can be classified according to the type of analyzed sequence and the applied methods of analysis which both determine their genetic behaviour, i.e. their codominant or dominant character.

The codominant markers, such as RFLP (Restriction Fragment Length Polymorphism), STS (Sequence tagged site), EST (Expressed Sequence Tag) and SSR (Single Sequence Repeats), are an important source of information because they allow the differentiation of homozygotes and heterozygotes being co-dominants, but each category also has a number of disadvantages.

Considering that the microsatellite markers have shown to be promising to evaluate the genetic diversity, Bredemeijer et al [98] constructed database comprising information about more than 500 tomato varieties cultivated in Europe evaluated with 20 SSR markers. The obtained results showed a relatively reduced variability of the studied tomato genotypes, with the average of allele per locus of 4.7, ranging between 2 and 8. Besides, the same test was performed in five different laboratories to emphasize the robustness of the marker system. It was concluded that the use of this set of 20 SSR markers lead to suitable results when homogeneous varieties were studied, but in the case of heterogeneous genotypes it is necessary to analyze a mixed DNA sample from 6 different individuals [98].

When Spanish landraces were analyzed, it was possible to differentiate cultivars only with a small number of SSR markers, even if they were phenotypic different, emphasizing a low level of variation within this species [99].

In an Italian study 50 tomato landraces originated from central of the country and other vintage and modern cultivars were analyzed with 29 SSR markers. The molecular data were associated with the study of 15 morpho-physiological traits. Two categories of markers were used – the markers from the first category were part of a linkage area where QTLs previously associated with the shape and size of the fruits were positioned and in the second category were markers from some chromosomal regions without any known linkage. Besides, DNA samples collected from plants grown in two different locations were analyzed. It was pointed out a high polymorphism of the tomato landraces compared to modern cultivars and many relations between the markers from the QTL region and the traits associated with fruit shape and size. These results are promising for the identification of SSR markers associated with traits of agronomic interest [100].

Later, 42 tomato varieties originated from different regions from China and Kenya were evaluated with SSR markers, emphasizing a high degree of diversity. The results analysis distributed the genotypes in different clusters without any relation with their origin [101]. In other study Italian local landraces were analyzed with 19 SSR markers generating a number of 60 alleles with moderate level of diversity but very different compared to the commercial varieties [102].

It was pointed out that the SSR markers could be used for the evaluation of tomato landraces variability, but it must be considered that their development is expensive and time consuming, therefore may be the markers which generated a high amount of data in only one analysis could be more efficient.

In 2000, species of wild tomato relatives originated from Peru (named PC – Peruvian complex) were evaluated with RAPD markers in comparison with cultured genotypes. A high diversity was shown, emphasizing the potential of the wild genotypes to be used as a source of genes for breeding [103].

In India, based on the molecular fingerprints generated by RAPD markers, the reduction of genetic diversity for tomato cultivars has been highlighted. This has been attributed to breeding processes that target plants with very similar traits [104].

The evaluation of the brasilian tomato landraces based on RAPD primers showed that most of them were part of a single cluster, different from the commercial cultivars [45]. Similar results were obtained when tomato landraces originated from Azerbaijan were analyzed [105].

ISSR markers were used to evaluate the genetic variability for 100 Brazilian tomato genotypes of different origin. Finally, a correlation between the fingerprints generated by ISSR markers and the origin of the genotypes was established [106].

In 2016, landraces originated from East Anatolian region of Turkey and North-West of Iran, along with three commercial cultivars were evaluated with ISSR markers. It turned out that the genotypes originating from the same region, often located in the same group or two adjacent groups [107].

The same markers were used to evaluate tomato genotypes with different antioxidant content. The obtained fingerprints were used to confirm the nature of the hybrids in breeding programs, thus accelerating the selection process [108].

The AFLP markers (Amplified Polymorphic DNA) were used in conjunction with SSR markers to characterize 48 traditional tomato cultivars collected from the south-east of Spain. The discrimination power was similar for both category of markers and the constructed dendrograms were grouped in the main types. The conclusion was that it would be more appropriate to use in combination the information obtained with several categories of markers [59].

In the early 2000's SRAP markers (Sequence-Related Amplified Polymorphism) were developed as a technique with low cost, simple, highly variable, with high reproducibility [109], based on a random amplification reaction. Considering that 3 'UTR region is usually polymorphic due to insertions and deletions the probability to identify polymorphism random in the coding regions is high. This marker had a widely use for diversity evaluation for different plant species.

Ruiz et al [99] studied the diversity of some traditional tomato cultivar from Spain based on SSR and SRAP markers. It was pointed out that SRAP markers clustered together the genotypes with the same origin. Comparable results were observed when SSR markers were used, but the level of resolution was lower [99].

Al Shaye et al [110] evaluated Saudi tomato landraces with SDS-PAGE and SRAP markers. It was shown that almost all of the landraces with the same origin were grouped in the same cluster emphasizing the usefulness of these markers in future breeding programs [110].

Similar to SRAP markers, which bind in the coding gene region, ScoT markers (Start Codon Targeted) involve the amplification with a single primer that anneal to the highly conserved region positioned next to start codon ATG of two close genes [111]. The ScoT primers were used in comparison with the ISSR to evaluate the variability for 8 Egyptian tomato genotypes. The genetic fingerprints were different for the two categories of markers and it was considered that ScoT ones were more related to the morphological traits compared to ISSR for evaluation of tomato diversity. Therefore, the use of more than one marker system is recommended for a higher resolution of the analysis [112]. Following the introduction of modern analytical techniques, they have also been applied in the area of diversity assessment.

Therefore, the sequencing system Illumina was used for evaluation of 75 landraces originated from Sothern Italy and distinguished a number of 152 single nucleotide polymorphisms (SNP). 30% variability was identified between local populations, the differences being associated especially with fruit-related traits. The developed SNP system was considered to be very useful for genetic characterization, effective conservation and application on tomato breeding process [58].

A complex research had been done in Italy to investigate 123 tomato genotypes originated from all over the world. A very wide range of genotypes has been analyzed in order to succeed in the polymorphism identification and its correlation with different 18 morphological traits, mainly related to fruits. A tomato array was used and a number of almost 8000 SNP were analyzed. The results showed that 36 of the SNP markers were correlated with 15 of the studied traits. These markers were mapped on chromosomes along with a number of 98 candidate genes as follows: 19 SNPs were located in six chromosomal regions in which candidate genes are positioned, and 17 SNPs in regions where no such genes are found. Thus, it can be stated that chromosomal regions have been identified where unknown genes related to the traits are positioned. Thus, new research lines are opened to identify genes of interest [61].

In the following years, considering the development of the SNP analysis system, point mutations associated with organoleptic characters and metabolites content were identified [113] and mutations in genes involved in drought tolerant and fruit maturation and quality [114].

Besides SNP identification, the whole genome sequencing was also applied to identify genes of interest involved in tolerance to drought, good quality and storage proprieties. Therefore, the whole genome of two landraces with the mentioned traits was sequenced. In their genome regions similar to *Solanum pimpinellifolium* and *S. pennellii* and candidate genes for the interest traits were identified [115].

Therefore, it can be said that over time several molecular marker systems have been used to assess variability in local tomato landraces. But it has rarely been possible to correlate with the phenotype, i.e. the genes determine certain characters. But these molecular markers have shown their importance in screening populations to determine the degree of similarity or to remove identical genotypes from the study and from the conservation. Instead, the development of SNP markers and sequencing of the entire genome is expected to be a strategy that will underpin the identification of all genes of interest in both biological and agricultural areas.

7. Conclusions

Vegetable landraces constitute a valuable genetic pool of genetic diversity, which can be exploited both in breeding programs for obtaining new commercial genotypes with targeted traits and as a valuable source of germplasm for traditional farmers.

Tomatoes are the most important vegetable with fruits and many landraces are preserved around the world as local varieties or farm varieties. Variability of chemical composition, plant morphology and agronomic performance have shown that cultivation technology has a major impact on the shelflife of tomato fruits.

The conservation of vegetable landraces is associated with their cultural value, geographical isolation of sites, aesthetic and organoleptic preferences of consumers and traditional farmers.

There is an optimistic outlook on harnessing landraces and traditional vegetable varieties in a quality-oriented sustainable horticultural system.

Acknowledgements

The authors would also like to thank the Centre for Consultancy and Euro-regional Rural Cooperation for financial support.

Author details

Radu-Liviu Sumalan[1*], Sorin-Ion Ciulca[2], Renata-Maria Sumalan[3] and Sorina Popescu[4]

1 Department of Plant Physiology, Faculty of Horticulture and Forestry, Banat University of Agricultural Sciences and Veterinary Medicine "King Michael I of Romania", Timisoara, Romania

2 Department of Quantitative Genetics, Statistics and Plant breeding, Faculty of Horticulture and Forestry, Banat University of Agricultural Sciences and Veterinary Medicine "King Michael I of Romania", Timisoara, Romania

3 Department of Soil Microbiology, Faculty of Horticulture and Forestry, Banat University of Agricultural Sciences and Veterinary Medicine "King Michael I of Romania", Timisoara, Romania

4 Department of Biotechnologies, Faculty of Horticulture and Forestry, Banat University of Agricultural Sciences and Veterinary Medicine "King Michael I of Romania", Timisoara, Romania

*Address all correspondence to: sumalanagro@yahoo.com

References

[1] Schreinemachers P, Simmons EB, Wopereis MCS. Tapping the economic and nutritional power of vegetables, Global Food Security. 2018;16: 36-45. DOI: 10.1016/j.gfs.2017.09.005

[2] Fellows P. Guidelines for small-scale fruit and vegetable processors. FAO Agricultural Services Bulletin – 127. 1997. Edited by Midway Technology Ltd. St Oswalds Barn, Clifford Hay on Wye, Hereford,United Kingdom

[3] Global Vegetable Seed Markets [Internet]. 2017. Available from: https://marketlitmus.com/report-store/agriculture/seeds-commodities/global-vegetable-seed-market/ [Accessed: 2020-11-12]

[4] World - Vegetable - Market Analysis, Forecast, Size, Trends and Insights [Internet]. 2020. Available from: https://www.researchandmarkets.com/reports/4828911/world-vegetable-market-analysis-forecast? [Accessed: 2020-12-17]

[5] van Rijswick C. Worl Vegetable Map 2018-More Than Just a Local Affair [Internet]. 2018. Available from: https://research.rabobank.com/far/en/sectors/regional-food-agri/world_vegetable_map_2018.html. [Accessed: 2020-11-29]

[6] Lipinski B, Hanson C, Lomax J, Kitinoja L, Waite R, Searchinger T. Reducing Food Loss and Waste. World Resources Institute Working Paper. Installment 2 of Creating a Sustainable Food Future. Washington, DC: World Resources Institute [Internet]. 2013. Available from: http://www.worldresourcesreport.org [Accessed:2020-12-15].

[7] Monier V, Mudgal S, Escalon V, O'Connor C, Gibon T, Anderson G, Morton G. Preparatory study on food waste across EU 27. 2010. Report for the European Commission (DG ENV – Directorate C) BIO Intelligence Service (BIOIS), Paris

[8] Stenmarck Å, Jensen C, Quested T, Moates G. Estimates of European foodwaste levels. FUSIONS Project, Wageningen. [Internet]. 2016. Available from: https://eu-fusions.org/phocadownload/Publications/Estimates%20of%20European%20food%20waste%20levels.pdf [Accessed: 2020-11-22]

[9] De Laurentiis V, Corrado S, Sala S. Quantifying household waste of fresh fruit and vegetables in the EU, Waste Management.2018;77:238-251. DOI:10.1016/j.wasman.2018.04.001

[10] Vilariño MV, Franco C, Quarrington C. Food loss and waste reduction as an integral part of a circular economy. Front Environ Sci. 2017;5:1-5. DOI:10.3389/fenvs.2017.00021

[11] Godfray HCJ, Garnett T. Food security and sustainable intensification. Philos. Trans. R. Soc. Lond. B. Biol. Sci. 2014:369, 20120273. DOI:10.1098/rstb.2012.0273

[12] United Nations. Sustainable development goals. 17 goals to transform our world. [Internet]. 2015. Available from: http://www.un.org/sustainabledevelopment/oceans/> [Accessed: 2020-11-08]

[13] European Commission. Closing the loop-An EU action plan for the Circular Economy. Communication from the Commission to the European Parliament, the Council, the European Economic and Social Committee and the Committee of the Regions COM. Brussels [Internet]. 2015. Available from: https://eur-lex.europa.eu/legal-content/EN [Accessed: 2020-11-23]

[14] Acquaah G. Principles of plant genetics and breeding.

Blackwell publishing Ltd. 2007.p 96, ISBN-13:978-1-4051-3046-4

[15] Weise S, Lohwasser U, Oppermann M. Document or Lose It—On the Importance of Information Management for Genetic Resources Conservation in Genebanks. Plants. 2020;9:1050. DOI:10.3390/plants9081050

[16] Wambugu PW, Ndjiondjop M-N, Henry RJ. Role of genomics in promoting the utilization of plant genetic resources in genebanks, Briefings in Functional Genomics.2020;17:198-206, DOI: 10.1093/bfgp/ely014

[17] FAO. The Second Report on the State of the World's Plant Genetic Resources for food and Agriculture. Rome [Internet]. 2010. Available from: http://www.fao.org/3/i1500e/i1500e.pdf [Accessed: 2020-11-23]

[18] Hammer K, Teklu Y. Plant Genetic Resources: Selected Issues from Genetic Erosion to Genetic Engineering. Journal of Agriculture and Rural Development in the Tropics and Subtropics. 2008;109/ 1:15-50

[19] Dulloo ME, Thormann I, Fiorino E, De Felice S, Rao VR, Snook L. Trends in Research using Plant Genetic Resources from Germplasm Collections: From 1996 to 2006. Crop Sci. 2013;53:1217-1227. DOI: 10.2135/cropsci2012.04.0219

[20] Sumalan RM, Ciulca SI, Poiana MA, Moigradean D, Radulov I, Negrea M, Crisan ME, Copolovici L, Sumalan RL. The Antioxidant Profile Evaluation of Some Tomato Landraces with Soil Salinity Tolerance Correlated with High Nutraceutical and Functional Value. Agronomy.2020;10/4:500. DOI: 10.3390/agronomy10040500

[21] Sonnino A. International Instruments for Conservation and Sustainable Use of Plant Genetic Resources for Food and Agriculture: An Historical Appraisal. Diversity.2017;9:50.DOI: 10.3390/d9040050

[22] Hammer K, Laghetti G. Genetic erosion – Examples from Italy. Genet Resour. Crop. Evol. 2005;52:629-634. DOI: 10.1007/s10722-005-7902-x

[23] Dias JCS. Biodiversity and vegetable breeding in the light of developments in intellectual property rights. In: O. Grillo and G. Venora (eds.) Ecosystems biodiversity. Intech, Rijeka, Croatia. 2011; 389-428 p. ISBN 978-953-307-417-7

[24] Dwivedi SL, Ceccarelli S, Blair MW, Upadhyaya HD, Are AK, Ortiz R. Landrace germplasm for improving yield and abiotic stress adaptation. Trends Plant Sci. 2016; 21(1):31-42. DOI: 10.1016/j.tplants.2015.10.012

[25] Ebert AW. The Role of Vegetable Genetic Resources in Nutrition Security and Vegetable Breeding. Plants. 2020;9:736. DOI: 10.3390/plants9060736

[26] Casañas F, Simó J, Casals J, Prohens J. Toward an Evolved Concept of Landrace. Front. Plant Sci. 2017;8:145. doi: 10.3389/fpls.2017.00145.

[27] Kiessling H. Die züchterische Bearbeitung der Landsorten in Bayern. Beiträge zur Pflanzenzücht 2:1912: 74-96.

[28] Zeven AC. Landraces: a review of definitions and classifications. Euphytica. 1998;104:127-139.

[29] Prospéri JM, Demarquet F, Angevain M, Mansat P. Evaluation agronomique de variétés de pays de sainfoin (*Onobrychis sativa* L.) originaires du sud-est de la France. Agronomie. 1994;14: 285-298.

[30] Camacho-Villa TC, Maxted N, Scholten M, Ford-Lloyd B. Defining

and identifying crop landraces. Plant Genetic Resources, Characterization and Utilization. 2005;3:373-384. DOI: 10.1079/PGR200591

[31] Boscaiu M, Fita A. Physiological and Molecular Characterization of Crop Resistance to Abiotic Stresses. Agronomy. 2020;10: 1308. DOI: 10.3390/agronomy10091308

[32] Terzopoulos PJ, Bebeli PJ. DNA and morphological diversity of selected Greek tomato (*Solanum lycopersicum* L.) landraces. Sci. Hortic. 2008;116(4):354-361. DOI: 10.1016/j.scienta.2008.02.010

[33] Byrne P, Richards C, Volk GM. From wild species to landraces and cultivars. In: Volk GM, Byrne P (Eds.) Crop wild relatives and their use in plant breeding. Fort Collins, Colorado: Colorado State University [Internet]. 2020. Available from: https://colostate.pressbooks.pub/cropwildrelatives/front-matter/crop-wild- relatives-and-their-use-in-plant-breeding/ [Accessed: 2020-11-28]

[34] Newton AC, Akar T, Baresel JP, Bebeli PJ, Bettencourt E, Bladenopoulos KV, Czembor JH, Fasoula DA, Katsiotis A, Koutis K, Koutsika-SotiriouM,KovacsG,LarssonH, de Carvalho MAAP, Rubiales D, Russell J, Dos Santos TMM, Patto MCV. Cereal landraces for sustainable agriculture. A review. Agron.Sustain. Dev. 2010; 30 (2): 237-269. DOI: 10.1051/agro/2009032

[35] van de Wouw M, Kik C, van Hintum T, van Treuren R, Visser B. Genetic erosion in crops: concept, research results and challenges. Plant Genet. Resour. Util. 2010;8:1-15. DOI: 10.1017/s1479262109 990062

[36] Missio JC, Rivera A, Figàs MR, Casanova C, Camí B, Soler S, Simó J. A comparison of landraces vs. modern varieties of lettuce in organic farming during the winter in the Mediterranean area: An approach considering the viewpoints of breeders, consumers, and farmers. Front. Plant Sci. 2018; 9: 1491. DOI: 10.3389/fpls.2018.01491

[37] Villa TCC, Maxted N, Scholten M, Ford-Lloyd B. Defining and identifying crop landraces. Plant Genet. Resour. 2005;3: 373-384. DOI: 10.1079/PGR200591

[38] da Silva Dias, J.C. Impact of improved vegetable cultivars in overcoming food insecurity. Euphytica. 2010;176:125-136 . Doi: 10.1007/s10681-010-0237-5

[39] Sumalan RL, Popescu I, Schmidt B, Sumalan R M, Popescu C, Gaspar S. Salt tolerant tomatoes local landraces from Romania – Preserving the genetic resources for future sustainable agriculture. Journal of Biotechnology; 8: S18. EUROPEAN BIOTECHNOLOGY CONGRESS, 20.08.2015.Bucharest.

[40] Zhu C, Gore M, Buckler ES, Yu J. Status and prospects of association mapping in plants. Plant Genome J. 2008;1:5-20. DOI: 10.3835/plantgenome2008.02.0089.

[41] Prada D. Molecular population genetics and agronomic alleles in seed banks: searching for a needle in a haystack? J. Exp. Bot. 2009;60:2541-2552. DOI: 10.1093/jxb/erp130

[42] Biasi R, Brunori E. The on-farm conservation of grapevine (*Vitis vinifera* L.) landraces assures the habitat diversity in the viticultural agro-ecosystem. Vitis J. Grapevine Res. 2015;54:265-269.

[43] Gascuel Q, Diretto G, Monforte AJ, Fortes A M, Granell A. Use of natural diversity and biotechnology to increase the quality and nutritional content of tomato and grape. Front. Plant Sci. 2017;8:652-660.DOI: 10.3389/fpls.2017.00652.

[44] Garcia-Martinez S, Andreani L, Garcia-Gusano M, Geuna F, Ruiz JJ.

Evaluation of amplified fragment length polymorphism and simple sequence repeats for tomato germplasm fingerprinting: utility for grouping closely related traditional cultivars. Genome. 2006;49(6):648-656. DOI:10.1139/g06-016

[45] Carelli PM, Gerald LTS, Grazziotin GF, Echeverrigaray S. Genetic diversity among Brazilian cultivars and landraces of tomato *Lycopersicon esculentum* Mill. Revealed by RAPD markers. Genet. Resour. Crop. Evol.2006;53: 395-400.DOI: 10.1007/s10722-004-0578-9

[46] Mazzucato A, Ficcadenti N, Caioni M, Mosconi P, Piccinini E, Sanampudi VR, Sestili S, Ferrari V. Genetic diversity and distinctiveness in tomato (*Solanum lycopersicum L.*) landraces: The Italian case study of 'A pera Abruzzese' 2010. Scientia Horticulturae.2010;25:55-62. DOI:10.1016/j.scienta.2010.02.021

[47] Di Paola-Naranjo RD, Otaiza S, Saragusti AC, Baroni V, Carranza AV, Peralta IE, Valle EM, Carrari F, Asis R. Data on polyphenols and biological activity analyses of an Andean tomato collection and their relationships with tomato traits and geographical origin. Data Brief. 2016;7:1258-1268. DOI:10.1016/j.dib. 2016.04.005.

[48] Quadrana L, Almeida J, Asis R, Duffy T, Dominguez PG, Bermudez L, et al. Natural occurring epialleles determine vitamin E accumulation in tomato fruits. Nat. Commun. 2014;5, 4027. DOI:10.1038/ncomms5027

[49] Fang C, Luo J, Wang S. The Diversity of Nutritional Metabolites: Origin, Dissection, and Application in Crop Breeding. Front Plant Sci. 2019;10:1028. DOI:10.3389/fpls.2019.01028

[50] Asprelli PD, Sance M, Insani M, Asis R, Valle EM, Carrari F, Galmarini CR, Peralta IE. Agronomic performance and fruit nutritional quality of an Andean tomato collection. Acta Horticulturae. 2017;1159:197-204. DOI:10.17660/ActaHortic.2017.1159.29

[51] Cortina PR, Asis R, Peralta IE, Asprelli PD, Santiago AN. Determination of volatile organic compounds in Andean tomato landraces by headspace solid phase microextraction-gas chromatography-mass spectrometry. Journal of the Brazilian Chemical Society, 2017;28(1):30-41. DOI:10.5935/0103-5053.20160142

[52] Giovannoni JJ. Prospects: the tomato genome as a cornerstone for gene discovery In: Causse M, Giovannoni J, Bouzayen M, Zouine M, eds. The tomato genome. Berlin, Heidelberg: Springer Berlin Heidelberg, 2016. p 257-259. DOI 10.1007/978-3-662-53389-5_13

[53] Sanchez E, Sifres A, Casanas F, Nuez F. The endangered future of organoleptically prestigious European landraces: Ganxet bean (*Phaseolus vulgaris* L.) as an example of a crop originating in the Americas. Genetic Resources and Crop Evolution.2008;55:45-52. DOI 10.1007/s10722-007-9213-x

[54] Hurtado M, Vilanova S, Plazas M, Gramazio P, Andújar I, Herraiz FJ, et al. Enhancing conservation and use of local vegetable landraces: the Almagro eggplant (Solanum melongena L.) case study. Genet.Resour.CropEvol. 2014;61:787-795. DOI:10.1007/s10722-013-0073-2.

[55] Maxim A, Strajeru S, Albu C, Sandor M, Mihaiescu L, Pauliuc SE. Conservation of vegetable genetic diversity in Transylvania-Romania. Sci. Rep. 2020;10:18416. DOI.org/10.1038/s41598-020-75413-x

[56] Sarli G, Tigan E, Bitonte D, Montemurro F, Montesano V,

Laghetti G, Hammer K. Collecting landraces of vegetable crop species in the South-West Romania. Journal of Environmental Science and Engineering B 5. 2016:17-25. DOI:10.17265/2162-5263/2016.01.003

[57] Solanke AU, Kumar PA. Phenotyping of tomatoes. In: Panguluri SK, Kumar AA , editors. Phenotyping for plant breeding: Applications of phenotyping methods for crop improvement. Springer Science+Business Media New York; 2013. p. 169-204. DOI: 10.1007/978-1-4614-8320-5_6

[58] Corrado G, Caramante M, Piffanelli P, Rao R. Genetic diversity in Italian tomato landraces: Implications for the development of a core collection. Sci. Hortic. 2014;168:138-144. DOI:10.1016/j.scienta.2014.01.027

[59] García-Martínez S, Andreani L, Garcia-Gusano M, Geuna F, Ruiz JJ. Evaluation of amplified fragment length polymorphism and simple sequence repeats for tomato germplasm fingerprinting: utility for grouping closely related traditional cultivars. Genome. 2006;49:648-656. DOI: 10.1139/g06-016

[60] Andreakis N, Giordano I, Pentangelo A, Fogliano V, Graziani G, Monti LM, Rao R. DNA fingerprinting and quality traits of Corbarino cherry-like tomato landraces. J. Agr. Food. Chem. 2004;52:3366-3377. DOI: 10.1021/jf049963y

[61] Sacco A, Ruggieri V, Parisi M, Festa G, Rigano MM, Picarella ME, Mazzucato A, Barone A. Exploring a Tomato Landraces Collection for Fruit-Related Traits by the Aid of a High-Throughput Genomic Platform. PLoS ONE. 2015;10 (9): e0137139. DOI:10.1371/journal.pone.0137139

[62] Manzano S, Navarro P, Martínez C, Megías ZM, Rebolloso MM, Jamilena M. Evaluation of fruit quality in tomato landraces under organic greenhouse conditions. Acta Hortic. 2015; 1099: 645-652. DOI:10.17660/ ActaHortic.2015.1099.79

[63] Lázaro A. Tomato landraces: An analysis of diversity and preferences. Plant Genetic Resources: Characterization and Utilization. 2018;16(4):315-324. DOI:10.1017/ S1479262117000351

[64] Hawkes JG, Maxted N, Ford-Lloyd BV. The ex situ conservation of plant genetic resources. Kluwer Academic Publishers, Dordrecht, 2000

[65] Cebolla-Cornejo J, Rosello S, Nuez F. Phenotypic and genetic diversity of Spanish tomato landraces. Sci. Hortic. 2013;162:150-164. DOI:10.1016/j. scienta.2013.07.044

[66] Tembe KO, Chemining'wa G, Ambuko J, Owinob W. Evaluation of african tomato landraces (*Solanum lycopersicum*) based on morphological and horticultural traits. Agriculture and Natural Resources. 2018;52(6):536-542.DOI:10.1016/j. anres.2018.11.014

[67] Renna M, Montesano FF, Signore A, Gonnella M, Santamaria P. BiodiverSO: A case study of integrated project to preserve the biodiversity of vegetable crops in Puglia (Southern Italy). Agriculture. 2018;8:128. DOI:10.3390/ agriculture8080128

[68] Rodriguez GR, Kim HJ, van der Knaap E. Mapping of two suppressors of OVATE (sov) loci in tomato. Heredity. 2013;111:256-264. DOI:10.1038/ hdy.2013.45

[69] Scarano A, Olivieri F, Gerardi C, Liso M, Chiesa M, Chieppa M, Frusciante L, Barone A, Santino A, Rigano MM. Selection of tomato landraces with high fruit yield and nutritional quality under elevated

temperatures. J Sci Food Agric 2020; 100: 2791-2799. DOI 10.1002/jsfa.10312

[70] Moreno-Ramírez YR, Hernández-Bautista A, Ramírez-Vallejo P, Castillo-Gónzalez F, Rocandio-Rodríguez M, Vanoye-Eligio V, Mora-Ravelo, S G. Social and environmental factors in the diversity of tomato landraces from the South-Central region of Mexico. Ciencia Rural. 2019; 49 (5): e20180514. DOI:10.1590/0103-8478 cr20180514

[71] Kouam EB, Dongmo JR, Djeugap JF. Exploring morphological variation in tomato (*Solanum lycopersicum*): A combined study of disease resistance, genetic divergence and association of characters. Agricultura Tropica et Subtropica. 2018;51(2):71-82. DOI:10.2478/ats-2018-0008

[72] Foolad MR. Genome mapping and molecular breeding of tomato. Int. J. Plant Genomics. 2007;64358. DOI:10.1155/2007/64358

[73] Carrillo-Rodríguez JC, Chávez-Servia JL, Lobato-Ortiz R, Perales-Segovia C. Generation and evaluation of heterogeneous genotypes of tomato for small-scale farmers. Journal of Plant Breeding and Crop Science. 2019;11(3):91-99. DOI:10.5897/JPBCS2018.0782

[74] Rocchi L, Paolotti L, Cortina C, Boggia A. Conservation of landrace: the key role of the value for agrobiodiversity conservation. An application on ancient tomatoes varieties. Agriculture and Agricultural Science Procedia. 2016;8:307-316. DOI:10.1016/j.aaspro.2016.02.025

[75] Martínez-Vázquez E, Hernández BA, Lobato OR, García ZJJ, Reyes LD. Exploring the breeding potential of Mexican tomato landraces. Sci. Hortic. 2017;220:317-325. DOI:10.1016/j.scienta.2017.03.031

[76] Massaretto IL, Albaladejo I, Purgatto E, Flores FB, Plasencia F, Egea-Fernández JM, Bolarin MC, Egea I. Recovering tomato landraces to simultaneously improve fruit yield and nutritional quality against salt stress. Front. Plant Sci. 2018;9:1778. DOI: 10.3389/fpls.2018.01778

[77] Figàs MR, Prohens J, Raigón MD, Pereira-Dias L, Casanova C, García-Martínez MD, Rosa E, Soler E, Plazas M, Soler S. Insights into the adaptation to greenhouse cultivation of the traditional Mediterranean long shelf-life tomato carrying the alc mutation: A multi-trait comparison of landraces, selections, and hybrids in open field and greenhouse. Front. Plant Sci. 2018;9:1774. DOI:10.3389/fpls.2018.01774

[78] Conesa MÀ, Fullana-Pericàs M, Granell A, Galmés J. Mediterranean long shelf-life landraces: An untapped genetic resource for tomato improvement. Front. Plant Sci. 2020;10:1651. DOI: 10.3389/fpls.2019.01651

[79] Zsögön A, Čermák T, Naves E R, Notini M M, Edel K H, Weinl S, Freschi l, Voytas DF, Kudla J, Pereira Peres LE. De novo domestication of wild tomato using genome editing. Nat. Biotechnol. 2018;36:1211-1216. DOI: 10.1038/nbt.4272

[80] Schouten HJ, Tikunov Y, Verkerke W, Finkers R, Bovy A, Bai Y and Visser RGF. Breeding has increased the diversity of cultivated tomato in The Netherlands. Front. Plant Sci. 2019;10:1606. DOI: 10.3389/fpls.2019.01606

[81] Mazzucato A, Papa R, Bitocchi E, Mosconi P, Nanni L, Negri V, Picarella ME, Siligato F., Soressi GP, Tiranti B, Veronesi F. Genetic diversity, structure and marker-trait associations in a collection of Italian tomato (*Solanum lycopersicum* L.) landraces. Theor. Appl. Genet.

2005;116(5):657-669. DOI:10.1007/s00122-007-0699-6.

[82] Mallor C, M.S. Arnedo-Andrés MS, Garcés-Claver A. Assessing the genetic diversity of Spanish *Allium cepa* landraces for onion breeding using microsatellite markers, Sci. Hortic. 2014;170:24-31. DOI:10.1016/j.scienta.2014.02.040.

[83] Scholten O E, Heusden AWV, Khrustaleva LI, Burger-Meijer K, Mank RA, Antonise RGC, Harrewijn JL, Van Haecke W, Oost EH, Peters RJ, Kik C. The long and winding road leading to the successful introgression of downy mildew resistance into onion. Euphytica. 2007;156:345—353. DOI:10.1007/s10681-007-9383-9

[84] Baldwin S, Revanna R, Pither-Joyce M, Shaw M, Wright K, Thomson S, Moya L, Lee R, Macknight R, McCallum J. Genetic analyses of bolting in bulb onion (*Allium cepa* L.). Theor. Appl. Genet. 2017;127:535—547. DOI:10.1007/s00122-013-2232-4

[85] Khosa JS, McCallum J, Dhatt AS, Macknight RC. Enhancing onion breeding using molecular tools. Plant Breed. 2016;135:9-20. DOI:10.1111/pbr.12330

[86] Lee R, Baldwin S, Kenel F, McCallum J, Macknight R. FLOWERING LOCUS T genes control onion bulb formation and flowering. Nat. Commun. 2013;4:2884. DOI:10.1038/ncomms3884

[87] Taylor A, Massiah AJ, Thomas B. Conservation of Arabidopsis thaliana photoperiodic flowering time genes in onion (*Allium cepa* L.). Plant Cell Physiol. 2010;51:1638-1647. DOI:10.1093/pcp/pcq120

[88] Hanci F, Gökçe AF. Molecular characterization of Turkish onion germplasm using SSR markers. Czech J. Genet. Plant Breed. 2016;52(2):71-76. DOI:10.17221/162/2015-CJGPB

[89] Mitrová K, Svoboda P, Ovesná J. The selection and validation of a marker set for the differentiation of onion cultivars from the Czech Republic. Czech J. Genet. Plant Breed. 2015;51:62-67. DOI:10.17221/16/2015-CJGPB

[90] González-Pérez S, Mallor C, Garcés-Claver A, Merino F, Taboada A, Rivera A, Pomar F, Perovic D, Silvar C. Exploring genetic diversity and quality traits in a collection of onion (Allium cepa L) landraces from north-west Spain. GENETIKA. 2015;47(3):885-900. DOI:10.2298/GENSR1503885G

[91] Khar A, Lawande KE, Negi KS. Microsatellite marker-based analysis of genetic diversity in short day tropical Indian onion and cross amplification in related Allium spp. Genet. Resour. Crop. Evol. 2011;58:741-752. DOI:10.1007/s10722-010-9616-y

[92] Bark OH, Havey MJ. Similarities and relationships among populations of the bulb onion as estimated by nuclear RFLPs. Theor. Appl. Genetics. 1995;90:407-414. DOI:10.1007/BF00221983

[93] McCallum J, Thomson S, Pither-Joyce M, Kenel F, Clarke A., Havey M J. Genetic diversity analysis and single-nucleotide polymorphism marker development in cultivated bulb onion based on expressed sequence tag–simple sequence repeat markers. J. Amer. Soc. Hort. Sci. 2010;133(6):810-818. DOI:10.21273/JASHS.133.6.810

[94] Simó J, Pascual L, Cañizares JF, Casañas F. Spanish onion landraces (Allium cepa L.) as sources of germplasm for breeding calçots: a morphological and molecular survey. Euphytica. 2014;195:287-300. DOI:10.1007/s10681-013-0995-y

[95] Rivera A, Mallor C, Garcés-Claver A, García-Ulloa A, Pomar F, Silvar C.

Assessing the genetic diversity in onion (Allium cepa L.) landraces from northwest Spain and comparison with the European variability. New Zealand Journal of Crop and Horticultural Science. 2016;44(2):103-120, DOI:10.1080/01140671.2016.1150308

[96] Shigyo M, Kik C. Onion. In: Prohens J, Nuez F, (eds). Vegetables II. Handbook of plant breeding. New York: Springer. 2008; p. 121-159.

[97] Porta B, Rivas M, Gutiérrez L, Galván GA. Variability, heritability, and correlations of agronomic traits in an onion landrace and derived S1 lines. Crop Breed. Appl. Biotechnol. 2014;14(1):29-35. DOI:10.1590/S1984-70332014000100005

[98] Bredemeijer GMM, Cooke RJ, Ganal MW, Peeters R, Isaac P, Noordijk Y, Rendell S, Jackson J, Röder MS, Wendehake K, Dijcks M, Amelaine M, Wickaert V, Bertrand L, Vosman B. Construction and testing of a microsatellite database containing more than 500 tomato varieties. Theor Appl Genet. 2002;105:1019-1026. DOI: 10.1007/s00122-002-1038-6

[99] Ruiz JJ, García-Martínez S, Picó B, Gao M, Quiros CF. Genetic variability and relationship of closely related Spanish traditional cultivars of tomato as detected by SRAP and SSR markers. J. Am. Soc. Hortic. Sci. 2005;130:88-94. DOI:10.21273/JASHS.130.1.88

[100] Mazzucato A, Papa R, Bitocchi E, Mosconi P, Nanni L, Negri V, Veronesi F. Genetic Diversity, Structure and Marker-Trait Associations in a Collection of Italian Tomato (*Solanum lycopersicum* L.) Landraces, Theor Appl Genet. 2008;116(5):657-669.DOI 10.1007/s00122-007-0699-6

[101] Korir NK, Diao W, Tao R, Li X, Kayesh E, Li A, Zhen W.2014, Genetic Diversity and Relationships among Different Tomato Varieties Revealed By EST-SSR Markers, Genet Mol Res, 2014;13(1):43-53. DOI: 10.4238/2014

[102] Castellana S, Ranzino L, Beritognolo I. Genetic characterization and molecular fingerprint of traditional Umbrian tomato (*Solanum lycopersicum* L.) landraces through SSR markers and application for varietal identification. Genet Resour Crop Evol. 2020;67: 1807-1820 (2020). DOI:10.1007/s10722-020-00942-3

[103] Egashira H, Ishihara H, Takashina T, Imanishi S. Genetic diversity of the 'Peruviamum-complex' (Lycopersicon peruviamum L. Mil and L. chilense dun.) revealed by RAPD analysis. Euphytica.2000;116: 23-31.

[104] Archak S, Karihaloo JL, Jain A. RAPD markers reveal narrowing genetic base of Indian tomato cultivars. Current Science. 2002;82:1139-1143

[105] Sharifova S, Mehdiyeva S, Theodorikas K, Roubos K. Assessment of Genetic Diversity in Cultivated Tomato (S. lycopersicum L.) Genotypes Using RAPD Primers. Journal of Horticultural Research. 2013;21(1):83-89. DOI: 10.2478/johr-2013-0012

[106] Aguilera JG, Pessoni LA, Rodrigues GB, Elsayed AY, Silva DJH, Barros EG. Genetic variability by ISSR markers in tomato (Solanum lycopersicum Mill.). Rev. Bras. Cienc. Agrar. 2011;6: 243-252. DOI:10.5039/agraria.v6i2a998.

[107] Henareh M, Dursun A, Mandoulakani BA, Haliloğlu K. Assessment of genetic diversity in tomato landraces using ISSR Markers. Genetika, 2016;48: 25- 35 DOI:10.2298/GENSR1601025H

[108] Angelov MB, Ivanova A, Pavlov D, Ganeva ZH, Danailov P, Bojinov BM. Development of ISSR markers for a Bulgarian tomato breeding collection aiming to improve antioxidant

compounds in fruits. Bulg. J. Agric. Sci. 2017;23(3):405-410

[109] Li G, Quiros CF. Sequence-related amplified polymorphism (SRAP), a new marker system based on a simple PCR reaction: its application to mapping and gene tagging in Brassica. Theor. Appl. Genet. 2001;103:455-461. DOI:10.1007/s001220100570

[110] Al Shaye N, Migdadi H, Charbaji A, Alsayegh S, Daoud S, Al-Anazi W. Alghamdi S2 Genetic variation among Saudi tomato (*Solanum lycopersicum* L.) landraces studied using SDS-PAGE and SRAP markers, Saudi Journal of Biological Sciences, 2018;25(6):1007-1015 DOI: 10.1016/j.sjbs.2018.04.014

[111] Collard BC, Mackill DJ. Start Codon Targted (SCoT) polymorphism: A simple novel DNA marker technique for generating gene targeted markers in plants. Plant Mol. Bio. 2009;27:86-93. DOI: 10.1007/s11105-008-0060-5

[112] Abdein MA, El-Moneim DA, Taha SS, Al-Juhani WSM, Mohamed SE. Molecular characterization and genetic relationships among some tomato genotypes as revealed by ISSR and SCoT markers. Egyptian Journal of Genetics and Cytology. 2018;47(1):140-159

[113] Baldina S, Picarella ME, Troise AD, Pucci A, Ruggieri V, Ferracane R, Barone A, Fogliano V, Mazzucato A. Metabolite Profiling of Italian Tomato Landraces with Different Fruit Types, Front Plant Sci. 2016;19(7):664. DOI: 10.3389/fpls.2016.00664

[114] Tranchida-Lombardo V, Mercati F, Avino M, Punzo P, Fiore, M C, Poma I. Genetic diversity in a collection of Italian long storage tomato landraces as revealed by SNP markers array. Plant Biosyst.2018;3504, 1-10. DOI: 10.1080/11263504.2018.1478900

[115] Tranchida-Lombardo V, Aiese-Cigliano R, Anzar I, Landi S, Palombieri S, Colantuono C. Whole-genome re-sequencing of two Italian tomato landraces reveals sequence variations in genes associated with stress tolerance, fruit quality and long shelf-life traits. DNA Res.2018;25:149-160. DOI: 10.1093/dnares/dsx045

Chapter 5

Genomic Tools to Accelerate Improvement in Okra (*Abelmoschus esculentus*)

Suman Lata, Ramesh Kumar Yadav and B.S. Tomar

Abstract

Okra (*Abelmoschus esculentus* L. Moench), is an important vegetable crop with limited studies on genomics. It is considered as an essential constituent for balanced food due to its dietary fibers, amino-acid and vitamins. It is most widely cultivated for its pods throughout Asia and Africa. Most of the okra cultivation is done exclusively in the developing countries of Asia and Africa with very poor productivity. India ranks first in the world with a production of 6.3 million MT (72% of the total world production). Cultivated okra is mostly susceptible to a large number of begomoviruses. Yellow vein mosaic disease (YVMD) caused by Yellow vein mosaic virus (YVMV) of genus Begomovirus (family Geminiviridae) results in the serious losses in okra cultivation. Symptoms of YVMD are chlorosis and yellowing of veins and veinlets at various levels, small size leaves, lesser and smaller fruits, and stunting growth. The loss in yield, due to YVMD in okra was found ranging from 30 to 100% depending on the age of the plant at the time of infection. Exploitation of biotechnological tools in okra improvement programmes is often restricted, due to the non availability of abundant polymorphic molecular markers and defined genetic maps. Moreover, okra genome is allopolyploid in nature and possess a large number of chromosomes (2n = 56–196) which makes it more complicated. Genomics tools like RNA- seq. for transcriptome analysis has emerged as a powerful tool to identify novel transcript/gene sequences in non-model plants like okra.

Keywords: Improvement, NGS, transcriptome, YVMV, bhindi, marker, orphan crop

1. Introduction

Okra (*Abelmoschus esculentus* L. Moench), which belongs to Malvaceae family, is an important fruit vegetable grown throughout the tropics and warmer parts of the temperate zone. It is cultivated in India, Nigeria, Europe, Turkey, Iran, West Africa, Afghanistan, Pakistan, Burma, Japan, Bangladesh, Brazil, China, Ethiopia, Cyprus, United States and all parts of tropics. It has 1.12 million hectare area and 8.7 million tonnes production in the world. It is one the most important traditional vegetable crops of India from production point of view, as India contributes around 73% of total worlds okra production. Okra is one of the important vegetables export from India. India produces annually over 63 lakhs metric tonnes of okra from an area of 5.24 lakh hectares which is valued at Rs. 534,037 lakhs at current market rates [1]. It is fourth most important crop after tomato, brinjal and chilli from seed industry viewpoints in India. Share of hybrid seed is more than 70% in nearly 6000 metric tonnes seed market in India. Okra is proved to be a very remunerative crop for farmers, but due to Bhendi Yellow Mosaic and Enation Leaf Curl Virus disease its successful production has become a challenge for the farmers all over the country, as most of the previously bred varieties like Parbhani Kranti, P-7, Arka Anamika and Arka have lost the resistance to YVMV and ELCV diseases [2]. Therefore, against viruses development of varieties/hybrids should the continuous process to enhance the crop productivity.

Tender pods of okra are used as delicious vegetable. To a limited extent, it is used in canned, dehydrated and frozen form. It removes constipation when 2-3 fresh pods are eaten regularly. It is often included in weight loss diet as it is both fat and cholesterol free and rich in fibre. It is rich source of protein, calcium, potassium and iodine. Fresh pod contains around 88% water, 0.1% fat, 8% carbohydrate, 1.8% protein and 0.9% fibre. Okra mucilage has potential for use as food, non-food products and medicine. Dried stems and roots of okra are used for cleaning sugarcane juice from which molasses is prepared. The dry seeds are rich source of oil (18-20%) and crude protein (20-23%). Edible oil of okra has pleasant taste and odour and it is high in unsaturated fats such as oleic and linoleic acid. Its ripe, roasted seeds are also used as coffee additive or substitute after grinding. It has a vast potential as one of the foreign exchange earning crops and accounts for about 60% of the export of fresh vegetables excluding potato, onion and garlic. Fresh okra is exported to Middle East, U.K., Western Europe, Singapore and USA. Frozen okra is also exported to U.K. cultivated okra, *A. esculentus* ($2n = 130$), is natural amphidiploid from chromosome doubling of cross between *A. tuberculatus* ($2n = 58$) as one parent and *A. ficulneus* ($2n = 72$) the other probable parent.

Genomic resources are practically absent in *Abelmoschus*, only two mRNA and few coding sequences of this genus are deposited in the public domain [3]. Okra genome is allopolyploid in nature and posses a large number of chromosomes ($2n = 56–196$) which makes it more complex for genome sequencing. Some, of the complexities in non model genomes like okra can be bypassed by sequencing the transcriptome rather than the genome [4]. mRNA sequencing also known as RNA-seq. Has emerged as a powerful tool to identify novel transcript/gene sequences and to develop molecular marker in non-model plants like okra.

2. Begomoviruses infecting okra

There are at least 27 begomoviruses which infects okra; of which, two viruses i.e. Yellow Vein Mosaic Virus (YVMV) and Okra Enation Leaf Curl Virus (OELCV) most severely affect quality of pod and lowers production by reducing yield [5]. The diseases of begomoviruses are mainly transmitted by insect vector whitefly (*Bemisia tabaci*). Begomoviruses can also be transmitted by grafting; but, seed-transmission or transmission through mechanical inoculation has not yet been established [6].

The yellow vein mosaic disease of okra (YVMD) caused by Bhendi yellow vein mosaic virus (BYVMV) was first reported in 1924 from the erstwhile Bombay Presidency [7]. The begomoviruses native to the New World have only bipartite genomes (having DNA-A and DNA-B components) whereas, of Old World are mostly monopartite (have DNA-A homolog and lacks DNA-B). In India, bhendi-infecting monopartite begomoviruses were mostly associated along with a specific betasatellite, Bhendi yellow vein betasatellite (BYVB). These BYVBs have been reported to be pathogenicity determinant and found responsible for the characteristic yellow vein mosaic symptoms. Minimum16 types of begomoviruses and 4 types of beta satellites are found associated with the YVMD in different combinations [8]. Begomoviruses isolated from okra throughout the world are of monopartite nature [9]. However, tomato leaf curl New Delhi virus (ToLCNDV), which is a bipartite begomovirus and bhendi yellow vein Delhi virus (BYVDeV) also a bipartite species were characterized from okra [10].

The typical symptom of YVMD is yellowing of veins with in green leaf and if infection becomes severe infected leaves turn completely yellowish. In case of early infection of YVMV there is significant reduction i.e. up to 94 to 100% in terms of yield [11]. Occurrence of infection after 50–65 days of germination reduces the damage and loss by 49–84% [12]. Moreover, the popular varieties of okra in India have become susceptible to YVMV [13], new biotypes of whitefly vectors have surfaced and vectors have become partly resistant to the insecticides [14]. All these factors lead to decrease in overall production of okra in India. Therefore, advance biotechnological and genomic tools like RNA interference (RNAi), genome editing and sequencing along with conventional methods like transfer of resistance from wild sources are required to enhance production of okra under YVMD condition.

3. Okra Enation Leaf Curl Virus

Okra Enation Leaf Curl Virus Disease (OELCD) was first identified from Bangalore during the early 1980s in India; OELCD can reduces yield up to 80– 90% in okra [15]. The typical symptoms of OELCD are curling in leaves, thick veins, and reduced leaf surface area. In addition, the disease bearing plants become severely stunted along with small and malformed fruits which make them unsuitable for consumers. This disease is going to be the future menace of okra cultivation and needs a strategic breeding program to evolve resistance against OELCV [8].

4. Molecular marker development and gene diversity studies in okra

Isolation of purified DNA is challenging in okra due to the presence of large amounts of mucilaginous acidic polysaccharides like polygalacturonic acid and polyphenols which reduces yield as well as purity of DNA [16]. Presence of impurities in the DNA hinder the downstream processing of DNA like PCR, digestion with restriction enzymes and labelling of DNA segment [17]. Molecular markers has emerged as a potential system for evaluation of genetic variations and associations at inter and intra species level [18, 19]. Most commonly used markers for genetic studies and marker assisted breeding programme are Random Amplified Polymorphic DNA (RAPD), Amplified Fragment Length Polymorphism and Simple Sequence Repeat (SSR) [20]. However, okra lacks information on molecular markers [21].

There are only few studies where markers were used to assess the genetic diversity in okra using general DNA markers [22, 23]. Gene diversity studies reported in okra is listed in **Table 1**. RAPD was initially used in genetic diversity studies between different accessions of okra [23, 26, 27, 32]. Sequence related amplified polymorphism (SRAP) [22] have also been used in okra for diversity analysis studies. SSR markers are an important marker tool in the application of plant genetics and breeding because of their high reproducibility, multiallelic nature, codominant inheritance and good genome coverage [33]. To develop the microsatellite markers in okra, Ravishankar et al. [34] has performed sequencing of genomic DNA employing Roche 454 Titanium pyrosequencing. A total of 979,806 bp data was generated and 61,722 reads were attained. Out of 3735 contigs

Species	No. of accessions/ genotypes	Type of markers	No. of primers	PIC	Reference
Abelmoschus escullentus	48	ISSRS	—	54.55%	[24]
Abelmoschus escullentus	24	ISSRS	22	0.531929	[19]
Abelmoschus escullentus	66	(iPBS)-retrotransposons and SSRs	83 iPBS, 9 SSRs	0.66 iPBS 0.62 SSRs	[25]
Abelmoschus caillei (50), *A. esculentus* *(43)*	93	RAPD	13	—	[26]
Abelmoschus spp	39	RAPD	31	—	[27]
Abelmoschus esculentus	50	AFLP	33	12	[28]
Abelmoschus esculentus	22	AFLP	8	0.26	[29]
Abelmoschus escullentus	23	SRAP	39	—	[22]
Abelmoschus esculentus, *A. moschatus*, *A. manihot*	65	SSR	19	0.49	[3]
Abelmoschus esculentus (21) *A. tuberculatus* (1), *A. moschatus* (1), *A. manihot* (1)	24	SSR	18	0.53	[30]
A. esculentus (92) *A. tuberculatus* (1), *A. moschatus* (1), *A. moschatus subspecies tuberosus* (1), and *A. manihot* (1)	96	SSR	40	0.52	[31]

Table 1.
Gene diversity studies in okra.

obtained from assembled reads, a total of 2708 contigs had microsatellites. Finally 402 microsatellites were used for selection of 50 SSR primers for amplification in okra. This is the first report on the development of genomic SSR markers in okra using next-generation sequencing technology.

5. Next generation sequencing (NGS) and transcriptomics studies in okra

The next-generation sequencing (NGS) technology has transformed the field of molecular breeding, particularly in the identification and development of SSR markers. The advantage of NGS techniques are cost efficiency and large number of SSR can be identified in shorter time [34]. There is limited literature available in okra related to studies using genomic approaches. Transcriptome analysis has appeared as a potential approach to identify the transcript/gene sequences in the crops like okra where limited or no genome sequence information is available. The first study on transcriptome assembly in okra was reported by [3]. Both leaf and pod tissues of okra were taken for RNA sequencing and short read assembly SRA accession no. SRX206126. They have identified more than 150,000 unigenes and 935 SSRs from unigenes (**Table 2**). These SSRs were used to study genetic diversity

Okra Species	Plant organ	Objective of study	Sequencing platform	Raw reads (M)	Final assembly	Marker discovery	N50	NCBI accession	Reference
(*Abelmoschus esculentus* (L.) Moench) CV. Mahnco Arka Abhhay	leaf and pod	Transcriptome assembly	Illumina HiSeq™ 2000	26,324,557 263	150,000 unigenes	935 SSRs	321 bp	SRX206126	[3]
Abelmoschus esculentus cv. Xianzhi	Roots, stems, and leaves	Transcriptome assembly	Illumina HiSeq X Ten platform	716,330,252 716	293,971 unigenes	—	1885 bp	SRP130180	[35]
Abelmoschus esculentus	Leaves	Transcriptome assembly	Illumina NextSeq 500	206.3 million	66,382 unigenes	9,578 SSRs	1,408 bp	SRX2995608, SRX2995609, SRX2995611, and SRX2995612	[36]
Abelmoschus esculentus cv. Arka Anamika	Leaves	Genome assembly	Roche 454 GS FLX Titanium	61,722	3735 contigs	402 SSRs markers		—	[37]
Abelmoschus esculentus	Leaves	miRNA identification	Illumina and Ion torrent	207,285,863	845 novel miRNAs			PRJNA352593	[38]

Table 2.
Description of transcriptomic and genomic studies published in okra.

in diverse okra germplasm by many workers and found informative for classification and understanding of okra germplasm. Ravishankar et al. [34] first reported development of genomic SSR markers in okra using Roche 454 Titanium pyrosequencing technology. A total of 61,722 reads were generated from 979,806 bp data. These reads were assembled into 3735 contigs of which 2708 had microsatellites. Primers were designed for 402 microsatellites, from which 50 randomly selected SSR primers were standardized for amplification of okra DNA.

MicroRNAs (miRNAs) are regulatory RNAs which plays a crucial role in regulating gene expressions at post-transcriptional levels in disease conditions. Vimala Kumar et al. [38] applied next generation sequencing approach for global profiling to characterize the miRNAs and their associated pre-miRNAs. Data analysis using miRPlant revealed 128 known and 845 novel miRNA candidates. They identified 57 known miRNAs of 15 families and 18 novel miRNAs. A total of 845 novel candidates were predicted when using cotton as a reference genome which is closely related to *A. esculentus.* In 2018, Zhang and co-workers used transcriptome approach to identify the transcripts involved in the synthesis of bioactive compounds like flavonoids and polysaccharides in various organs like roots, stems, leaves, flowers, and fruits. They have identified 293,971 unique unigene sequences, 931 unigenes related to enzymes of flavonoids biosynthesis were identified and quantified. 691 unigenes encoded 13 key enzymes related to fructose and mannose metabolism. The transcriptome data will be useful for the gene expression analysis study of the genes encoding bioactive compounds in okra. Priyavathi et al. [36] reported high quality leaf transcriptome of *A. esculentus* from leaf samples. 16,307 unigenes, 76 transcription factor, 9,578 potential SSRs have been identified from *A. esculentus* leaf transcriptome. The *A. esculentus* sequence information presented in this study will be a valuable resource for further molecular genetics and functional genomics studies for the improvement of this crop plant.

6. Proteomic studies in okra

Proteomics analysis is a tool to facilitate the study of global protein expression, and to provide a wealth of information on the role of individual proteins in specific biological processes. Due to the complex allopolyploid genome of okra little attention has been paid to the genetic improvement of this crop until recently. Soil salinity is one of the main abiotic stresses limiting plant growth and agricultural productivity. Understanding the mechanisms that protect plants from salt stress will help in the development of salt-stress-tolerant crop. Using TMT-based proteomic technique in 2019, Zhan and associates analyzed the differentially expressed proteins between the NaCl-treated seedlings and control. They have identified a total of 7179 proteins, there were 317 differentially expressed proteins (DEPs), of which 165 proteins were upregulated and 152 proteins downregulated in the presence of NaCl.

7. Discussion

The molecular markers can be effectively used to enhance okra breeding programme through marker-assisted selection (MAS). Marker assisted breeding allows selection of desired trait at early stage which leads to accelerated development of improved varieties. Although, molecular markers have been broadly employed for DNA fingerprinting, gene mapping and gene tagging, seed purity testing and to know the molecular basis of heterosis in various crops, but in okra its use is still limited, therefore, it is the need of hour to use these approaches to accelerate the okra breeding programme at faster pace. The genomics and bioinformatics should also be well integrated into the programme for effective application of markers to okra breeding. A comparative genomics approach of other crop should also be applied for breeding programmes of those crops where the genome information is not available. Development of cost-effective genotyping technologies should always be the integral part of any improvement programme. There is need to use SSR and SNP based genotyping technologies as well as advanced technologies such as next generation sequencing.

Resilient resistance to begomoviruses like Yellow vein mosaic virus (YVMV) poses a serious challenge to both breeders and pathologists as these viruses are highly diverse, and constantly generate new forms via recombination. Biotechnological tools to generate resistant cultivar against Yellow vein mosaic disease (YVMD) are limited due to the lack of informative polymorphic markers, genetic maps and genome sequence information. Therefore, use of novel molecular and genomic tools will help in the accomplishing resistance against YVMV in okra. Identification of markers linked to the YVMV resistance gene/s and its pyramiding for combining multiple disease resistance genes in various backgrounds will help in okra crop improvement. In addition, genomic tools will help in elucidating the metabolic pathway governing disease resistance.

8. Conclusion

Okra is considered as a non model crop with a complex genome. Genomic studies like genome sequencing and transcriptome sequencing will help in identification of genes/transcripts for important agronomic traits like disease resistance in okra. Tools like RNAi and CRISPR/Cas9 genome editing can be employed for imparting resistance as well as functional characterization of genes. Identification of genes/transcripts and markers linked to the resistance genes will help in breeding for resistance varieties. Also, there is requirement to bred durable/stable resistance against multiple diseases. ELCV is emerging as new havoc for okra along with YVMV which may be more difficult for production of okra in future. Therefore, gene pyramiding for combating multiple disease resistance genes in various genetic backgrounds should be done. There is also need to breed varieties/hybrids tolerant to abiotic stresses like cold, moisture and salt stress in the changing climatic scenario.

Author details

Suman Lata*, Ramesh Kumar Yadav and B.S. Tomar
Division of Vegetable Science, ICAR-Indian Agricultural Research Institute, New Delhi, India

*Address all correspondence to: sumanlata3@gmail.com

References

[1] NHB databse 2018. National Horticulture Board. Ministry of Agriculture & farmer welfare, Govt. of India. http://www.nhb.gov.in.

[2] Padidam, M., Sawyer, S., & Fauquet, C. M. (1999). Possible emergence of new geminiviruses by frequent recombination. Virology, 265(2), 218-225.

[3] Schafleitner, R., Kumar, S., Lin, C. Y., Hegde, S. G., & Ebert, A. (2013). The okra (*Abelmoschus esculentus*) transcriptome as a source for gene sequence information and molecular markers for diversity analysis. Gene, *517*(1), 27-36.

[4] Hirsch, C. N., & Robin Buell, C. (2013). Tapping the promise of genomics in species with complex, non model genomes. Annual review of plant biology, 64, 89-110.

[5] Swanson, M. M., and Harrison, B. D. (1993). Serological relationships and epitope profiles of isolates of okra leaf curl geminivirus from Africa and the Middle East. Biochimie 75, 707-711. doi: 10.1016/0300-9084(93)90101-W

[6] Brown, J. K., Fauquet, C. M., Briddon, R. W., Zerbini, M., Moriones, E., and Navas-Castillo, J. (2012). "Geminiviridae," in Virus Taxonomy-Ninth Report of the International Committee on Taxonomy of Viruses, eds A. M. Q. King, M. J. Adams, E. B. Carstens, and E. J. Lefkowitz (San Diego, CA: Elsevier Academic Press), 351-373.

[7] Kulkarni, C. S. (1924). Mosaic and Other Related Diseases of Crops in the Bombay Presidency. Poona Agriculture College Magazine, Pune

[8] Singh, B., Sanwal, S., Venkataravanappa, V., and Halder, J. (2013). "Breeding strategies for biotic stresses of okra: prospects and potential," in Abstract Book of National Symposium on Abiotic and Biotic Stress Management in Vegetable Crops, (Varanasi), 32-33.

[9] Jose, J., and Usha, R. (2003). Bhendi yellow vein mosaic disease in India is caused by association of a DNA ß satellite with a begomovirus. Virology 305, 310-317. doi: 10.1006/viro.2002.1768

[10] Venkataravanappa, V., Reddy, C. N. L., Jalali, S., and Reddy, M. K. (2012). Molecular characterization of distinct bipartite begomovirus infecting bhendi (Abelmoschus esculentus L.) in India. Virus Genes 44, 522-535. doi: 10.1007/s11262-012-0732

[11] Pun KB, Doraiswamy S (1999) Screening of plant species for the presence of antiviral principles against Okra yellow vein mosaic virus. Indian Phytopath 52:221-223.

[12] Sastry KSM, Singh SJ (1974) Effect of yellow vein mosaic virus infection on growth and yield of okra crop. Indian Phytopath 27:294-297.

[13] Borah GC, Saikia AK, Shadeque A (1992) Screening of okra genotypes for resistance to yellow vein mosaic virus disease. Indian J Virol 8:55-57.

[14] Rashida P, Sultan MK, Khan MA, Noor-Ul-Islam (2005) Screening of cotton germplasm against cotton leaf curl Begomovirus (CLCuV). J Agri Soc Sci 3:35-238.

[15] Singh, S. J. (1996). Assessment of losses in okra due to enation leaf curl virus. Indian J. Virol. 12, 51-53.

[16] Aljanabi, S. M., Forget, L., & Dookun, A. (1999). An improved and rapid protocol for the isolation of polysaccharide-and polyphenol-free

sugarcane DNA. Plant Molecular Biology Reporter, 17(3), 281-282.

[17] Sahu, S. K., Thangaraj, M., & Kathiresan, K. (2012). DNA extraction protocol for plants with high levels of secondary metabolites and polysaccharides without using liquid nitrogen and phenol. International Scholarly Research Notices, 2012.

[18] Chakravarthi, B.K. and Naravaneni, R. (2006) SSR Marker Based DNA Fingerprinting and Diversity Study in Rice (Oryza sativa. L.). African Journal of Biotechnology, 5, 684-688.

[19] Yuan C. Y., Zhang C., Wang P., Hu S., Chang H. P., Xiao W. J. et al. 2014 Genetic diversity analysis of okra (*Abelmoschus esculentus* L.) by inter-simple sequence repeat (ISSR) markers. Genet. Mol. Res.13, 3165-3175.

[20] Sawadogo, M., Ouedraogo, J. T., Balma, D., Ouedraogo, M., Gowda, B. S., Botanga, C., et al. (2009). The use of cross species SSR primers to study genetic diversity of okra from Burkina Faso. Afr. J. Biotechnol. 8, 2476-2482. doi: 10.5897/AJB08.1126.

[21] Sunday E., Aladele O. J. Ariyo L. and Robert D. L. 2008 Genetic relationship among West African Okra (Abelmo schuscaillei) and Asian genotypes (Abelmo schusesculentus) using RAPD. Afri. J. Biotechnol. 7, 1426-1431.

[22] Gulsen, O., Karagul, S., Abak, K., (2007) Diversity and relationships among Turkish okra germplasm by SRAP and phenotypic marker polymorphism. Biol. Bratislava 62, 41-45.

[23] Prakash,K., Pitchaimuthu,M. and Ravishankar, K.(2011)Assessment of genetic relatedness among okra genotypes[Abelmoschus esculentus(L) Moench] using RAPDMarkers. Electronic J Plant Breed2: 80-86.

[24] Yuan, C. Y., Wang, P., Chen, P. P., Xiao, W. J., Zhang, C., Hu, S., ... & Guo, X. H. (2015). Genetic diversity revealed by morphological traits and ISSR markers in 48 Okras (Abelmoschus escullentus L.). Physiology and Molecular Biology of Plants, *21*(3), 359-364.

[25] Yıldız, M., Koçak, M., & Baloch, F. S. (2015). Genetic bottlenecks in Turkish okra germplasm and utility of iPBS retrotransposon markers for genetic diversity assessment. Genetics and Molecular Research, *14*(3), 10588-10602.

[26] Aladele, S.E., Ariyo, O.J., de La Pena, R., (2008). Genetic relationships among West African okra (Abelmoschus caillei) and Asian genotypes (Abelmoschus esculentus) using RAPD. Afr. J. Biotechnol. 7, 1426-1431.

[27] Martinello, G.E., Leal, N.R., Amaral Jr., A.T., Pereira, M.G., Daher, R.F., 2001. Comparison of morphological characteristics and RAPD for estimating genetic diversity in *Abelmoschus* spp. Acta Hort. 546, 101-104.

[28] Kyriakopoulou, O. G., Arens, P., Pelgrom, K. T., Karapanos, I., Bebeli, P., & Passam, H. C. (2014). Genetic and morphological diversity of okra (Abelmoschus esculentus [L.] Moench.) genotypes and their possible relationships, with particular reference to Greek landraces. Scientia Horticulturae, *171*, 58-70.

[29] Akash, M. W., Shiyab, S. M., & Saleh, M. I. (2013). Yield and AFLP analyses of inter-landrace variability in okra (Abelmoschus esculentus L.). Life Science Journal, *10*(2), 2771-2779.

[30] Fougat, R. S., Purohit, A. R., Kumar, S., Parekh, M. J., & Kumar, M. (2015). SSR based genetic diversity in Abelmoschus species. Indian J. Agr. Sci, *85*, 1223-1228.

[31] Kumar, S., Parekh, M. J., Fougat, R. S., Patel, S. K., Patel, C. B., Kumar, M., & Patel, B. R. (2017). Assessment of genetic diversity among okra genotypes using SSR markers. Journal of plant biochemistry and biotechnology, 26(2), 172-178.

[32] Nwangburuka C. C., Kehinde O. B., Ojo D. K., Denton O. A. and Popoola A. R. 2011 Morphological classification of genetic diversity in cultivated okra, Abelmoschu sesculentus (L.) Moench using principal component analysis (PCA) and single linkage cluster analysis (SLCA). Afr. J. Biotechnol. 10, 54

[33] Bertini C. D., Schuster I., Sediyama T., de Barros E. G. and Moreira M. A. 2006 Characterization and genetic diversity analysis of cotton cultivars using microsatellites. Genet. Mol. Biol. 29, 321-329.

[34] Ravishankar, K. V., Muthaiah, G., Mottaiyan, P., & Gundale, S. K. (2018). Identification of novel microsatellite markers in okra (Abelmoschus esculentus (L.) Moench) through next-generation sequencing and their utilization in analysis of genetic relatedness studies and cross-species transferability. *Journal of genetics*, *97*(1), 39-47.

[35] Zhang, C., Dong, W., Gen, W., Xu, B., Shen, C., & Yu, C. (2018). De novo transcriptome assembly and characterization of the synthesis genes of bioactive constituents in Abelmoschus esculentus (L.) moench. *Genes*, *9*(3), 130.

[36] Priyavathi P, Nagesh S, Kavitha VVK, Johnson C, Gopal P (2018) Comprehensive Leaf Transcriptome of a Non-model Plant, *Abelmoschus esculentus* for the Functional Genomics Studies. J Genet Genome Res 5:036. doi. org/10.23937/2378-3648/1410036

[37] Ravishankar, K. V., Muthaiah, G., Mottaiyan, P., & Gundale, S. K. (2018). Identification of novel microsatellite markers in okra (*Abelmoschus esculentus* (L.) Moench) through next-generation sequencing and their utilization in analysis of genetic relatedness studies and cross-species transferability. Journal of genetics, 97(1), 39-47.

[38] Velayudha Vimala Kumar, K., Srikakulam, N., Padbhanabhan, P., & Pandi, G. (2017). Deciphering microRNAs and Their Associated Hairpin Precursors in a Non-Model Plant, Abelmoschus esculentus. Non-coding RNA, *3*(2), 19.

Chapter 6

Nutritional Value and Phytochemical Content of Crop Landraces and Traditional Varieties

Inés Medina-Lozano and Aurora Díaz

Abstract

Over the years, crops have been improved through breeding, mainly to increase production and, secondly, to introduce resistance to diseases and to achieve tolerance to abiotic stresses, these two latter by resorting to Crop Wild Relatives (CWR). This has resulted, in most cases, in homogeneous and nutritionally poor commercial varieties. Landraces and traditional varieties, barely taken into account, are key resources as they retain nutrients frequently "washed away" in the commercial varieties and also harbour a great genetic variability. They could represent a shortcut when compared to CWR in breeding, saving time and resources. The consumer's growing interest in health and food quality has caused breeders to redirect their attention toward them. This chapter provides information about the content in compounds with health benefits, such as phenolics, minerals, vitamins, etc., of landraces and traditional varieties of the most important crops, which could help to obtain healthier and more nutritious products.

Keywords: biofortification, carotenoids, micronutrients, health-promoting compounds, minerals, plant breeding, phenolic compounds, vitamins

1. Introduction

1.1 Landraces and traditional varieties: similarities, differences and comparison with wild species and commercial varieties

In the wide spectrum of plant material in terms of domestication and/or breeding, the concepts seem to be clear in both extremes, wild forms and commercial varieties. On one hand, the wild plants (either the Crop Wild Relatives, CWR, or those belonging to more distant gene pools) are those that have not been domesticated or subject to processes of artificial (human) selection and breeding. They do not exhibit traits typically present in cultivated plants, like uniform seed germination and homogeneous fruit ripening, or desirable characteristics present in those plants destined to human consumption, mainly related to quality (**Figure 1**). On the other hand, commercial varieties are those obtained by a breeding programme aimed to improve certain traits of the crop and that differ from other existing varieties by distinctive properties, which are uniformly expressed, and transferred in a stable way to the subsequent generations (**Figure 1**).

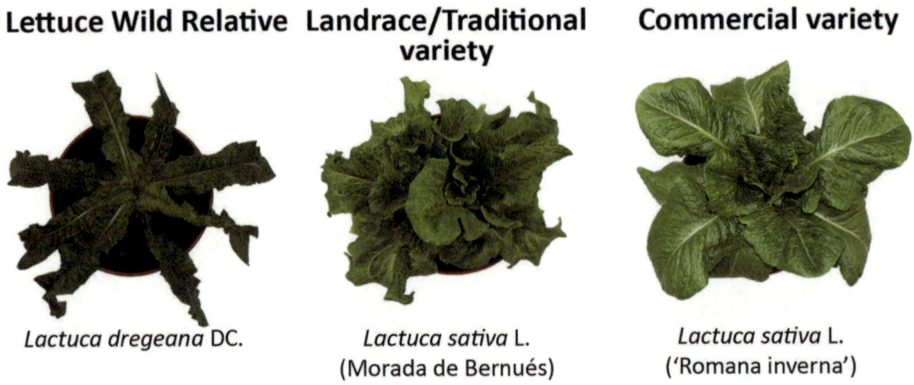

Figure 1.
Examples of phytogenetic resourses within the genus Lactuca: A lettuce wild relative (Lactuca dregeana DC.*) and two cultivated forms (Lactuca sativa* L.*), a landrace/traditional variety (Morada de Bernués), and a commercial variety ('Romana inverna').*

In between those two ends, a wide plethora of intermediate forms can be found. That is a grey area with blurred boundaries, what explains the general lack of consensus in even defining the plant material. In many cases, different terms have been used to refer to the same (or similar) type of plant, like ecotype, landrace, race, farmer variety, folk variety, local variety, traditional cultivar, etc. [1]. Even if some definitions are contradictory, there seems to be some recurrent ideas when authors refer to landraces and traditional varieties.

Landraces are profusely described in the literature as autochthonous cultivars or, at least, cultivars that have been grown in a certain area since ancestral times and, hence, are adapted to local growing conditions and uses through natural selection but without any active intervention from farmers. There are several terms difficult to verify in that definition. It does not seem easy to trace back the origin of the cultivars, especially if we take into account that the crop dispersals and the human migrations are inseparable. Besides, even if they have been cultivated in a region for a long period of time and, hence, they are adapted to the predominant environmental conditions, that does not imply that they exhibit a great tolerance to adverse conditions, biotic and abiotic stresses as stated before [1, 2]. Actually, the adverse edaphic, climatic and phytosanitary conditions would be mitigated even by the most traditional low input agricultural systems in comparison to those that the wild plants would have to face in the same region. Finally, it is difficult to defend the idea of farmers growing a cultivar for generations without carrying out any type of selection of the outstanding individuals, even if it is not fully conscious, as stated mainly in the earliest definitions [3, 4]. In fact, the agriculture procedures (seeding, harvesting ...) exert an artificial selection under which the most suitable genotypes for those cultural practises, prosper (and they probably rely on them for their survival, in return). Furthermore, in a scenario in which only the natural selection is acting, the resulting plants would probably be more similar to their wild relatives and less to the bred cultivars (and that is not the case with the landraces). In more recent definitions, the idea of a more or less directed human selection has been embraced [5–7], even if it cannot be considered a formal breeding programme [8].

In contrast with landraces, traditional varieties (also called folk varieties or farmer-bred varieties), have usually been defined like those that have been maintained by active selection and/or breeding by farmers. And, if this is the main difference between landraces and traditional varieties (as the latter are also cultivated locally and are well adapted to the particular climatic and growing conditions), is it really possible to distinguish them? How do we determine if a certain variety is the product of merely natural selection or the human intervention has also played a role on it? Is it actually possible to separate both processes? It could be that the question is nowadays irrelevant and the important aspect is that, either if we call them landraces or traditional varieties (**Figure 1**), they consist of dynamic populations that harbour enough genetic variability to show adaptability to local conditions and plasticity to overcome eventual changes, even if they can be fairly uniform for the selected traits. That broad genetic base would explain that, under eventual adverse conditions, they are still able to yield stably (though moderately), as some genotypes within the population will possibly show a better performance. These aspects were early emphasised by the plant breeder Harlan [9] when stated that some of the most important characteristics of landraces are their genetic diversity and dynamism, what has also been adopted in more modern times by other authors [10, 11]. Harlan also pointed out that they are the result of millennia of natural and artificial selection, as a way of integrating these two indiscernible processes. Another approach to overcome this thorny aspect consisted of eliminating the type of selection undergone in the definition of the landraces [12]. Any realistic and updated definition of this type of plant material will have to include the impact of agriculture and, hence, the human influence in their evolution as proposed recently [13].

Another aspect that blurs the lines between landraces and traditional varieties is the gene flow between them. With the availability of molecular markers and Next-Generation Sequencing (NGS) techniques, it is possible to trace the allele introgression from cultivated (all types) to wild plants and *vice versa*. Even if there were landraces exclusively product of natural selection and traditional varieties obtained by men selection, obviously, gene transfer could have also happened between them, especially taking into account that exchanging plant material is a common practise among farmers.

In any case, the main differences when compared to commercial varieties are that landraces and traditional varieties do not always have a traceable origin, they exhibit a great diversity and, precisely for that reason, most of their traits are less uniform within them and less stable through the descendants.

1.2 Importance and conservation of landraces and traditional varieties in germplasm banks worldwide

The great variability harboured by the landraces and traditional varieties is one of their most outstanding characteristics. Historically, all this richness had been preserved and used (a vicious circle of cause and effect) by the agriculturalists. That situation started to change when the erosion of the plant genetic resources became patent for scientists and breeders, not only in the case of landraces and traditional varieties, but also concerning the wild species. Since then, the germplasm banks have assumed a principal role in safeguarding this plant biodiversity [14]. The strategy has revealed itself so successful that, according to the World Information and Early Warning System (WIEWS) on Plant Genetic Resources for Food and Agriculture (PGRFA), approximately 5.4 million accessions are being conserved in over 710 genebanks from 103 countries and 17 international/regional centres [15]. Landraces and traditional varieties represent the heart of the collections, what becomes obvious when the numbers of the different types of plant resources are consulted. As an example, in Genesys, which is a portal that supplies not only seeds, but also characterisation and evaluation data about PGRFA from germplasm banks around the world [16], landraces and traditional varieties account for the highest proportion of accessions (37%), followed by breeding and research material (27%), advanced and improved cultivars (19%), and finally, wild forms (17%) (**Figure 2**).

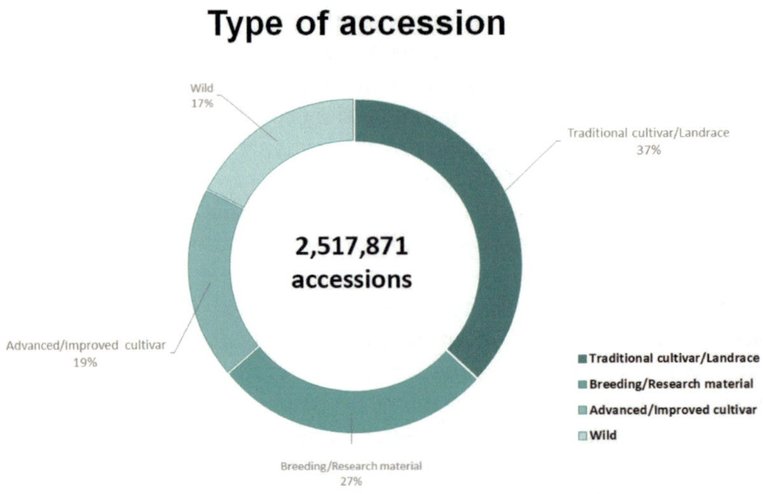

Figure 2.
Relative amount of the different types of accessions attending to their biological status (excluding the "not specified" material) hold at Genesys [16], the online platform which harbour information about PGRFA conserved in genebanks worldwide.

The high genetic variability exhibited by landraces and traditional varieties obviously translates into characteristics that could be desirable in modern varieties, particularly those related to their nutritional value and content of health-promoting compounds, which is the subject under discussion in this chapter. In modern breeding programmes, flavour selection has prevailed over nutritional quality. That explains why, for instance, modern lettuce varieties have almost lost their ancestral bitterness. That is a direct consequence of the drastic decrease in the content of sesquiterpene lactones, which are not only responsible for the bitter taste but have also beneficial properties for the plant itself and for the animals that feed on it [17]. In other cases, the main objective has not been to eliminate flavours detrimental to the taste but to enhance the pleasant ones. This is the case for sweet corn. Its sugar content has been escalating over the last decades by the selection of varieties with an increasing polysaccharide content: sugar-enhanced varieties, supersweet or extrasweet varieties, high sugar varieties... [18]. The side effect has been the disappearance of non-sweet dark-grain primitive varieties rich in anthocyanins, which happen to be powerful antioxidants with an important role for health by preventing cardiovascular diseases and by presenting anti-cancer activity [19, 20]. The landraces and traditional varieties were shaped under very different criteria, what does not necessary implies that they are better, for instance, from a nutritional perspective, than any commercial variety, though they contribute to increase the agrodiversity and to enrich the diet. In this sense, the germplasm banks can act as gene reservoir to improve crops, allowing us to dive for valuable characteristics to obtain all types of plant material (coming from crosses between different traditional varieties, between traditional varieties and CWR, between traditional varieties and breeding material ...), using both conventional and biotechnological tools.

2. Essential micronutrients

Essential micronutrients are nutrients that must be obtained in the diet as they cannot be synthesised by the human body. They are required in small quantities and usually consist of vitamins and minerals. Micronutrients play vital roles in human health, so their deficiencies can be devastating. These deficiencies, also known as hidden hunger, are mainly consequence of micronutrient low concentrations in the daily diet, resulting in malnutrition that is considered an important global problem of public health, especially in developing countries. In addition, the impact is more serious in women of reproductive age (especially pregnant women) and under-five children due to their higher micronutrient requirements. In fact, maternal and child malnutrition or micronutrient deficiencies affect approximately half of the world's population [21]. Nevertheless, hidden hunger also affects developed countries due to low quality food or bad habits, like extreme diets to lose weight or alcohol and drug abuse.

Generally, fruits and vegetables are the sources of vitamins and minerals, but their concentrations in most plant foods are not sufficient to reach the daily dietary requirements, even if the recommended amounts are consumed [22]. Besides, micronutrient content usually depends on the plant genotype, among other factors like environmental conditions, time of harvest, etc. Cases in which landraces and traditional varieties of important crops exhibit higher contents of micronutrients than commercial and modern cultivars are described here. They actually could play a key role in human health by supplying an enhanced nutrition.

2.1 Organic micronutrients: vitamins

Vitamins are a diverse group of organic molecules that are essential in trace quantities for a proper metabolism of all living organisms and are mainly synthesised by plants and microorganisms. Vitamins can be classified into fat-soluble (A, D, E and K) or water-soluble (vitamin B-complex, C and H) compounds. Their main function is to act as cofactors for many enzymes and as natural antioxidants, both in plants and animals. In addition, some vitamins play specific roles, for example, in human vision (vitamin A) [23] or as hormones implied in calcium and phosphorus homeostasis in the blood stream (vitamin D) [24], and many of them are indispensable to prevent chronic diseases [19, 20].

Plants, mostly fruits and vegetables, are the main source of vitamins for humans. However, their concentration in the edible portions of most important crops is usually below the recommended daily intake, which entails important implications for global human health [24]. Interestingly, some landraces exceed these minimal requirements or, at least, they are richer than commercial cultivars in these micronutrients, especially for vitamins A, C and E.

2.1.1 Vitamin A

Vitamin A is a fat-soluble vitamin group that includes retinol and its derivatives, like retinoic acid and retinal, among many others [25]. Besides, among the large group of compounds known as carotenoids, there are some that can act as precursors of vitamin A, known generically as provitamin A, such as α-carotene, β-cryptoxanthin and β-carotene, the most abundant and nutritionally active within them all. The richest sources of vitamin A are from animal origin, whereas carotenoids are synthesised mainly by plants, but also by some fungi and microorganisms.

Carotenoids play important roles in plant metabolism: acting as pigments in different tissues, mediating plant–animal interaction for pollination or seed dispersal, participating in cell photoprotection against photooxidative damage and heat stress, and protecting membranes from lipid peroxidation thanks to their antioxidant activity [26].

In humans, provitamin A is involved in vision, immune responses, cellular growth, development and reproduction [23]. Vitamin A deficiency is one of the micronutrient deficiencies with more devastating consequences for health. It is the main cause of preventable blindness in children and pregnant women, especially in low-income countries, and raises the risk of suffering several diseases and of dying as a result of severe infections. Between 250,000 and 500,000 vitamin A deficient children become blind every year, half of them dying 12 months later [27]. Therefore, it is a question of the utmost importance to know what plant-based foods contain high levels of provitamin A.

The β-carotene content was measured in two Spanish landraces of tomato (*Solanum lycopersicum* L.) and in the commercial variety 'Moneymaker' [28]. A higher concentration of this carotenoid was found in green fruits of the two landraces when compared to 'Moneymaker', whereas in ripe fruits, only the landrace Negro Yeste had a higher amount, even more than double. Also in comparison with the commercial variety 'Moneymaker', three tomato landraces, two from Italy and one from Guatemala, showed a significantly higher β-carotene content [29]. In other study carried out in melon (*Cucumis melo* L.), landraces of different origins exhibited the highest levels of β-cryptoxanthin and β-carotene compared with commercial melons [30]. In an analysis of the β-carotene content of mungbean (*Vigna radiata* L. Wilzeck), the landrace VI000323 B-G happened to have grains significantly richer than two improved mungbean lines at their maturity stage [31]. Though modern wheat (*Triticum* spp.) varieties were not analysed, old varieties (from the 1900–1960 breeding period) were included as reference, and the average values obtained for β-carotene and β-cryptoxanthin were significantly higher in the wheat landraces than in the old cultivars [32]. Also in landraces of pepino (*Solanum muricatum* Ait.) from the Andean region [33] and in the landrace G-4615 of sweet potato (*Ipomoea batatas* (L.) Lam.) from Solomon Islands [34], higher contents of β-carotene than in improved varieties have been obtained.

2.1.2 Vitamin C

Vitamin C is a water-soluble vitamin that comprises ascorbic acid (AA), the main biologically active form, and its oxidation product, dehydroascorbic acid (DHAA), easily convertible into AA in the human body [35]. In plants, vitamin C plays relevant roles in metabolic and defence processes, as it is an important antioxidant in the ascorbate-glutathione pathway, it protects enzymes with prosthetic metal ions, it is a cofactor for many enzymes (including those involved in cell wall synthesis), it is involved in photosynthesis and respiration, etc. [36].

In humans, it is crucial in some metabolic processes as it participates in collagen formation and inorganic iron absorption, and contributes to a healthy state by reducing the cholesterol levels, preventing chronic diseases and enhancing the immune system by its antioxidant action [37]. The main consequence of vitamin C deficiency is scurvy and, although relatively few people suffer this deficiency, the benefits of the micronutrient are evident, so it is important to find vitamin C-rich plant food.

Some studies have reported a higher content in vitamin C in crop landraces with respect to commercial varieties. For example, 17 Spanish melon traditional varieties were evaluated and most of them had significantly higher values of AA when compared with 10 commercial accessions of reference, in some cases even doubling the AA values of the commercial variety within the same market class (Piel de Sapo, Yellow, Ananás...) [38]. Traditional varieties of lettuce (*Lactuca sativa* L.) from Aragón (Spain) have also been reported to have higher average contents in vitamin C than commercial varieties, especially AA content [39, 40]. Some Spanish landraces of eggplant (*Solanum melongena* L.) had also a higher concentration of both, AA and DHAA, than commercial hybrids [41]. In other experiment, four to seven traditional varieties of tomato contained higher concentrations of vitamin C than the commercial variety 'Baghera', with significant differences for the traditional varieties CIDA-62 and BGW-004123. In addition, CIDA-62 fruits showed the highest antioxidant activity, whereas the lowest was observed in the commercial variety [42]. Other authors also reported 10 indeterminate tomato landraces that exhibited significantly higher AA contents than the commercial variety 'Moneymaker' [29]. In analyses of the AA content in accessions of garlic (*Allium sativum* L.) from Plugia region (Italy), the six landraces evaluated had a higher content than the commercial cultivar used as reference [43]. Higher contents of total vitamin C have also been obtained in grains of the mungbean landrace VI000323 B-G from Taiwan [31], in the Greek onion (*Allium cepa* L.) landrace Vatikiotiko [44] and in two rare landraces of Italian turnip (*Brassica rapa* L. subspecies *rapa*) [45] when compared with commercial and improved varieties.

2.1.3 Vitamin E

Vitamin E is a fat-soluble vitamin group made up of tocopherols and tocotrienols, a group of lipid-soluble compounds. Both tocopherols and tocotrienols can present four different methylated forms, α, β, γ and δ, and although all of them are antioxidants, α-tocopherol is the most abundant form and has the highest activity [46].

In plants, the main function of vitamin E is as antioxidant, quenching and scavenging singlet oxygen, controlling the extent of lipid peroxidation, preserving the integrity of the membranes, and protecting against photoinhibition and photo-oxidative stresses [36].

In humans, vitamin E also acts as a potent antioxidant and it is involved in multiple physiological processes, such as regulation of gene expression and cognitive performance. Besides, vitamin E plays a key role in maintaining a healthy state by controlling the inflammation, enhancing the immune function and preventing light-induced pathologies of the skin and eyes, and degenerative disorders like cardiovascular diseases, atherosclerosis and cancer. Its deficiency is common in developing countries and affects mainly children and the elderly, and can cause haemolytic anaemia in premature babies and neurological and ophthalmological disorders as well as myopathy in children. In developed countries it is rare and only appears in some stages of development, such as in premature babies, and specific conditions, like in digestive and genetic pathologies [24].

A total of 28 Korean accessions of soybean (*Glycine max* L.) were evaluated and the highest total tocopherol contents were observed in the seeds of the 7 landraces analysed, especially in HaNagari, in comparison with the modern cultivars developed by cross-breeding, in which paradoxically the content decreased gradually with the year of registration [47]. Furthermore, a strong negative correlation between tocopherol contents and lipid peroxidation was observed (what demonstrates the vitamin E role in oxidative stress tolerance), with the soybean landraces showing the lowest lipid peroxidation. In wheat, higher contents of tocopherols and tocotrienols were obtained for some landraces in comparison with modern cultivars when individual genotypes were analysed [48]. Hazelnut (*Corylus avellana* L.) is also a good source of vitamin E and an Argentinian landrace has been reported to have the highest total tocopherol content in comparison with different commercial cultivars [49]. The total contents of tocopherols and tocotrienols, as well as total vitamin E, were higher in traditional red rice (*Oryza sativa* L.) varieties than in three light brown new-improved varieties [50].

2.2 Mineral micronutrients

Mineral micronutrients are inorganic elements required in small quantities to play vital functions in both, plants and animals. The nutrient classifications are dynamic and, sometimes, even contradictory. That is because, on one hand, the limit between small and big quantities that determine the inclusion of an element in the micronutrient or macronutrient list can result arbitrary. On the other hand, new discoveries about the participation of some elements in important physiological mechanisms cause their transfer from the "nonessential" to the "essential" list. Magnesium (Mg) is a clear example of discrepancies on the first criteria as, depending on the author, is described as micronutrient or macronutrient as ranks in an intermediate position in terms of recommended daily allowances [51]. Regarding the second criteria, some minerals like boron (B) have been known to be essential for plant nutrition for a long time but it has not been until a few decades ago that its important effect on human nutrition was noted [52].

In plants, mineral micronutrients participate in different physiological processes of primary and secondary metabolism, like photosynthesis, electron transfer, activation of enzymes, cell defence, hormone perception, gene regulation... So, mineral deficiencies affect the plant life cycle seriously, causing even plant death in the severest cases [53].

In humans, more than 22 mineral elements (altogether micro- and macronutrients) are essential and they can be obtained with an appropriate diet [51]. Nevertheless, mineral deficiencies are very common, especially in developing countries, and their consequences, such as learning disabilities in children, increased morbidity and mortality, low productivity at work..., are detrimental for humans. Iron (Fe), zinc (Zn) and iodine (I) are the mineral elements most frequently lacking in the diet and their deficiencies, together with vitamin A deficiency, are responsible for about 12% of the deaths among under-five children globally [21]. Fe is important for oxygen transport and haemoglobin formation, and its deficiency is the main cause of preventable iron-deficiency anaemia, poor cognitive development, and maternal and child deaths [54, 55]. Zn plays a central role in growth, development and in the normal functioning of the immune system, so its deficiency hampers growth, alters immunity and also causes diarrhoea among children [56, 57]. Moreover, both deficiencies are also associated with childhood stunting. I is a component of the thyroid hormones and a strong antioxidant. Its deficiency can also cause growth impairments and, in addition, thyroid enlargement (goitre), hypothyroidism, pregnancy loss, infant mortality and cognitive and neuron psychological impairments [58]. On the other hand, manganese (Mn), copper (Cu) or selenium (Se) deficiencies are not a global issue, but they are common in some populations of developing countries, specifically in parts of China, India and Africa [51].

Many landraces of horticultural crops are reported to present higher contents of minerals and oligoelements than commercial varieties. In a study carried out with seeds of Turkish lentils (*Lens culinaris* Medik.), the average values of all the micro-minerals quantified (Cu, Fe, Mn, and Zn) were higher in the landraces than in the commercial cultivars, being Kahmar1 the richest in Zn and Cu, Diykub in Fe, and Kahmar2 in Mn [22]. Also in Turkey, higher contents of Zn and Se have been observed in common bean (*Phaseoulus vulgaris* L.) landraces than in modern varieties, especially in the landrace LR05 [59]. The Greek onion landrace Vatikiotiko [44] and a Greek garlic landrace [60] were both richer in Zn, Mn and Fe than well-established onion cultivars and hybrids commercialised in Greece and a garlic commercial cultivar used as control, respectively. In addition, the mineral content of the onion landrace was even higher than the suggested by USDA (United States Department of Agriculture) for raw onions as a standard reference, especially for

Fe. Results obtained in chickpea (*Cicer arietinum* L.) revealed that landraces from Kyrgyzstan presented higher average values for Fe, Mn, Cu and Zn compared with a breeding line [61]. In Andean landraces of pepino [33], in several eggplant landraces from Spain and Cuba [62], and in landraces of mungbean [31], higher contents in Fe and Zn than in commercial and modern varieties have also been reported.

For cereal crops there are also several studies in which landraces are reported to be richer in mineral micronutrients than commercial varieties. Wheat is one of the most important cereal crops worldwide and there are many studies on wheat landraces. The maximum contents in Fe, Cu, Zn, Mn, and Se were observed in wheat landraces from Canary Islands in comparison with the commercial cultivar 'Vitrón' [63]. Other authors [64] also reported landraces and old cultivars of wheat with a higher average concentration of B, Cu, Fe, and Zn, and of Cu, Fe, Zn, and Mn, respectively, than modern cultivars. Similarly, the average content in Fe, Zn, Mn, Cu, and strontium (Sr) in wheat grain was reported to be significantly higher in 12 Sicilian landraces than in 3 modern cultivars [65]. Other study found seven Afghan wheat landraces with higher content in Fe than reference lines in three different locations [66]. In the case of rice, two Indian landraces showed a higher content in Zn in brown and even polished (considered a poor source of micronutrients) grains than the commercial variety 'BPT 5204' ('Samba Mahsuri'), very appreciated for its high yield and quality [67].

3. Health-promoting phytochemicals

Plant-based foods are rich in different phytochemicals with health-promoting properties for the human body, in spite of not being essential nutrients. Polyphenols and carotenoids are the most important ones among these plant phytochemicals. Unlike micronutrients, their deficiencies in humans are not devastating, but their health benefits are very significant.

3.1 Polyphenols

Phenolic compounds (monophenols and polyphenols) are one of the most abundant and widely distributed groups of chemicals in plants, with more than 8,000 structurally-different compounds currently identified [68]. Particularly, polyphenols are characterised by the presence of aromatic rings with one or more hydroxyl groups and, depending on the basic chemical structure, they are classified in at least 10 different types. However, there is a growing tendency to group them in 2 main categories: flavonoid (flavones, flavonols, flavanols, flavanones, isoflavones, and anthocyanins) and non-flavonoid (phenolic acids, stilbenes, lignans, xanthones, and tannins) compounds. In plants, polyphenols are involved, on one hand, in crucial biological processes, such as cell division, development, hormonal regulation, reproduction, photosynthesis, pigmentation and pollinator attraction, and, on the other hand, in protection mechanisms against oxidative damage due to radiation or biotic stress (pathogens), among other causes, thanks to their antioxidant properties [69].

Polyphenols seems to be the main contributor to the total antioxidant activities of fruits, with flavonoids being the most abundant in human diets. The health-promoting effects associated with phenolic compounds include the elimination of free radicals, as well as the prevention of chronic diseases, such as cancer, diabetes and cardiovascular and neurodegenerative diseases [68].

There are a number of studies in which different polyphenols are more abundant in horticultural crop landraces than in commercial cultivars. This could be because some polyphenols contribute to the bitterness and astringency of the food, what could have been negatively selected in modern breeding programmes. Tomato is one of the most important crops worldwide and it is very rich in polyphenols. Several Italian and Spanish landraces have been reported to have higher contents of total phenolic compounds than the commercial varieties 'Brigade' and 'Moneymaker', with significant higher levels of the flavonoids quercetin-3-rutinoside, kaempferol-*O*-rutinoside and kaempferol-*O*-glucoside in the case of the Spanish landraces [29, 70]. Nevertheless, polyphenols are abundant in many other crops. For example, different Spanish landraces of eggplant exhibited the highest average and individual contents of total phenolic compounds when compared with several commercial cultivars in two independent studies [41, 62]. Other study found higher levels of chlorogenic acid in three Italian landraces of carrots (*Daucus carota* L.) in comparison with a commercial cultivar used as reference [71]. Landraces of pepino from the Andean region also exhibited a higher average content of total phenolics than commercial cultivars [33]. Two rare Italian landraces of turnip showed similar concentration of total phenols between them, which was up to a 61% higher than in the commercial genotype also included in the study [45]. An Ecuadorian landrace of sweet potato showed the highest content in two particular anthocyanins (peonidin and cyaniding glucosides) when compared with several improved varieties [34]. Regarding phenolic acids and flavonoids, significant higher contents were observed in landraces of mungbean [9], garlic [43], and apple (*Malus domestica* Borkh.) [72], in comparison with improved lines and commercial varieties. Finally, in winery by-products from Majorcan landraces of grape (*Vitis vinifera* L.), the highest values of total anthocyanins, tannins, and total phenolic compounds were observed in the Escursac red landrace, with the commercial variety 'Cabernet Sauvignon' used as reference [73].

In the case of cereals, also some landraces have been reported to be richer in polyphenols than commercial cultivars. In extracts of wheat bread flour, the landrace Biancola showed higher contents of flavonoids and total phenolic compounds than three modern cultivars, as well as higher reducing power and lipid peroxidation inhibition levels [74]. Similarly, the landrace Gentil Rosso had a much higher amount of total, free, and bound polyphenols than three modern and five old cultivars [75]. In extracts of wheat grains, the highest contents of the 13 phenolic compounds identified were found in landraces when compared with commercial cultivars, especially in Tumminia SG3, Tripolino, Scavuzza, and Urria [76]. In maize (*Zea mays* L.), several Mexican landraces have been reported to have the highest content of phenylpropanoids in comparison with two commercial genotypes, especially Sinaloa 35, which contained exceptionally high levels of diferulates [77]. Also in maize, the Italian landrace Rostato Rosso contained a higher concentration of anthocyanins than an inbred line and a hybrid assayed [78]. Finally, in rice, traditional red-grained varieties of Sri Lanka exhibited significantly stronger antioxidant activity and higher total phenolic content in both, bran and grains, than light brown-grained newly improved varieties, with proanthocyanidins and phenolic acids among the most abundant phenolic compounds identified [50].

3.2 Carotenoids

Carotenoids are the second most abundant natural pigments, behind only chlorophyll, with more than 750 different structures known until now. They are synthesised by photosynthetic organisms (bacteria, algae and plants) and by some non-photosynthetic bacteria and fungi. They can be classified in two main groups: carotenes, composed of carbon and hydrogen atoms, such as α-carotene, β-carotene, and lycopene, among others; and xanthophylls, that are oxygenated hydrocarbon derivatives, like lutein, cryptoxanthin, violaxanthin, zeaxanthin, etc. [79]. Carotenoids play key roles in several biological processes in plants. Apart from some of them being vitamin A precursors (as mentioned above), they are also precursors of the plant hormones abscisic acid (ABA) and strigolactones (SLs), they are one of the most important attractants to pollinators thanks to their pigmentation and fragrances (provided by volatile carotenoids), and they also participate in development, photosynthesis, photomorphogenesis and photoprotection processes [26].

The antioxidant potential of carotenoids is very important in human health due to their ability to reduce and, sometimes, prevent the development of various ROS (reactive oxygen species)-mediated disorders, such as cardiovascular diseases, cancer and neurological and photosensitive pathologies [80]. As humans are not able to synthesise these compounds, it is interesting to find crops rich in carotenoids. Vitamin A precursors (α-carotene, β-carotene and β-cryptoxanthin) have been described previously, so they are not dealt with here. Lycopene is the carotenoid responsible for tomato's red colour and it has been reported to be more abundant in two Spanish traditional varieties of tomato than in the commercial variety 'Baghera' [42]. In addition, one of these traditional varieties showed the strongest antioxidant activity. In two other studies carried out in tomato, not only lycopene, but also lutein content were significantly higher in a Spanish landrace and in three Italian landraces, respectively, than in the commercial variety 'Moneymaker' [28, 29]. Higher levels of lutein were also found in three Italian landraces of carrot, especially in the Tiggiano Yellow-Purple landrace [71], and in the melon landrace Casca de Carvalho [30] in comparison with commercial varieties. Cereal grains are also rich in carotenoids, especially lutein and zeaxanthin [81]. In this sense, several landraces of wheat exhibited higher levels of both compounds than old cultivars used as reference [32]. Finally, higher contents of lutein were also found in kernels of some maize traditional varieties from Italy, especially in Storo, in comparison to the hybrid B73/MO17, used as control [82].

4. Applications

As we all know, malnutrition is a public health problem with global dimensions. In 2019, almost 690 million people, 8.9% of the world population, were undernourished, mostly in developing countries. Beside this, about 2 billion people in the world suffered moderate or severe food insecurity, i.e. they did not have regular access to safe, nutritious, and sufficient food that year [83]. Overweight is also a growing matter of concern. In addition, since Green Revolution, the main objective of crop improvement programmes has been yield increase, what has resulted in a nutrient decrease in foodstuffs, contributing to malnutrition. However, quality has started to receive higher priority and agriculture objectives are undergoing changes from yield gains to the production of nutrient-rich food crops in sufficient amounts.

A search for crop landraces and traditional varieties with an enhanced nutritional value could be an interesting approach to combat nutrient deficiencies because, as seeing above, some of them are richer in micronutrients and health-promoting phytochemicals. However, they do not always cover minimal nutrient requirements and they are usually adapted to local environmental conditions. Therefore, a more feasible measure could be developing nutritionally enhanced foods with an increased bioavailability of nutrients for the human population. These efforts are normally directed toward raising the levels of minerals, vitamins, amino acids, and antioxidant compounds, as well as improving fatty acid composition in the edible portion of crop plants [84]. Crops with a higher nutritional value can be obtained by agronomic practices, conventional plant breeding, and modern biotechnological techniques.

4.1 Fortification

Fortification through agronomic practices or traditional fortification consists of the physical addition of micronutrients to the plants to improve their nutritional quality. It is generally achieved by using mineral fertilisers to increase their content, bioavailability and/or transport from the soil to the edible portion of the plant. Plant growth-promoting soil microorganisms can also be used [85]. This approach is simple and fast but requires regular applications in every crop season, what can increases costs, and also needs supervision in order not to reach toxicity levels, both in the environment and for humans.

One example of this approach is the Se fortification through foliar application in different wheat genotypes [86]. The greater Se accumulations were obtained in the grains of the landrace Timilia and the obsolete variety 'Cappelli' when compared with modern varieties, with an increase of up to 35-fold in mineral grain concentration at the maximum Se application. In another study, fortification with I was carried out in the carrot landrace Carota di Polignano through foliar fertilisation in open field experiments and through both, foliar fertilisation and fertigation with nutrient solution, in greenhouse experiments [87]. In open field, the root content in I increased a 51% and a 194% with low and high levels of the fertiliser, respectively, when compared with untreated carrots, whereas in greenhouse, the I content increased a 9% and only with the fertigation.

4.2 Biofortification

Quite the opposite that the fortification, the biofortification consists of developing crops with a higher nutritional value *per se*, either through conventional breeding or through genetic engineering, without the need of external micronutrient addition. That means that the plants are able to synthesise greater amounts of the particular micronutrients.

Biofortification is a one-time investment and offers a long-term and cost-effective approach to prevent malnutrition: once a crop has been biofortified, no more costs, like adding fertilisers to the soil or fortificants to the processed food are needed. In addition, low-income countries could develop biofortified crops through traditional practices, so in theory, low cultivation and production costs are feasible [88]. Reducing the amount of fertilisers required to obtain a more nutritious crop has also unarguable environmental benefits. Nevertheless, biofortification is not the final solution but an additional tool to combat malnutrition.

4.2.1 Biofortification through conventional plant breeding

Biofortification through conventional plant breeding requires crosses between parent lines rich in nutrients and recipient lines that present desirable agronomic traits during several generations. This is a time-consuming method, though sustainable. However, this conventional biofortification relies on genetic variability, which is usually limited in commercial cultivar gene pools, especially of staple crops. Landraces and traditional varieties are an adequate alternative here, thanks to their wide genetic diversity. This approach has been applied to a wide variety of crops, especially since HarvestPlus Challenge Programme was launched in 2003 to develop biofortified staple food crops with enhanced essential micronutrients through plant conventional breeding [89].

DOI: http://dx.doi.org/10.5772/intechopen.95514

Technique	Crop	Landrace or traditional variety	Enhanced trait	Method	Achievement	Reference
Agronomic practices	Wheat	Landrace Timilia; obsolete variety 'Capelli'	Se	Foliar fertilisation	↑ [Se] (up to 35-fold)	[86]
	Carrot	Landrace Carota di Polignano	I	Foliar fertilisation	↑ 51% and 194% with low and high levels of fertiliser, respectively	[87]
				Fertigation with nutrient solution	↑ 9%	
Conventional plant breeding	Rice	Traditional variety Zawa Bonday	Fe	Modern variety ('IR72') × traditional variety	Improved line with ↑ [Fe] (about 21 ppm in brown rice)	[90]
	Rice	Landrace Chittimuthyalu	Zn	Modern variety ('IR64') × landrace	Hybrid with ↑ [Zn] (26.9 mg/kg)	[91]
	Maize	Landrace ITA0370005	Carotenoids	Single cross: landrace × landrace (same population)	Hybrid with a ↑ [carotenoid] already commercialised	[92]
	Tomato	Landrace San Marzano	Polyphenols, tannins, flavonoids	Multiple crosses: landrace × landrace (same population)	Hybrid ('Torpedino di Fondi') with ↑ [polyphenols] and ↑ antioxidant activity in pink ripeness stage	[93]
	Eggplant	Nine landraces from Spain (ANS24, ANS26, ANS6, IVIA25, IVIA371, IVIA400, IVIA604, MUS8, VS22, VS9), one from China (ASIS1), and one from Cuba (SUDS5)	Polyphenols, Fe, Zn	Multiple crosses: landrace × landrace (different landraces)	Collection of hybrids with ↑ [phenolic compounds], ↑ [Fe], and ↑ [Zn]	[62]
	Eggplant	Landrace Almagro	Reduced prickliness	Backcrosses: three non-prickly commercial varieties × landrace	Improved pure line (H15) with nutritional properties of Almagro and ↓ prickliness	[94]
Modern biotechnology	Rice	Landrace Krabe	Seed yield	CRISPR-Cas9	Mutants with Krabe nutritional propierties and ↑ seed yield	[95]

Table 1.
Fortified and biofortified crops through different approaches by using landraces and traditional varieties.

Nevertheless, there is not a large number of studies carried out in landraces (**Table 1**). For example, in the International Rice Research Institute (IRRI) programme, an improved line (IR68144-3B-2-2-3) with a high concentration of Fe in the grain was obtained through a cross between a high-yield variety ('IR72') and a traditional variety (Zawa Bonday) from India. This new variety was reported to have about 80% more Fe than the commercial variety 'IR64' [90]. Useful information have been collected about the Zn content of different mapping populations of rice including wild germplasm, landraces and varieties, as well as hybrids [91]. Using 'IR64' as one of the parents, the hybrid with the highest Zn content (26.9 mg/kg) resulted from a cross with the landrace Chittimuthyalu. A collection of 14 hybrids between different landraces of eggplant has also been characterised [62]. These hybrids exhibited a higher average content of phenolics, as well as Fe and Zn, than commercial varieties. Zn average concentration was also higher in the hybrids than in the landraces tested. A maize hybrid with a high carotenoid content has also been identified [92]. It is a single-cross hybrid developed from the landrace ITA0370005 and it is currently being used by an Italian beer brewer. The metabolite profile and the antioxidant activity of the tomato hybrid Torpedino di Fondi (TF), developed from the landrace San Marzano (SM), has been characterised in two ripening stages, pink and red, both considered ideal for fresh consumption. In comparison with SM, pink TF tomatoes exhibited the highest content of total polyphenols, tannins, and flavonoids besides the greatest antioxidant activity [93]. Within a breeding programme, the eggplant landrace Almagro, known to contain higher values of vitamin C and total phenolics than regular varieties, but also having higher prickle presence, was used as recurrent parent in a backcross, whereas three non-prickly eggplant accessions were used as donors of this desirable trait [94]. Finally, an improved pure line (H15) with the Almagro eggplant ideotype and reduced prickliness was developed.

4.2.2 Biofortification through modern biotechnological techniques

Biofortification can be tackled through the genetic transformation of crops to express desirable genes from a plant species, independently of their taxonomic status, or even from other type of organisms, in the plant of interest to increase their nutrient content and bioavailability. This approach overcomes the limitation of the availability of genetic variability, allows the transfer of several genes simultaneously, and makes possible to biofortify crops with particular nutrients that are not naturally produced by themselves. Biofortification through transgenesis implies large investment of time, resources and researching: it is necessary to identify and characterise gene functions previously, and then, use these genes to transform crops. However, once the crop has been biofortified, it becomes a cost-effective approach [96].

The cisgenesis is a very interesting alternative to the transgenesis. With this approach, only genetic material from either the same species, or close relatives that hybridise naturally with it, is introduced [97]. In this way, the pool of genes available is exactly the same than when classical breeding methods are used. Cisgenic crops are subject to the same regulation than transgenic crops, but the EFSA (European Food Safety Authority) have concluded that cisgenics pose similar risks than plants obtained by conventional breeding [98]. Furthermore, the consumer's acceptance of cisgenics is greater than of transgenics [99].

Furthermore, the application of modern biotechnological techniques to landraces also allows the development of crops with higher yield, as it has been achieved recently [95]. The CRISPR-Cas9 technique was applied to the African rice landrace Kabre, considered a valuable resource, obtaining mutants with significantly improved seed yield and low lodging by disrupting genes known to control seed size and/or yield (**Table 1**).

5. Conclusion

In spite of not having been widely used in fortification and biofortification, especially with modern biotechnological approaches, crop landraces and traditional varieties could be key to improve the nutritional quality of food crops, as they can provide the desired genetic variability without sexual incompatibility barriers to overcome. Hopefully, in the near future there could be less restrictive regulations about the use of these biotechnological tools in crop breeding.

Acknowledgements

This work was funded by the projects RTA2017-00093-00-00 from the National Institute for Agricultural and Food Research and Technology (INIA) and LMP164_18 from the Government of Aragón; and by the Operational Programme FEDER Aragón 2014-2020 and the European Social Fund from the European Union [Grupo Consolidado A12-17R: "Grupo de investigación en fruticultura: caracterización, adaptación y mejora genética"]. We gratefully acknowledge the Vegetable Germplasm Bank of Zaragoza (BGHZ-CITA, Spain) for supplying the seeds used for this work. I. M.-L. was granted with a predoctoral contract for training doctors from the Spanish Ministry of Science, Innovation and Universities (MCIU) and the Spanish State Research Agency (AEI).

Conflict of interest

The authors declare no conflict of interest.

Author details

Inés Medina-Lozano[1,2] and Aurora Díaz[1,2]*

1 Department of Horticulture, Agrifood Research and Technology Centre of Aragon (CITA), Zaragoza, Spain

2 AgriFood Institute of Aragon – IA2 (CITA-University of Zaragoza), Zaragoza, Spain

*Address all correspondence to: adiazb@cita-aragon.es

References

[1] Zeven, A.C. Landraces: A review of definitions and classifications. Euphytica. 1998;104:127-139. DOI: 10.1023/A:1018683119237

[2] Mansholt, U.J. Van Pesch Plantenteelt, beknopte handleiding tot de kennis van den Nederlandschen landbouw. In Plantenteelt; 3rd ed. Zwolle; 1909. 228 p.

[3] von Rümker, K. Die systematische Einteilung und Benennung der Getreidesorten für praktische Zwecke. Jahrb. der Dtsch. landwirtschafts-Gesellschaft. 1908;23:137-167

[4] Fruwirth, C.; Roemer, T. Einführung in die landwirtschaftlichen Pflanzenzüchtung. Berlin; 1921. 150 p.

[5] Bellon, M.R.; Brush, S.B. Keepers of maize in Chiapas, Mexico. Econ. Bot. 1994;48:196-209. DOI: 10.1007/BF02908218

[6] Prospéri, J.; Demarquet, F.; Angevain, M.; Mansat, P. Évaluation agronomique de variétés de pays de sainfoin (*Onobrychis sativa* L.) originaires du sud-est de la France. Agronomie. 1994;14:285-298. DOI: 10.1051/agro:19940502

[7] Louette, D.; Charrier, A.; Berthaud, J. *In situ* conservation of maize in Mexico: Genetic diversity and maize seed management in a traditional community. Econ. Bot. 1997;51:20-38. DOI: 10.1007/BF02910401

[8] Teshome, A.; Baum, B.R.; Fahrig, L.; Torrance, J.K.; Arnason, T.J.; Lambert, J.D. Sorghum [*Sorghum bicolor* (L.) Moench] landrace variation and classification in North Shewa and South Welo, Ethiopia. Euphytica. 1997;97:255-263. DOI: 10.1023/A:1003074008785

[9] Harlan, J.R. Our vanishing genetic resources. Science. 1975;188:618-621. DOI: 10.1126/science.188.4188.617

[10] Villa, T.C.C.; Maxted, N.; Scholten, M.; Ford-Lloyd, B. Defining and identifying crop landraces. Plant Genet. Resour. 2005;3:373-384. DOI: 10.1079/pgr200591

[11] Del Greco, A.; Negri, V.; Maxted, N. Report of a task force on on-farm conservation and management. In Proceedings of the Second Meeting; Stegelitz, Germany: Rome: Biodiversity International; 2007; p. 19-20

[12] Negri, V.; Maxted, N.; Veteläinen, M. European landrace conservation: an introduction. In Veteläinen, M., Negri, V., Maxted, N., editors. European Landraces: On-farm Conservation, Management and Use: Biodiversity Technical Bulletin No. 15. Rome, Italy: Biodiversity International; p. 275-282

[13] Casañas, F.; Simó, J.; Casals, J.; Prohens, J. Toward an evolved concept of landrace. Front. Plant Sci. 2017;8:145. DOI: 10.3389/fpls.2017.00145

[14] Mallor, C.; Díaz, A. Melon germplasm characteristics, diversity, preservation and uses. In Walton M, editors. Germplasm: Characteristics, Diversity and Preservation. New York: Nova Science Publishers; 2016. p. 1-26

[15] FAO. Food and Agriculture Organization of United Nations [Internet]. 2020. Available from: http://www.fao.org/ [Accessed: 2020-11-17]

[16] Genesys. Gateway to genetic resources [Internet]. 2020. Available from: https://www.genesys-pgr.org/a/overview [Accessed: 2020-11-19]

[17] Chadwick, M.; Trewin, H.; Gawthrop, F.; Wagstaff, C. Sesquiterpenoids lactones: Benefits to plants and people. Int. J. Mol. Sci. 2013;14:12780-12805. DOI: 10.3390/ijms140612780

[18] Lertrat, K.; Pulam, T. Breeding for increased sweetness in sweet corn. Int. J. Plant Breed. 2007;1:27-30

[19] He, J.; Monica Giusti, M. Anthocyanins: Natural colorants with health-promoting properties. Annu. Rev. Food Sci. Technol. 2010;1:163-187. DOI: 10.1146/annurev.food.080708.100754

[20] Yousuf, B.; Gul, K.; Wani, A.A.; Singh, P. Health benefits of anthocyanins and their encapsulation for potential use in food systems: A review. Crit. Rev. Food Sci. Nutr. 2016;56:2223-2230. DOI: 10.1080/10408398.2013.805316

[21] Ahmed, T.; Hossain, M.; Sanin, K.I. Global burden of maternal and child undernutrition and micronutrient deficiencies. Ann. Nutr. Metab. 2012;61:8-17. DOI: 10.1159/000345165

[22] Karaköy, T.; Erdem, H.; Baloch, F.S.; Toklu, F.; Eker, S.; Kilian, B.; Özkan, H. Diversity of macro-and micronutrients in the seeds of lentil landraces. Sci. World J. 2012;2012:710412. DOI: 10.1100/2012/710412

[23] Haskell, M.J.; Brown, K.H. Maternal vitamin A vutriture and the vitamin A content of human milk. J. Mammary Gland Biol. Neoplasia. 1999;4:243-257. DOI: 10.1023/A:1018745812512

[24] Fitzpatrick, T.B.; Basset, G.J.C.; Borel, P.; Carrari, F.; DellaPenna, D.; Fraser, P.D.; Hellmann, H.; Osorio, S.; Rothan, C.; Valpuesta, V.; Caris-Veyrat, C. Fernie, A.R. Vitamin deficiencies in humans: Can plant science help? Plant Cell. 2012;24:395-414. DOI: 10.1105/tpc.111.093120

[25] Olson, J.A. Vitamin A. In Decker M, editors. Handbook of Vitamins. New York: Eastern Hemisphere Distribution; 2001. p. 1-50

[26] DellaPenna, D.; Pogson, B.J. Vitamin synthesis in plants: Tocopherols and carotenoids. Annu. Rev. Plant Biol. 2006;57:711-738. DOI: 10.1146/annurev.arplant.56.032604.144301

[27] WHO. World Health Organization [Internet]. 2020. Available from: https://www.who.int/nutrition/topics/vad/en/ [Accessed: 2020-10-30]

[28] Massaretto, I.L.; Albaladejo, I.; Purgatto, E.; Flores, F.B.; Plasencia, F.; Egea-Fernández, J.M.; Bolarin, M.C.; Egea, I. Recovering tomato landraces to simultaneously improve fruit yield and nutritional quality against salt stress. Front. Plant Sci. 2018;871:1778. DOI: 10.3389/fpls.2018.01778

[29] Scarano, A.; Olivieri, F.; Gerardi, C.; Liso, M.; Chiesa, M.; Chieppa, M.; Frusciante, L.; Barone, A.; Santino, A.; Rigano, M.M. Selection of tomato landraces with high fruit yield and nutritional quality under elevated temperatures. J. Sci. Food Agric. 2020;100:2791-2799. DOI: 10.1002/jsfa.10312

[30] Esteras, C.; Rambla, J.L.; Sánchez, G.; López-Gresa, M.P.; González-Mas, M.C.; Fernández-Trujillo, J.P.; Bellés, J.M.; Granell, A.; Picó, M.B. Fruit flesh volatile and carotenoid profile analysis within the *Cucumis melo* L. species reveals unexploited variability for future genetic breeding. J. Sci. Food Agric. 2018;98:3915-3925. DOI: 10.1002/jsfa.8909

[31] Ebert, A.W.; Chang, C.H.; Yan, M.R.; Yang, R.Y. Nutritional composition of mungbean and soybean sprouts compared to their adult growth stage. Food Chem. 2017;237:15-22. DOI: 10.1016/j.foodchem.2017.05.073

[32] Hussain, A.; Larsson, H.; Kuktaite, R.; Olsson, M.E.; Johansson, E. Carotenoid content in organically produced wheat: Relevance for human nutritional health on consumption. Int. J. Environ. Res. Public Health. 2015;12:14068-14083. DOI: 10.3390/ijerph121114068

[33] Herraiz, F.J.; Raigón, M.D.; Vilanova, S.; García-Martínez, M.D.; Gramazio, P.; Plazas, M.; Rodríguez-Burruezo, A.; Prohens, J. Fruit composition diversity in land races and modern pepino (*Solanum muricatum*) varieties and wild related species. Food Chem. 2016;203:49-58. DOI: 10.1016/j.foodchem.2016.02.035

[34] Drapal, M.; Rossel, G.; Heider, B.; Fraser, P.D. Metabolic diversity in sweet potato (*Ipomoea batatas*, Lam.) leaves and storage roots. Hortic. Res. 2019;6:2. DOI: 10.1038/s41438-018-0075-5

[35] Lee, S.K.; Kader, A.A. Preharvest and postharvest factors influencing vitamin C content of horticultural crops. Postharvest Biol. Technol. 2000;20:207-220. DOI: 10.1016/S0925-5214(00)00133-2

[36] Ishikawa, T.; Dowdle, J.; Smirnoff, N. Progress in manipulating ascorbic acid biosynthesis and accumulation in plants. Physiol. Plant. 2006;126:343-355. DOI: 10.1111/j.1399-3054.2006.00640.x

[37] Carr, A.C.; Frei, B. Toward a new recommended dietary allowance for vitamin C based on antioxidant and health effects in humans. Am. J. Clin. Nutr. 1999;69:1086-1107. DOI: 10.1093/ajcn/69.6.1086

[38] Escribano, S.; Lázaro, A. Physicochemical and nutritional evaluation of Spanish melon landraces. Plant Genet. Resour. 2017;15:177-186. DOI: 10.1017/S1479262115000507

[39] Medina-Lozano, I.; Bertolín, J.R.; Zufiaurre, R.; Diaz, A. Improved UPLC-UV method for the quantification of vitamin C in lettuce varieties (*Lactuca sativa* L.) and crop wild relatives (*Lactuca* spp.). J. Vis. Exp. 2020;160:e61440. DOI: 10.3791/61440

[40] Medina-Lozano, I.; Bertolín, J.R.; Díaz, A. (in press). Nutritional value of commercial and traditional lettuce (*Lactuca sativa* L.) and wild relatives: vitamin C and anthocyanin content. Food Chem.

[41] San José, R.; Sánchez-Mata, M.-C.; Cámara, M.; Prohens, J. Eggplant fruit composition as affected by the cultivation environment and genetic constitution. J. Sci. Food Agric. 2014;94:2774-2784. DOI: 10.1002/jsfa.6623

[42] Gonzalez-Cebrino, F.; Lozano, M.; Ayuso, M.C.; Bernalte, M.J.; Vidal-Aragon, M.C.; Gonzalez-Gomez, D. Characterization of traditional tomato varieties grown in organic conditions. Spanish J. Agric. Res. 2011;9:444-452. DOI: 10.5424/sjar/20110902-153-10

[43] Bonasia, A.; Conversa, G.; Lazzizera, C.; Loizzo, P.; Gambacorta, G.; Elia, A. Evaluation of garlic landraces from Foggia province (Puglia region; Italy). Foods. 2020;9:850. DOI: 10.3390/foods9070850

[44] Petropoulos, S.A.; Fernandes, Â.; Barros, L.; Ferreira, I.C.F.R.; Ntatsi, G. Morphological, nutritional and chemical description of "Vatikiotiko", an onion local landrace from Greece. Food Chem. 2015;182:156-163. DOI: 10.1016/j.foodchem.2015.03.002

[45] Conversa, G.; Lazzizera, C.; Bonasia, A.; Rotonda, P. La; Elia, A. Nutritional characterization of two rare landraces of turnip (*Brassica rapa*. var. *rapa*) tops and their on-farm conservation in Foggia province. Sustainability. 2020;12:3842. DOI: 10.3390/su12093842

[46] Fryer, M.J. The antioxidant effects of thylakoid Vitamin E (α-tocopherol). Plant. Cell Environ. 1992;15:381-392. DOI: 10.1111/j.1365-3040.1992.tb00988.x

[47] Lee, Y.Y.; Park, H.M.; Hwang, T.Y.; Kim, S.L.; Kim, M.J.; Lee, S.K.; Seo, M.J.; Kim, K.J.; Kwon, Y.U.; Lee,

S.C.; Kim, Y.H. A correlation between tocopherol content and antioxidant activity in seeds and germinating seeds of soybean cultivars. J. Sci. Food Agric. 2014;95:819-827. DOI: 10.1002/jsfa.6963

[48] Hussain, A.; Larsson, H.; Olsson, M.E.; Kuktaite, R.; Grausgruber, H.; Johansson, E. Is organically produced wheat a source of tocopherols and tocotrienols for health food? Food Chem. 2012;132:1789-1795. DOI: 10.1016/j.foodchem.2011.11.141

[49] Cittadini, M.C.; Martín, D.; Gallo, S.; Fuente, G.; Bodoira, R.; Martínez, M.; Maestri, D. Evaluation of hazelnut and walnut oil chemical traits from conventional cultivars and native genetic resources in a non-traditional crop environment from Argentina. Eur. Food Res. Technol. 2020;246:833-843. DOI: 10.1007/s00217-020-03453-8

[50] Gunaratne, A.; Wu, K.; Li, D.; Bentota, A.; Corke, H.; Cai, Y.Z. Antioxidant activity and nutritional quality of traditional red-grained rice varieties containing proanthocyanidins. Food Chem. 2013;138:1153-1161. DOI: 10.1016/j.foodchem.2012.11.129

[51] White, P.J.; Broadley, M.R. Biofortifying crops with essential mineral elements. Trends Plant Sci. 2005;10:586-593. DOI: 10.1016/j.tplants.2005.10.001

[52] Bolt, H.M.; Duydu, Y.; Başaran, N.; Golka, K. Boron and its compounds: current biological research activities. Arch. Toxicol. 2017;91:2719-2722. DOI: 10.1007/s00204-017-2010-1

[53] Vatansever, R.; Ozyigit, I.I.; Filiz, E. Essential and beneficial trace elements in plants, and their transport in roots: a review. Appl. Biochem. Biotechnol. 2016;181:464-482. DOI: 10.1007/s12010-016-2224-3

[54] Subramaniam, G.; Girish, M. Iron deficiency anemia in children. Indian J. Pediatr. 2015;82:558-564. DOI: 10.1007/s12098-014-1643-9

[55] Abbaspour, N.; Hurrell, R.; Kelishadi, R. Review on iron and its importance for human health. J. Res. Med. Sci. 2014;19:164-174

[56] Prasad, A.S. Discovery of human zinc deficiency: Its impact on human health and disease. Adv. Nutr. 2013;4:176-190. DOI: 10.3945/an.112.003210.176

[57] Roohani, N.; Hurrell, R.; Kelishadi, R.; Schulin, R. Zinc and its importance for human health: An integrative review. J. Res. Med. Sci. 2013;18:144-157. DOI: 10.1016/j.foodpol.2013.06.008

[58] Zimmermann, M.B.; Jooste, P.L.; Pandav, C.S. Iodine-deficiency disorders. Lancet. 2008;372:1251-1262. DOI: 10.1016/S0140-6736(08)61005-3

[59] Celmeli, T.; Sari, H.; Canci, H.; Sari, D.; Adak, A.; Eker, T.; Toker, C. The nutritional content of common bean (*Phaseolus vulgaris* L.) landraces in comparison to modern varieties. Agronomy. 2018;8:166. DOI: 10.3390/agronomy8090166

[60] Petropoulos, S.A.; Fernandes, Â.; Ntatsi, G.; Petrotos, K.; Barros, L.; Ferreira, I.C.F.R. Nutritional value, chemical characterization and bulb morphology of Greek garlic landraces. Molecules. 2018;23:319. DOI: 10.3390/molecules23020319

[61] Torutaeva, E.; Asanaliev, A.; Prieto-Linde, M.L.; Zborowska, A.; Ortiz, R.; Bryngelsson, T.; Garkava-Gustavsson, L. Evaluation of microsatellite-based genetic diversity, protein and mineral content in chickpea accessions grown in Kyrgyzstan. Hereditas. 2014;151:81-90. DOI: 10.1111/hrd2.00042

[62] Raigón, M.D.; Prohens, J.; Muñoz-Falcón, J.E.; Nuez, F. Comparison of eggplant landraces and commercial

varieties for fruit content of phenolics, minerals, dry matter and protein. J. Food Compos. Anal. 2008;21:370-376. DOI: 10.1016/j.jfca.2008.03.006

[63] Hernández Rodríguez, L.; Afonso Morales, D.; Rodríguez Rodríguez, E.; Díaz Romero, C. Minerals and trace elements in a collection of wheat landraces from the Canary Islands. J. Food Compos. Anal. 2011;24:1081-1090. DOI: 10.1016/j.jfca.2011.04.016

[64] Hussain, A.; Larsson, H.; Kuktaite, R.; Johansson, E. Mineral composition of organically grown wheat genotypes: Contribution to daily minerals intake. Int. J. Environ. Res. Public Health. 2010;7:3442-3456. DOI: 10.3390/ijerph7093442

[65] Sciacca, F.; Allegra, M.; Licciardello, S.; Roccuzzo, G.; Torrisi, B.; Virzì, N.; Brambilla, M.; Romano, E.; Palumbo, M. Potential use of Sicilian landraces in biofortification of modern durum wheat varieties: evaluation of caryopsis micronutrient concentrations. Cereal Res. Commun. 2018;46:124-134. DOI: 10.1556/0806.45.2017.056

[66] Kondou, Y.; Manickavelu, A.; Komatsu, K.; Arifi, M.; Kawashima, M.; Ishii, T.; Hattori, T.; Iwata, H.; Tsujimoto, H.; Ban, T.; Matsui, M. Analysis of grain elements and identification of best genotypes for Fe and P in Afghan wheat landraces. Breed. Sci. 2016;66:676-682. DOI: 10.1270/jsbbs.16041

[67] Neeraja, C.N.; Kulkarni, K.S.; Babu, P.M.; Rao, D.S.; Surekha, K.; Babu, V.R. Transporter genes identified in landraces associated with high zinc in polished rice through panicle transcriptome for biofortification. PLoS One. 2018;13:e0192362. DOI: 10.1371/journal.pone.0192362

[68] Lima, G.P.P.; Vianello, F.; Corrêa, C.R.; Campos, R.A. da S.; Borguini, M.G. Polyphenols in fruits and vegetables and its effect on human health. Food Nutr. Sci. 2014;5:1065-1082. DOI: 10.4236/fns.2014.511117

[69] Parr, A.J.; Bolwell, G.P. Phenols in the plant and in man. The potential for possible nutritional enhancement of the diet by modifying the phenols content or profile. J. Sci. Food Agric. 2000;80:985-1012. DOI: 10.1002/(sici)1097-0010(20000515)80:7<985::aid-jsfa572>3.3.co;2-z

[70] Siracusa, L.; Patanè, C.; Avola, G.; Ruberto, G. Polyphenols as chemotaxonomic markers in Italian "long-storage" tomato genotypes. J. Agric. Food Chem. 2011;60:309-314. DOI: 10.1021/jf203858y

[71] Scarano, A.; Gerardi, C.; D'Amico, L.; Accogli, R.; Santino, A. Phytochemical analysis and antioxidant properties in colored Tiggiano carrots. Agriculture. 2018;8:102. DOI: 10.3390/agriculture8070102

[72] Jakobek, L.; Barron, A.R. Ancient apple varieties from Croatia as a source of bioactive polyphenolic compounds. J. Food Compos. Anal. 2016;45:9-15. DOI: 10.1016/j.jfca.2015.09.007

[73] Garau, M.C.; González-Centeno, M.R.; Luna, J.M.; Negre, A.; Rosselló, C.; Femenia, A. Potential of landrace winery byproducts (*Vitis vinifera* L.) as a source of phenolic compounds with antioxidant properties. J. Int. des Sci. la Vigne du Vin. 2015;49:241-252. DOI: 10.20870/oeno-one.2015.49.4.45

[74] Falcinelli, B.; Calzuola, I.; Gigliarelli, L.; Torricelli, R.; Polegri, L.; Vizioli, V.; Benincasa, P.; Marsili, V. Phenolic content and antioxidant activity of wholegrain breads from modern and old wheat (*Triticum aestivum* L.) cultivars and ancestors enriched with wheat sprout powder. Ital. J. Agron. 2018;13:297-302. DOI: 10.4081/ija.2018.1220

[75] Migliorini, P.; Spagnolo, S.; Torri, L.; Arnoulet, M.; Lazzerini, G.; Ceccarelli, S. Agronomic and quality characteristics of old, modern and mixture wheat varieties and landraces for organic bread chain in diverse environments of northern Italy. Eur. J. Agron. 2016;79:131-141. DOI: 10.1016/j.eja.2016.05.011

[76] Bianco, M. Lo; Siracusa, L.; Dattilo, S.; Venora, G.; Ruberto, G. Phenolic fingerprint of sicilian modern cultivars and durum wheat landraces: A tool to assess biodiversity. Cereal Chem. 2017;94:1045-1051. DOI: 10.1094/CCHEM-06-17-0125-R

[77] Bily, A.C.; Burt, A.J.; Ramputh, A.I.; Livesey, J.; Regnault-Roger, C.; Philogène, B.R.; Arnason, J.T. HPLC-PAD-APCI assay of phenylpropanoids in cereals. Phytochem. Anal. 2004;15:9-15. DOI: 10.1002/pca.735

[78] Bernardi, J.; Stagnati, L.; Lucini, L.; Rocchetti, G.; Lanubile, A.; Cortellini, C.; De Poli, G.; Busconi, M.; Marocco, A. Phenolic profile and susceptibility to fusarium infection of pigmented maize cultivars. Front. Plant Sci. 2018;9:1189. DOI: 10.3389/fpls.2018.01189

[79] Nisar, N.; Li, L.; Lu, S.; Khin, N.C.; Pogson, B.J. Carotenoid metabolism in plants. Mol. Plant. 2015;8:68-82. DOI: 10.1016/j.molp.2014.12.007

[80] Fiedor, J.; Burda, K. Potential role of carotenoids as antioxidants in human health and disease. Nutrients. 2014;6:466-488. DOI: 10.3390/nu6020466

[81] Panfili, G.; Fratianni, A.; Irano, M. Improved normal-phase high-performance liquid chromatography procedure for the determination of carotenoids in cereals. J. Agric. Food Chem. 2004;52:6373-6377. DOI: 10.1021/jf0402025

[82] Puglisi, D.; Landoni, M.; Cassani, E.; Toschi, I.; Lucchini, G.; Cesari, V.; Borlini, G.; Pilu, R. Traditional farmers' varieties: a valuable source of genetic variability for biofortification programs. Maydica. 2018;63:1-10

[83] FAO; IFAD; UNICEF; WFP; WHO. The state of food security and nutrition in the world 2020. Transforming food systems for affordable healthy diets; Rome, Italy; 2020. 320 p. DOI: 10.4060/ca9692en

[84] Hirschi, K.D. Nutrient biofortification of food crops. Annu. Rev. Nutr. 2009;29:401-421. DOI: 10.1146/annurev-nutr-080508-141143

[85] Rengel, Z.; Batten, G.D.; Crowley, D.E. Agronomic approaches for improving the micronutrient density in edible portions of field crops. F. Crop. Res. 1999;60:27-40. DOI: 10.1016/S0378-4290(98)00131-2

[86] De Vita, P.; Platani, C.; Fragasso, M.; Ficco, D.B.M.; Colecchia, S.A.; Del Nobile, M.A.; Padalino, L.; Di Gennaro, S.; Petrozza, A. Selenium-enriched durum wheat improves the nutritional profile of pasta without altering its organoleptic properties. Food Chem. 2017;214:374-382. DOI: 10.1016/j.foodchem.2016.07.015

[87] Signore, A.; Renna, M.; D'Imperio, M.; Serio, F.; Santamaria, P. Preliminary evidences of biofortification with iodine of "Carota di Polignano", an Italian carrot landrace. Front. Plant Sci. 2018;9:170. DOI: 10.3389/fpls.2018.00170

[88] Nestel, P.; Bouis, H.E.; Meenakshi, J. V; Pfeiffer, W. Biofortification of staple food crops. J. Nutr. 2006;136:1064-1067. DOI: 10.1093/jn/136.4.1064.

[89] Bouis, H.E.; Saltzman, A. Improving nutrition through biofortification: A review of evidence from HarvestPlus, 2003 through 2016. Glob. Food Sec. 2017;12:49-58. DOI: 10.1016/j.gfs.2017.01.009

[90] Gregorio, G.B.; Senadhira, D.; Htut, H.; Graham, R.D. Breeding for trace minerals in rice. Food Nutr. Bull. 2000;21:382-386. DOI: 10.1177/156482650002100409

[91] Sanjeeva Rao, D.; Neeraja, C.N.; Madhu Babu, P.; Nirmala, B.; Suman, K.; Rao, L.V.S.; Surekha, K.; Raghu, P.; Longvah, T.; Surendra, P.; Kumar, R.; Babu, V.R.; Voleti, S.R. Zinc biofortified rice varieties: Challenges, possibilities, and progress in India. Front. Nutr. 2020;7:26. DOI: 10.3389/fnut.2020.00026

[92] Berardo, N.; Mazzinelli, G.; Valoti, P.; Laganà, P.; Redaelli, R. Characterization of maize germplasm for the chemical composition of the grain. J. Agric. Food Chem. 2009;57:2378-2384. DOI: 10.1021/jf803688t

[93] Ingallina, C.; Maccelli, A.; Spano, M.; Matteo, G. Di; Sotto, A. Di; Giusti, A.M.; Vinci, G.; Giacomo, S. Di; Rapa, M.; Ciano, S.; Fraschetti, C.; Filippi, A.; Simonetti, G.; Cordeiro, C.; Silva, M.S.; Crestoni, M.E.; Fornarini, S.; Mannina, L. Chemico-biological characterization of torpedino di fondi® tomato fruits: A comparison with san marzano cultivar at two ripeness stages. Antioxidants. 2020;9:1027. DOI: 10.3390/antiox9101027

[94] Hurtado, M.; Vilanova, S.; Plazas, M.; Gramazio, P.; Andújar, I.; Herraiz, F.J.; Castro, A.; Prohens, J. Enhancing conservation and use of local vegetable landraces: the Almagro eggplant (*Solanum melongena* L.) case study. Genet. Resour. Crop Evol. 2014;61:787-795. DOI: 10.1007/s10722-013-0073-2. The

[95] Lacchini, E.; Kiegle, E.; Castellani, M.; Adam, H.; Jouannic, S.; Gregis, V.; Kater, M.M. CRISPR-mediated accelerated domestication of African rice landraces. PLoS One. 2020;15:1-12. DOI: 10.1371/journal.pone.0229782

[96] Garg, M.; Sharma, N.; Sharma, S.; Kapoor, P.; Kumar, A.; Chunduri, V.; Arora, P. Biofortified crops generated by breeding, agronomy, and transgenic approaches are improving lives of millions of people around the world. Front. Nutr. 2018;5:12. DOI: 10.3389/fnut.2018.00012

[97] Schouten, H.J.; Krens, F.A.; Jacobsen, E. Cisgenic plants are similar to traditionally bred plants. EMBO Rep. 2006;7:750-753. DOI: 10.1038/sj.embor.7400769

[98] Panel, E. Scientific opinion addressing the safety assessment of plants developed using Zinc Finger Nuclease 3 and other Site-Directed Nucleases with similar function. EFSA J. 2012;10:10. DOI: 10.2903/j.efsa.2012.2943

[99] Holme, I.B.; Wendt, T.; Holm, P.B. Intragenesis and cisgenesis as alternatives to transgenic crop development. Plant Biotechnol. J. 2013;11:395-407. DOI: 10.1111/pbi.12055

Chapter 7

Interactive Effect of Organic and Inorganic Amendments along with Plant Growth Promoting Rhizobacteria on Ameliorating Salinity Stress in *Maize*

Sajid Rashid Ahmad, Sana Ashraf and Humaira Nawaz

Abstract

Saline soil is one of the common environmental issues that negatively affects the soil quality of agricultural lands. It reduces the plant growth and productivity worldwide. Soil Salinity and sodicity affecting land about 1128 million hectares globally determined by recent researches. The most important salt-sensitive cereal crops in the world are Maize (*Zea mays* L.) For food security, its need of hour to securing attainable production of maize crop in the salt affected soils. To reduce negative impacts of saline soil on plant growth, sustainable approaches such as organic amendments like press mud and inorganic amendments like silicon can be applied. For increasing crop productivity, plant growth promoting rhizobacteria (PGPR) which are salt-tolerant in saline agriculture can also be applied. In this book chapter interactive effect of different organic and inorganic amendments and plant growth-promoting rhizobacteria to reduce salinity stress on maize has been discussed.

Keywords: Salinity stress, Maize, Food security, Organic amendments, Inorganic amendments

1. Introduction

In the Arid and semi-arid areas salt affected soil poses immense threats to the agriculture industry worldwide [1]. Researchers have reported that about 1128 million ha of land is affected by salinity and sodicity globally [2]. Soil salinization at global level has caused food insecurity in several countries during last decade. In Pakistan approximately 6.8 million hectares of land is affected by salinity [3]. Saline soil is characterized by the presence of high level of sodium and its chlorides and sulphates [4]. Soil having 4 dS m^{-1} or more electrical conductivity of the saturation soil paste is considered as salt affected soil [5–7]. Due to high concentrations of Na^+ and Cl^- in the plant, Sodium chloride can reduce crop productivity by making the roots water uptake more difficult that can cause plant toxicity [8]. Research studies proved that approximately 20% of the world's cultivated land is affected by salinity [9].

Due to inappropriate management of irrigation and drainage, soil salinity is gradually increasing in irrigated lands [10]. Global warming as an environmental issue has greatly affected arid regions of the world that are at highest risk of soil salinization. Salt redistribution in the soil profile is due to climatic factor such as precipitation. Agricultural productivity is affected when salt is added by wind to coastal agricultural lands [11]. Soil Biodiversity and microbial activity is affected by the high salt concentration [12].

Under saline soil plant growth is negatively affected by osmotic effects and hormonal imbalance. It also causes nutritional disorder and specific ion toxicity [13]. The adverse effects of saline soil on plants include: (1) Osmotic potential is decreases due to excessive soluble salts in the soil solutions. It also causes physiological drought by decreasing plants ability to absorb water (2) toxicity due to salt ions inside the plant cells. The Growth inhibition is caused by sodium and chloride ions as sodium ions are retained in the roots and stems and only chloride ions become concentrated in the shoot in some plants which is causing negative affects to the plants [14, 15]. (3) Secondary stresses which is mainly caused by osmotic and ionic pressure. It includes high concentration of toxic compounds such as ROS and nutrient imbalance in plants. Sodium ions compete with potassium ions under saline condition and causing reproductive disorders by calcium ions in the cell membrane [16, 17]. The maize (*Zea mays* L). is a major food crop in the world food. The productivity of maize crop is declined as it moderately salt-sensitive plant [18].

In this scenario, it is the need of time that agronomists and environmentalists should develop eco-friendly, cost effective and sustainable methods to reclaim saline soils [19]. Currently, various physico-chemical processes are in practice for the reclamation of saline soils. To some extent these methods are unsustainable and inefficient at high salt concentration [20]. The traditional breeding and biotechnological methods for the production of salinity-tolerant crops is a time-consuming process. By using chemical neutralizers and sustainable approaches sustainable crop yield in saline soils must be secured. It can also be secured by using salt-tolerant varieties or amelioration methods.

For plant growth and development, microorganisms play an important role under different environmental conditions [21, 22]. For enhancing crop productivity in saline soil, the application of plant growth-promoting rhizobacteria has become sustainable approach [23, 24]. Inoculation with PGPR leads towards abiotic stress regulation which can cause systemic tolerance directly or indirectly [25]. Many PGPR have been applied for their positive role in improving plant-water relations and for ion homeostasis. It is also used for photosynthetic efficiency in plants under salt stress. Plants can effectively protect from many stresses by PGPR that produce IAA and ACC deaminase. IAA accumulation increase transcription of ACC synthase genes. It is resulted an increases ACC concentration that can lead to the production of ethylene. Excess ACC are broken by PGPR that produce ACC deaminase. It also decreases plant ethylene levels under harsh environmental conditions. It permits IAA to encourage the growth of the plants [26].

Bacteria secrete exopolysaccharides which can bind soil particles into aggregates. These are helpful in regulating soil structures. It also increases water holding as well as cation exchange capacity of soil [27]. An enclosed matrix of microcolonies is formed by EPS which provide protection against environmental changings. It also leads towards water as well as nutrient retention and epiphytic colonization [28]. The exopolysaccharide secretion by PGPR binds sodium ions and reduces its uptake in plants which is determined by researches studies [29].

In saline soil a diversity of salt-tolerant PGPR such as *Azospirillum, Burkholderia, Rhizobium*, *Pseudomonas*, *Acetobacter* and *Bacillus* have been applied. These are also tested for promoting plant growth under salt stress [30, 31]. Thus, it has been demonstrating by different researches that use of PGPR is a beneficial approach to increase plant performance in saline soil [32, 33]. The physiological drought is caused by salt in soil environment which reduces the ability of plants to remove water. Many biotic and physical stresses on plants can be reduced by application of Si fertilizer which can change the negative effects of saline soil [34, 35]. By improving sodium ions and potassium ions homeostasis, silicon may increase salinity tolerance in plants. It also improves nutritional status and photosynthetic efficiency of plants under stress conditions [36–38]. Many laboratory and greenhouse experiments have determined that under saline conditions, Si reduced the uptake of sodium ions and chloride ions [39, 40]. The use of organic matter increases the physico-chemical and biological properties of salt-affected soils [41]. Organic matter also plays an important role by improving roots to grow more uniformly. The soil CO_2 concentration is increased by decaying organic matter. It also releases H^+ and enhances $CaCO_3$ dissolution. It can release more calcium for sodium exchange [42] Application of press mud is very effective in reclaiming saline sodic soils [43, 44].

2. Impacts of soil salinity

2.1 Impact of soil salinity on plant growth and development

Saline soil affects plant growth, development and process of photosynthesis. It also affects protein synthesis and lipid metabolism [45]. Osmotic stress reduces photosynthetic efficiency which is resulted in partial closure of stomata [46]. The nutrient imbalance and membrane destabilization are caused by soil salinity [47]. The cell growth and development are decreased in plants in responses to osmotic stress. It resulted in decreased leaf area and chlorophyll content [48].

The nutritional imbalances are also caused by decrease in the uptake of calcium ions and potassium ions in leaves and an increase in the uptake of sodium ions. In some cases, there is a requirement of low sodium ions and high potassium ions or calcium ions are required for optimum function, but increased sodium ions resulted in metabolic disturbances. Cell swelling in plants is caused by accumulation of sodium and chloride which can affect plant enzymes. It can also result in physiological changes and reduced energy production [49]. The photosynthetic function is disturbed by nitrate reductase activity due to chloride ions [50]. There are competitive interactions with nutrient ions for binding sites. It can also affect transfer of protein in root cells under excessive sodium and chloride ions in rhizosphere. It also affects processes like movement of material, deposition, and partitioning within plants [51]. Salts can increase in intercellular spaces resulted in cell dehydration [52]. Oxidative stress increases due to the accumulation of reactive oxygen species which has negative impact on cell membranes, proteins, enzymes, and nucleic acids [53] Both antioxidant enzymes and non-enzymatic antioxidants are produced by plants to protect against oxidative stress [54].

2.2 Impacts of soil salinity on rhizosphere microbial diversity

Microbial biomass is an important parameter as it functions as an agent transformation and plays its role as the recycling of the organic matter by providing soil nutrients. In the first few centimeters of the soil surface, there are microbial biomass and organic matter. Microbiological activity is affected by the salinization process [55]. Microbial diversity, functions, and compositions are negatively affected by salinity [56]. Total bacteria and actinobacteria are reduced by a 5% increase in salinity. The attachment of *Azospirillum brasilense* to maize roots was observed to reduce due to salinity [57]. Due to increase salinity in the rhizosphere, the plant root secretion and organic matter decomposition by microorganisms are adversely affected [58].

3. Application of organic amendments to agricultural crops to mitigate salinity effects

3.1 Organic matter

There is an excess of salts in water which is used for irrigation purposes. It can reduce the crop yield due to its increased salt concentration [59]. Soil electrical conductivity is being increased due to the continuous increase of salts in it [60]. Water which is used for irrigation having excess salts in it resulted in negative impacts on plant physiology, soil water plant relationships, and limits the production of crops [61]. By application of organic manure in soil, the toxicity of salts can be minimized, and soil properties can be improved as cost-effective approaches [62].

There are agricultural practices that are used for the management of salt-affected soil [63]. Addition of organic martial is beneficial as a fertilizer which can modify and improve the soil characteristic. For recovery of saline soil, organic amendments like organic manure and compost are being tested as efficient methods [64]. Application of organic matter for the reclamation of sandy soil is an effective method to improve the physical properties of soil [65]. Researchers determined that poultry manure, farmyard manure (FYM), crop residues as compost are being used for the addition of nutrients in the soil. It is beneficial for improving plants' health. It can also modify physiochemical properties of plants [66]. Farmyard manure is the most commonly and easily available source of organic matter. There are different factors which can affect the efficiency of farmyard manure such as nature of feed consumed by the animal, type of animal and waste management methods [67].

3.2 Biochar

There are different long-term and short-term methods for reclamation of salt-affected soil, but short-term management approaches are useful as a management strategy that are cost effective and high-income generating methods [68]. The biochar is an effective method for organic amendment of salt-affected soil that results in

- Soil physicochemical and biological properties are improved
- Stomatal conductance and phytohormones can be regulated
- Reduction in oxidative stress
- Increase in mineral nutrient uptake
- Effects on plant growth, photosynthesis and biomass
- Na ion toxicity in plants is reduced

4. Application of inorganic amendments to agricultural crops to mitigate salinity effects

4.1 Exogenous application of sulfur

Salt affected soil has many salts in it and each salt has a differential contribution to salt stress. There are different salts such as Na_2SO4, NaCl, Na2CO3, CaSO4, MgCl2, KCl but the most important of these is NaCl [69–73]. For the regulation of cell metabolism and hormone signaling pathway, Sulfur plays a very important role. For regulating seed germination its acts as a biochemical agent [74, 75]. For the synthesis of protein, chlorophyll, vitamins, and glutathione which are helpful to tolerate various stresses, sulfur plays a very important role [76]. Sulfur compounds are also present in many amino acids and their composition changes by the application of sulfur [77]. To improve plant growth by improving its cellular functions especially in saline soil, the addition of sulfur is beneficial [78]. Different approaches are being applied to mitigate the deleterious effects of salinity on health of plants. The exogenous application of inorganic salts and osmo-protectants are cost efficient approach to reduce the negative effects of salt stress on plant growth [79, 80].

4.2 Use of silicon nutrition to alleviate the salt stress in maize

In contrast to Na^+ and Cl^- toxicity, silicon (Si) has ameliorative features. It can help plants to grow on saline soil. For industrialized counties, it can prove cost-effective. Under biotic stress, silicon can improve plant growth also reduces radiation effects on it. It is helpful in reducing water loss up to 30% [81]. The exogenous application of Si for different salt-tolerant plant species has been reported [82, 83]. Under saline environment, Si uptake by plats increases root activity and inhibits transpiration. But in the plasma membrane, it increases the activity of ATPase and PPase. This can result in decrease in Na uptake and an increase in K uptake [84, 85].

4.2.1 Silicon-mediated mechanisms underlying increased crops tolerance to salinity

Si application can directly influence growth of plants by diminishing the transport of Na^+ ions while indirectly activating physiological processes under saline conditions.

4.2.1.1 Reduced Na^+ uptake by plant roots due to Si application

Due to high concentration of Cl^- and Na^+ and low concentration of K^+ and Ca^{+2} in the saline environment, Na^+/K^+ ratio vary in plants [84]. Due to elevated level of Na^+ and overproduction of ROS, plant metabolism is being changed [85] Research studies demonstrated that Si can reduce ion toxicity which is resulted from the saline condition. It is also helpful in increasing K^+ and decreasing Na^+ uptake [86]. Thus, research studies determined that Si application resulted in reduced Na^+ buildup in the roots [86]. Si as phytolith, accumulates different parts of plant bodies. Si deposits underneath cell walls of roots to bind the Na^+ and reduces Na^+ toxicity by decreasing the Na^+ transport in upper regions and increasing the K^+ uptake.

4.2.1.2 Stimulation of antioxidant defense system in crops

Under the saline conditions, studies have determined the enhanced production of antioxidant due to the application of Si [87]. Effects of Si on the antioxidants depend upon different factors like the severity of saline stress, time, plant species, and the concentration of Si. Thus, studies determined that application of Si can regulate antioxidant defense system by reducing salinity effects. This also resulted in decrease lipid peroxidation and regulate membrane integrity. It also can decrease permeability of plasma membrane. The research studies determine that non-Si-treated and Si-treated plants show different responses under saline conditions. Application of Si plays a protective role to improve antioxidant activity.

5. Role of PGPRs in alleviation of salinity stress in maize crop

In the semi-arid environment, salinity pose negative effects on the growth and production of various crops. It also affects aggregate stability of soil. Soil structure stability has important for improvement of soil properties. The soil microbial communities such as free-living or symbiotic organisms play an immense role to improve soil structure. It is proved that the activities which microbes performed to soil aggregate stability are very advantageous [88]. It can efficient solution for saline soil and make it fit for agricultural practices. PGPRs can help in inducing plant tolerance to various abiotic stresses including salt stress. In saline environments, PGPR-crop interactions improved the plant growth. It can also promote plant survival in adverse conditions [89]. PGPR promote the growth and development of plants by providing nitrogen, phytohormones soluble phosphates, and iron [90]. The plant is being protected against various soil-borne diseases, and it is known that most of these diseases are caused by pathogenic fungi [91].

5.1 Various attributes of PGPsRs in mitigating negative effects of salt stress in maize crop

5.1.1 Enhanced root proliferation and plant vigor

PGPRs can promote the growth of the plants by means of PGPRs which colonize the rhizosphere [92]. The co-inoculation of seeds of different PGPR species is a beneficial strategy to remediate salt-stressed soil. This approach has improved the plant tolerance towards abiotic stresses and the structure of root hairs.

5.1.2 Phytohormones produced by bacteria

The physiological response in plants is increased by phytohormones produced by microbes in root zone. Production of indoleacetic acid and gibberellins promote the root length. It also increases number of tips, surface area of roots and uptake of nutrients thus promoting the plant vigor exposed to saline conditions [93–96]. Indole acetic acid production is a common characteristic of PGPR. This bacterium is observed to reduce salinity stress in plants.

5.1.3 Role of PGPR as a sink for 1-aminocyclopropane-1-carboxylate (ACC)

Increase in ACC levels can result in higher ethylene production under saline environment. It can also increase plant injuries [97, 98]. Cobalt ions and amino ethoxy vinyle glycine as chemical inhibitors of ethylene synthesis is often used to control salinity problems. These chemicals are expensive and have harmful effects on the environment. PGPR play a role of sink for ACC which can be hydrolyzed to generate a-ketobutyrate and ammonia to reduce the ethylene production.

5.1.4 PGPR-mediated ion homeostasis

Plants inoculated with PGPR have showed high concentration of K^+ which led to high Na^+/K^+ ratio and ultimately improved tolerance towards salt stress [99–101]. Salinity can damage the cell-membrane in plants which can enhance its permeability and electrolyte leakage. In maize, Lower the electrolyte leakage has been determined the inoculation with Rhizobium [102–104].

5.1.5 Accumulation of osmolytes

The functioning of photosynthetic structures and maintaining water homeostasis are essential for reducing salinity impact on plants. Excessive production of various compatible organic solutes (such as glycine betaine and proline) has been observed as stress responses in plants [105]. Accumulation of proline is a physiological response of plants to saline conditions [106]. It also maintains high leaf water potential and protects the plants from negative effects of oxidative stress. Researchers have determined that PGPRs contribute to accumulation of osmolytes to increase plant tolerance towards stress.

5.1.6 Antioxidative enzymes

Reactive oxygen species (ROS) damage the nucleic acids, proteins and lipids. Limited photosynthetic activity under salinity promotes the excessive production of ROS [107]. Antioxidants have been found to greatly reduce the oxidative damage. Under saline conditions, the activities of the enzymatic antioxidants such as guaicol peroxidase, catalase and superoxide dismutase increased [108]. Researchers determined that the application of PGPR caused a significant increase in polyphenol oxidase, superoxide dismutase and other enzymes involved in plant defense system. It also increases in enzymes such as peroxidase, phenyl alanine ammonia-lyase, catalase, phenolics and lipoxygenase [109–111]. These PGPR-stimulated enzymes are playing important role in removing hydrogen peroxide from stressed roots [112].

5.1.7 Ameliorating effects of bacterial extracellular polymeric substances (EPS)

Researchers determined that inoculation with EPS-producing PGPR have significantly increased the volume of soil macropores, rhizospheric soil aggregation, improved fertilizer as well as water availability. This approach can help plants to survive in salt-stressed soils. Different studies have shown positive effects of EPS-producing PGPR on the rhizospheric soil aggregation [113]. As bacterial EPS can sequester the cations, there may be an opportunity to eliminate the salinity stress by increasing the EPS-producing PGPR strains [114].

5.1.8 Enhancement of plant nutrient uptake

It is obvious that PGPR can regulate the availability of plant nutrients. So, employing PGPR can cut down the use of chemical fertilizers. Various PGPR strains are involved in solubilizing the inorganic phosphate and mineralization of organic phosphate, thus providing nutrients to plants [115]. However, the former activity of PGPR is the key role of PGPR in providing nutrients to plants.

5.1.9 PGPR-mediated disease suppression

Many rhizobacteria are known to produce antifungal metabolites like phenazines, HCN, pyrrolnitrin, tensin, pyoluteorin, 2,4-diacetylphloroglucinol, and viscosinamide [116]. However, various PGPR strains can control the pathogen of plants grown under salt stress.

6. Interactive techniques to ameliorate salinity stress in maize

6.1 Silicon and PGPR to mitigate salt stress in maize

An environment-friendly and cost-effective approach for lessening salinity in crop plants is the co-application of silicon and PGPR [117]. Different studies have shown that by improving photosynthetic efficiency, and scavenging enzyme activity soil salinity tolerance can be enhanced. It also determined that this approach can improve the plant tolerance towards salinity, ROS and Na^+/K^+ ratio [118]. PGPR promote the growth of plants via synthesis of phytohormones, exopolysaccharides, volatile organic compounds and different other mechanisms [118]. Recently, it has been found that both Si and PGPR can enhance plants tolerance to saline environment to improve growth and yield of plants [118].

6.2 Combined effects of biochar and plant growth-promoting bacterial endophytes on alleviating salt stress in maize

Employing the salt tolerant PGPR to enhance crop productivity has been a sustainable and efficient method [119–122]. Researchers have documented that PGPR produced the exopolysaccharide (EPSs) that prevent the uptake of Na + ions by sequestering these ions [123, 124]. Studies demonstrated that few PGPR have an important enzyme, ACC- deaminase, which can reduce ethylene production by metabolizing ACC into ammonia. ACC is the precursor of ethylene and a-ketobutyrate [125–127]. Unlike PGPR, plant growth-promoting bacterial endophytes colonize the internal tissues of plants without causing any harm to the plants [128]. It can lead to several physiological modifications that contribute to plant growth and development [129–131]. These, plant growth-promoting bacterial endophytes may promote plant growth by adopting the similar mechanisms as observed in PGPR [132]. Thus, it is proved that plant growth-promoting bacterial endophytes are more effective in promoting plant growth even under severe stresses as compared to PGPR. Different researchers have demonstrated that for reducing soil salinity addition of biochar along with endophytic bacteria is an efficient and environment friendly approach [133].

For enhancing crop growth and yield, use of biochar is cost effective and eco-friendly option to boost water and nutrient-holding capacity of soil [134–137]. Application of biochar has positive effects on physicochemical properties of soil. Moreover, Biochar can also improve a variety of soil microbes by providing them a favorable habitat and nourishment [138]. Thus, it is an excellent solution for recycling organic waste and solution to environmental pollution.

There are three important mechanisms underlying biochar-mediated reduction of salt stress in plants. These include:

a. High adsorption of Na^+ on biochar resulting in reduced availability of Na^+ in soil solution

b. Regulation of ions concentration in soil solution by liberating mineral nutrients

c. Dilution of soil solution via increasing available moisture contents of soil to reduce the osmotic stress [139].

7. Conclusion

Reclamation of saline soils is mainly achieved by employing various physicochemical processes. However, these processes are not sustainable and considered inefficient in the case of high salt concentration. PGPR contain a vital enzyme, 1-aminocyclopropane-1-carboxylate deaminase that can decrease salinity induced ethylene production. Silicon and elemental sulfur can also be applied to reduce the negative effects of soil salinity on plants. The organic matter such as press mud usually contains about 70% lime, 15–20% organic matter and 23% sugar. This organic matter is highly soluble and readily available to the microbial activity and soil. Due to microbial activity more carbon dioxide is produced that may increase the solubility of lime and hence reclaim the saline soils. Hence, the combine application of organic amendments (like press mud), inorganic amendments (like silicon and elemental sulfur) and PGPR can ameliorate the saline soil in an environmentally sustainable way.

Conflicts of interest

The authors declare no conflict of interest.

Author details

Sajid Rashid Ahmad*, Sana Ashraf* and Humaira Nawaz
College of Earth and Environmental Sciences, University of the Punjab, Lahore, Pakistan

*Address all correspondence to: sajidpu@yahoo.com and sana.cees@pu.edu.pk

References

[1] Muchate, N.S., G.C. Nikalje, N.S. Rajurkar, P. Suprasanna and T.D. Nikam. 2016. Plant salt stress: adaptive responses, tolerance mechanism and bioengineering for salt tolerance. Bot Rev. doi:10.1007/s12229-016-9173-y.

[2] Wicke, B., E. Smeet, V. Dornburg, B. Vashev, T. Gaiser, W. Turkenburg and A. Faaij. 2011. The global technical and economic potential of bioenergy from salt-affected soils. Energy Environment Science, 4: 2669-2681.

[3] QURESHI, R.H., M. Aslam and A. Javaid. 2003. Productivity enhancement in the salt affected lands of Joint Satiana Pilot Project Area of Pakistan. J Crop Prod 7, 277-297.

[4] Rajaravindran, M. and S. Natarajan. 2012. Effects of salinity stress on growth and antioxidant enzymes of the halophyte *Sesuvium portulacastrum*. International Journal Research Plant Science, 2: 23-28.

[5] Munns, R. and M. Tester. 2008. Mechanism of salinity tolerance. Annu. Rev. Plant Biol., 59: 651-681 Tester M, Davenport R (2003) Na+ tolerance and Na+ transport in higher plants. Ann Bot 91(5):503-27. doi:10.1093/aob/mcg058.

[6] Tester M, Davenport R (2003) Na+ tolerance and Na+ transport in higher plants. Ann Bot 91(5):503-27. doi:10.1093/aob/mcg058.

[7] Hanin, M., C.Ebel, M. Ngom, L. Laplaze and K. Masmoudi. 2016. New insights on plant salt tolerance mechanisms and their potential use for breeding. Front Plant Sci. 7:1787

[8] Deinlein, U., A.B. Stephan, T. Horie, W. Luo, G. Xu and J.I. Schroeder. 2014. Plant salt-tolerance mechanisms. Trends Plant Sci. 19:371-379.

[9] FAO. 2000. Global network on integrated soil management for sustainable use of salt-affected soils. Available in: http://www.fao.org/ag/AGL/agll/spush/intro.htm (28. Jan. 2015)

[10] WWAP. 2012. World Water Assessment Programme. The United Natins World Water Development Report 4: Managing Water under Uncertainity and Risk. Paris: UNESCO.

[11] FAO. 2008. Land and Plant Nutrition Management Service. http://www.fao.org/ag/agl/agll/. Accessed on November/ 15/2012.

[12] Schirawski, J., and M.H. Perlin. 2018. Plant- microbe interaction 2017-the good, the bad and the diverse. Int. J. Mol. Sci. 19:1374.

[13] Panuccio, M.R., S.E. Jacobsen, S.S. Akhtar and A. Muscolo. 2014. Effect of saline water irrigation on seed germination and early seedling growth of the halophyte quinoa. AoB Plants 6, plu047

[14] Mager, P., M. Gerth and J.I. Schreoeder. 2002. Molecular mechanisms of potassium and sodium uptake in plant. Plant Soil 247, 43-54.

[15] Tester, M., and R. Davenport. 2003. NaC tolerance and NaC transport in higher plants. Ann. Bot. 91, 503-527.

[16] Zhu, J.K. 2016. Abiotic stress signaling and responses in plants. Cell, 167, 313-324.

[17] Yang, Y. and Y. Guo. 2018. Elucidating the molecular mechanisms mediating plant salt-stress responses. New Phytol. 217:523-539.

[18] Fu, Q., C. Liu, N. Ding, Y. Lin and B. Guo. 2010. Ameliorative effects of inoculation with the plant growth promoting rhizobacterium Pseudomonas sp. DW1 ongrowth of eggplant (*Solanum melongena* L.)

seedlingsunder salt stress, Agr. Water Manage., 97: 1994-2000.

[19] Ma, Y., M. Rajkumar, C. Zhang and H. Freitas. 2016. Beneficial role of bacterial endophytes in heavy metal phytoremediation. J. Environ. Manag. 174, 14-25.

[20] Ayyam, V., S. Palanivel and S. Chandrakasan. 2019. "Approaches in land degradation management for productivity enhancement," in Coastal Ecosystems of the Tropics – Adaptive Management, eds V. Ayyam, S. Palanivel, and S. Chandrakasan (Singapore: Springer)

[21] ugtenberg, B. and F. Kamilova. 2009. Plant-growth-promoting rhizobacteria. Annu. Rev. Microbiol. 63, 541-556.

[22] Qin, Y., I.S. Druzhinina, X. Pan and Z. Yuan. 2016. Microbially mediated plant salt tolerance and microbiome-based solutions for saline agriculture. Biotechnol Adv. 34:1245-1259.

[23] Mayak, S., T. Tirosh and B.R. Glick. 2004. Plant growth-promoting bacteria confer resistance in tomato plants to salt stress. Plant Physiology and Biochemistry 42, 565-572.

[24] Ahmad, M., Z.A. Zahir, M. Khalid, F. Nazli and M. Arshad. 2013. Efficacy of Rhizobium and Pseudomonas strains to improve physiology, ionic balance and quality of mung bean under salt-affected conditions on farmer's fields. Plant Physiology and Biochemistry 63, 170-176.

[25] Yang, J., J.W. Kloepper and C.M. Ryu. 2009. Rhizosphere bacteria help plants tolerate abiotic stress. Trends Plant Sci. 14:1-4.

[26] Glick BR, Penrose DM, Li J (1998) A model for the lowering of plant ethylene concentrations by plant growth-promoting bacteria. Journal of Theoretical Biology 190, 63-68. doi:10.1006/jtbi.1997.0532.

[27] Upadhyay SK, Singh JS, Singh DP (2011) Exopolysaccharide-producing plant growth-promoting rhizobacteria under salinity condition.

[28] Balsanelli, E., V.A. de Baura, F.D. Pedrosa, E.M. de Souza and R.A. Monteiro. 2014. Exopolysaccharide biosynthesis enables mature biofilm formation on abiotic surfaces by Herbaspirillum seropedicae. PLOS ONE 9: 110-392.

[29] Ashraf, M., S.H. Berge and O.T. Mahmood. 2004. Inoculating wheat seedlings with exopolysaccharide-producing bacteria restricts sodium uptake and stimulates plant growth under salt stress. Biol. Fertil. Soils 40, 157-162.

[30] Chatterjee, P., S. Samaddar, R. Anandham, Y. Kang, K. Kim and G. Selvakumar. 2017. Beneficial soil bacterium Pseudomonas frederiksber-sgensis OS261augments salt tolerance and promotes red pepper plant growth. FrontPlant Sci. 8:1-9.

[31] Egamberdieva, D., D. Jabborova and A. Hashem. 2015. Pseudomonas induces salinity tolerance in cotton (*Gossypium hirsutum*) and resistance to Fusariumroot rot through the modulation of indole-3-acetic acid. Saudi J BiolSci. 22(6):773-9.

[32] Glick BR. 2014. Bacteria with ACC deaminase can promote plant growth and help to feed the world. Microbiol Res. 169:30-39.

[33] Numan, M., S. Bashir, Y. Khan, R. Mumtaz, Z.K. Shinwari, A.L. Khan, A. Khan and A. Al-Harrasi. 2018. Plant growth promoting bacteria as an alternative strategy for salt tolerance in plants: a review. Microbiol Res. 209:21-32.

[34] Alzahrani, Y., A. Kuşvuran, H.F. Alharby, S. Kuşvuran and M.M. Rady.

2018. The defensive role of silicon in wheat against stress conditions induced by drought, salinity or cadmium. Ecotoxicol Environ Saf.154:187-196.

[35] Etesami, H. 2018. Can interaction between silicon and plant growth promoting rhizobacteria benefit in alleviating abiotic and biotic stresses in crop plants? Agric Ecosyst Environ. 253:98-112.

[36] Garg, N. and P. Bhandari. 2016. Interactive effects of silicon and arbuscular mycorrhiza in modulating ascorbate-glutathione cycle and antioxidant scavenging capacity in differentially salt-tolerant *Cicer arietinum* L. genotypes subjected to long-term salinity. Protoplasma. 253:1325-1345.

[37] Li, Y.T., W.J. Zang, J.J. Cui, D.Y. Lang, M. Li, Q.P. Zhao and X.H. Zhang. 2016. Silicon nutrition alleviates the lipid peroxidation and ion imbalance of Glycyrrhiza uralensis seedlings under salt stress. Acta Physiol Plant. 38:96-105.

[38] Rios, J.J., M.C. Martínez-Ballesta, J.M. Ruiz, B. Blasco and M. Carvajal. 2017. Silicon-mediated improvement in plant salinity tolerance: the role of aquaporins. Front Plant Sci. 8:948.

[39] Abbas, T., R.M. Balal, M.A. Shahid, M.A. Pervez, C.M. Ayyub, M.A. Aqueel and M.M. Javaid. 2015. Silicon-induced alleviation of NaCl toxicity in okra (*Abelmoschus esculentus*) is associated with enhanced photosynthesis, osmoprotectants and antioxidant metabolism. Acta Physiol Plant. 37:6.

[40] Garg, N. and P. Bhandari. 2015. Silicon nutrition and mycorrhizal inoculations improve growth, nutrient status, K+/Na+ ratio and yield of *Cicer arietinum* L. genotypes under salinity stress. Plant Growth Regul. 78, 371-387.

[41] Clark, G.J., N. Dodgshun, P.W.G. Sale and C. Tang. 2007. Changes in chemical and biological properties of a sodic clay subsoil with addition of organic amendments. Soil Biology and Biochemistry, 39: 2806-2817.

[42] Ghafoor, A., G. Murtaza, B. Ahmad and T.M. Boers. 2008. Evaluation of amelioration treatments and economic aspects of using saline-sodic water for rice and wheat production on salt-affected soils under arid land conditions. Irrigation and Drainage, 57: 424-434.

[43] Wong, V.N., R.C. Dalal and R.S. Greene. 2009. Carbon dynamics of sodic and saline soils following gypsum and organic material additions: a laboratory incubation. Applied Soil Ecology, 41: 29-40.

[44] Cha-um, S. and C. Kirdmanee. 2011. Remediation of salt-affected soil by the addition of organic matter: an investigation into improving glutinous rice productivity. Scientia Agricola, 68: 406-410.

[45] Parida AK, Das AB (2005) Salt tolerance and salinity effects on plants: a review. Ecotox Environ Safe 60(3):324-49. doi:10.1016/j.ecoenv.2004.06.010

[46] Meloni DA, Oliva MA, Martinez CA, Cambraia J (2003) Photosynthesis and activity of superoxide dismutase, peroxidase and glutathione reductase in cotton under salt stress. Environ Exp Bot 49(1):69-76.doi:10.1016/S0098-8472(02)00058-8

[47] Hasegawa PM, Bressan RA, Zhu JK, Bohnert HJ (2000) Plant cellular and molecular responses to high salinity. Annu Rev Plant Physiol Plant Mol Biol 51:463-99. doi:10.1146/annurev.arplant.51.1.463

[48] Shannon MC, Grieve CM (1999) Tolerance of vegetable crops to salinity. Sci Hortic 78(1-4):5-38. doi:10.1016/S0304-4238(98)00189-7

[49] Larcher W. (1980) Physiological plant ecology: ecophysiology and stress

physiology of functional groups, 2nd edn. Springer-Verlag, Berlin.

[50] Xu ZH, Saffigna PG, Farquhar GD, Simpson JA, Haines RJ, Walker S et al (2000) Carbon isotope discrimination and oxygen isotope composition in clones of the F (1) hybrid between slash pine andCaribbean pine in relation to tree growth, water-use efficiency and foliar nutrient concentration. Tree Physiol 20(18):1209-17. doi:10.1093/treephys/20.18.1209

[51] Tester, M., and R. Davenport. 2003. NaC tolerance and NaC transport in higher plants. Ann. Bot. 91, 503-527

[52] White PJ, Broadley MR (2001) Chloride in soils and its uptake and movement within the plant: a review. Ann Bot 88(6):967-88. doi:10.1006/anbo.2001.1540.

[53] Ruiz-Lozano, J. M., Porcel, R., Azcón, C., & Aroca, R. (2012). Regulation by arbuscular mycorrhizae of the integrated physiological response to salinity in plants: new challenges in physiological and molecular studies. *Journal of Experimental Botany*, *63*(11), 4033-4044.

[54] Hasegawa PM, Bressan RA, Zhu JK, Bohnert HJ (2000) Plant cellular and molecular responses to high salinity. Annu Rev Plant Physiol Plant Mol Biol 51:463-99. doi:10.1146/annurev.arplant.51.1.463.

[55] ietz DN, Haynes RJ (2003) Effects of irrigation-induced salinity and sodicity on soil microbial activity. Soil Biol Biochem 35(6):845-54.doi:10.1016/S0038-0717(03)00125-1

[56] Borneman J, Skroch PW, O'Sullivan KM, Palus JA, Rumjanek NG, Jansen JL et al (1996) Molecular microbial diversity of an agricul-tural soil in Wisconsin. Appl Environ Microbiol 62(6):1935-43.

[57] Jofre E, Fischer S, Rivarola V, Balegno H, Mori G (1998) Saline stress affects the attachment of Azospirillum brasilense Cd to maize and wheat roots. Can J Microbiol 44(5):416-22. doi: 10.1139/w98-024.

[58] Ondrasek G, Rengel Z, Romic D, Savic R (2010) Environmental salinisation processes in agro-ecosystem of neretva river estuary. Novenytermeles 59:223-226.

[59] Fuller, M.P., J.H. Hamza, H.Z. Rihan and M. Al-Issawi. 2012. Germination of primed seed under NaCl stress in wheat. Int. Sch. Res. Netw. Bot.,12: 1-5. https://doi.org/10.5402/2012/167804

[60] Kim, H., H. Jeong, J. Jeon and S. Bae. 2016. Effects of irrigation with saline water on crop growth and yield in greenhouse cultivation. Water, 8(4): 127-135. https://doi.org/10.3390/w8040127.

[61] Plaut, Z., M. Edelstein and M. Ben-Hur. 2013.Overcoming salinity barriers to crop production using traditional methods. Crit. Rev. Plant Sci.,.32(4): 250291.https://doi.org/10.1080/07352689.2012.752236.

[62] Shaaban, M., M. Abid and R.A.I. Abou-Shanab. 2013. Amelioration of salt affected soils in rice paddy system by application of organic and inorganic amendments. Plant Soil Environ., 59(5): 227-233. https://doi.org/10.17221/881/2012-PSE

[63] Amezketa, E.A., R. Aragues and R. Gazol. 2005.Efficiency of sulfuric acid, mined gypsum and two gypsum by-products in soil crusting prevention and sodic soil reclamation. Agron.J., 97: 983-989. https://doi.org/10.2134/agronj2004.0236

[64] Wahid, A., S. Akhtar, I. Ali and E. Rasul. 2015.Amelioration of saline-sodic soils with organicmatter and their use for wheat growth. Commun. Soil Sci. Plant Anal., 29(15-16):

2307-2318.https://doi.org/10.1080/0010362980937011

[65] Mamo, M., J.F. Moncrief, C.J. Rosen and T.R.Halbach. 2000. Municipal solid waste compost application on soil water and water stress in irrigated corn. Compost Sci. Util., 8(3):236-246. https://doi.org/10.1080/1065657X.2000.10701996

[66] Ahmad, M., Z.A. Zahir, M. Khalid, F. Nazli and M. Arshad. 2013. Efficacy of Rhizobium and Pseudomonas strains to improve physiology, ionic balance and quality of mung bean under salt-affected conditions on farmer's fields. Plant Physiology and Biochemistry 63, 170-176.

[67] Iqbal, M., A. Hassan and M. Ibrahim. 2008. Effects of tillage systems and mulch on soil physical quality parameters and maize (*Zea mays* L.) yield in semi-arid Pakistan. Biol. Agric. Hortic.,25(4): 311-325. https://doi.org/10.1080/01448

[68] Qadir M, Quillérou E, Nangia V, Murtaza G, Singh M, Thomas RJ, Drechsel P, Noble AD (2014) Economics of salt-induced land deg-radation and restoration. Nat Res Forum 38:282-295

[69] Rengasamy, P. 2002. Transient salinity and subsoil constraints to dryland farming in Australian sodic soils: An overview.Aust. J. Exp. Agric., 42: 351-61.

[70] Munns, R. and M. Tester. 2008. Mechanisms of salinity tolerance. Annu. Rev. Plant Biol. 59, 651-681.

[71] Tavakkoli, E., P. Rengasamy and G.K. McDonald. 2010. High concentrations of Na+ and Cl– ions in soil solution have simultaneous detrimental effects on growth of faba bean under salinity stress. J. Exp. Bot., 61: 4449-4459.

[72] Shahzad, M., K. Witzel, C. Zörb, and K.H. Mühling. 2012. Growth-related changes in subcellular ion patterns in maize leaves (*Zea mays* L.) under salt stress. J. Agron. Crop Sci.,198: 46-56.

[73] Abbasi, G.H., J. Akhtar, M. Anwar-ul-Haq, S. Ali, Z. Chen, andW. Malik. 2014. Exogenous potassium differentially mitigates salt stress in tolerant and sensitive maize hybrids. Pak. J. Bot., 46: 135-146.

[74] Lauchli, A. and E. Epstein. 1990. Plant responses to saline and sodic conditions, in: (Ed.): Tanji K.K. agricultural salinity assessment and management, American Society of Civil Engineering, New York, p. 113-137.

[75] Johnson HE, Broadhurst D, Goodacre R, Smith AR (2003) Metabolic fingerprinting of salt-stressed tomatoes. Phytochemistry 62(6):919– 28. doi:10.1016/S0031-9422(02)00722-7.

[76] Spadaro, D., B.W. Yun, S.H. Spoel, C. Chu, Y.Q. Wang and G.J. Loake. 2010. The redox switch: dynamic regulation of protein function by cysteine modifications. Physiol. Plant.,138: 360-371.

[77] Singh, B.R. 2003. Sulfur and crop quality-agronomical strategies for crop improvement. Abstracts of COST Action 829 Meetings, Braunschweig, Germany. p. 35-36.

[78] Taiz, L. and E. Zeiger. 2006. Plant Physiology. 4th Edition.Sinauer Associates Inc. Sunderland, Massachusetts.

[79] Ashraf M, Afzal M, Ahmed R, Mujeeb F, Sarwar A, Ali L (2010). Alleviation of detrimental effects of NaCl by silicon nutrition in salt Esensitive and Etolerant genotypes of sugarcane (*Saccharum officinarum* L.). Plant Soill 326(12):381-391.

[80] Ashraf M, Ozturk M, Ahmad MSA, Aksoy A (2012). Crop production for

agricultural improvement. Springer Science+Business Media, NY.

[81] Dionisio-Sese, M. L., & Tobita, S. (1998). Antioxidant responses of rice seedlings to salinity stress. *Plant Science*, *135*(1), 1-9.

[82] Tuna, A. L., Kaya, C., Higgs, D., Murillo-Amador, B., Aydemir, S., and Girgin, A.R. (2008). Silicon improves salinity tolerance in wheat plants. Environ. Exp. Bot. 62, 10-16. doi: 10.1016/j.envexpbot.2007.06.006

[83] Liang Y, Sun W, Zhu YG, Christie P (2007). Mechanisms of silicon-mediated alleviation of abiotic stresses in higher plants: A review.Environmental Pollution 147(2):422-428.

[84] Khan, M.A. and I.A. Ungar. 1997. Effect of thermo period on recovery of seed germination of halophyte from saline conditions. Am. J. Bot., 84: 279-283.

[85] Mahajan, S., & Tuteja, N. (2005). Cold, salinity and drought stresses: an overview. *Archives of biochemistry and biophysics*, *444*(2), 139-158.

[86] Tuna, A. L., Kaya, C., Higgs, D., Murillo-Amador, B., Aydemir, S., and Girgin, A.R. (2008). Silicon improves salinity tolerance in wheat plants. Environ. Exp. Bot. 62, 10-16. doi: 10.1016/j.envexpbot.2007.06.006

[87] Li YQ, Zhao HL, Yi XY, Zuo XA, Chen YP (2006) Dynamics of carbon and nitrogen storages in plant–soil system during desert-ification process in horqin sandy land. Huan Jing Ke Xue 27(4):635-40.

[88] Jastrow JD, Miller RM (1991) Methods for assessing the effects of biota on soil structure. Agric Ecosyst Environ 34(1-4):279-303. doi:10.1016/0167-8809(91)90115-E.

[89] Dimkpa C, Weinand T, Asch F (2009) Plant-rhizobacteria interactions alleviate abiotic stress conditions. Plant Cell Environ 32(12):1682-94. doi: 10.1111/j.1365-3040.2009.02028.x.

[90] Hayat R, Ali S, Amara U, Khalid R, Ahmed I (2010) Soil beneficial bacteria and their role in plant growth promotion: a review. Ann Microbiol 60(4):579-98. doi:10.1007/s13213-010-0117-1.

[91] Lugtenberg B, Kamilova F (2009) Plant-growth-promoting rhizobacteria. Annu Rev Microbiol 63:541-56. doi:10.1146/annurev.micro.62.081307.162918.

[92] Diby P, Bharathkumar S, Sudha N (2005a) Osmotolerance in biocontrol strain of pseudomonas pseudoalcaligenes MSP-538: a study using osmolyte, protein and gene expression profiling. Ann Microbiol55(4):243-47

[93] Diby P, Sarma YR, Srinivasan V, Anandaraj M (2005b) *Pseudomonas fluorescens* mediated vigour in black pepper (*Piper nigrum* L.) undergreen house cultivation. Ann Microbiol 55(3):171-74.

[94] Egamberdieva D, Kucharova Z (2009) Selection for root colonizing bacteria stimulating wheat growth in saline soils. Biol Fertil Soils45(6):563-71. doi:10.1007/s00374-009-0366-y

[95] Egamberdieva D (2012) Pseudomonas chlororaphis: a salt-tolerant bacte-rial inoculant for plant growth stimulation under saline soil conditions.Acta Physiol Plant 34(2): 751-56. doi:10.1007/s11738-011-0875-9

[96] Egamberdieva D (2011) Survival of pseudomonas extremorientalis TSAU20 and P. Chlororaphis TSAU13 in the rhizosphere of com-mon bean (Phaseolus.

[97] Botella MA, del Amor FM, Amoros A, Serrano M, Martinez V, Cerda A(2000) Polyamine, ethylene and

other physico-chemical parametersin tomato (*Lycopersicon esculentum*) fruits as affected by salinity. Physiol Plant 109(4):428-34. doi:10.1034/j.1399-3054.2000.100409.x

[98] Botella MA, Martinez V, Pardines J, Cerdá A (1997) Salinity induced potassium deficiency in maize plants. J Plant Physiol 150(1-2):200-05. doi:10.1016/S0176-1617(97)80203-9.

[99] Serraj R, Sinclair TR (2002) Osmolyte accumulation: can it really help increase crop yield under drought conditions? Plant Cell Environ 25(2):333-41. doi:10.1046/j.1365-3040.2002.00754.x

[100] Nadeem SM, Shaharoona B, Arshad M, Crowley DE (2012) Population density and functional diversity of plant growth promoting rhizobacteria associated with avocado trees in saline soils. ApplSoil Ecol 62:147-54. doi:10.1016/j.apsoil. 2012.08.005

[101] Nadeem SM, Zahir ZA, Naveed M, Arshad M (2007) Preliminary investigations on inducing salt tolerance in maize through inoculation with rhizobacteria containing ACC deaminase activity. Can J Microbiol 53(10):1141-9. doi:10.1139/W07-081.

[102] Kohler J, Caravaca F, Roldan A (2010) An AM fungus and a PGPR intensify the adverse effects of salinity on the stability of rhizosphere soil aggregates of *Lactuca sativa*. Soil Biol Biochem 42(3):429-34.doi:10.1016/j. soilbio.2009.11.021

[103] Kohler J, Hernandez JA, Caravaca F, Roldan A (2009) Induction of antioxidant enzymes is involved in the greater effectiveness of a PGPR versus AM fungi with respect to increasing the tolerance of lettuce to severe salt stress. Environ Exp Bot 65(2-3):245-52. doi:10.1016/j.envexpbot. 2008.09.008.

[104] Rojas-Tapias D, Moreno-Galvan A, Pardo-Diaz S, Obando M, Rivera D, Bonilla R (2012) Effect of inoculation with plant growth-promoting bacteria (PGPB) on amelioration of saline stress in maize (Zeamays). Appl Soil Ecol 61:264-72. doi:10.1016/j.apsoil.2012.01.006.

[105] Bano A, Fatima M (2009) Salt tolerance in *Zea mays* (L) following inoculation with Rhizobium and Pseudomonas. Biol Fertil Soils 45(4):405-13. doi:10.1007/s00374-008-0344-9.

[106] Peng YL, Gao ZW, Gao Y, Liu GF, Sheng LX, Wang DL (2008) Eco-physiological characteristics of alfalfa seedlings in response to var- ious mixed salt-alkaline stresses. J Integr Plant Biol 50(1):29-39.doi:10.1111/j.1744-7909.2007.00607.x

[107] Johnson HE, Broadhurst D, Goodacre R, Smith AR (2003) Metabolic fingerprinting of salt-stressed tomatoes. Phytochemistry 62(6):919– 28. doi:10.1016/S0031-9422(02)00722-7.

[108] Mittova V, Tal M, Volokita M, Guy M (2003) Up-regulation of the leaf mitochondrial and peroxisomal antioxidative systems in response to salt-induced oxidative stress in the wild salt-tolerant tomato species Lycopersicon pennellii. Plant Cell Environ 26(6):845-56. doi:10.1046/j.1365-3040.2003.01016.x.

[109] Liang Y, Sun W, Zhu YG, Christie P (2007). Mechanisms of silicon-mediated alleviation of abiotic stresses in higher plants: A review.Environmental Pollution 147(2):422-428.

[110] Nautiyal CS, Govindarajan R, Lavania M, Pushpangadan P (2008) Novel mechanism of modulating natural antioxidants in functional foods: Involvement of plant growth promoting rhizobacteria NRRL B-30488. J Agr Food Chem 56(12):4474-81. doi:10.1021/jf073258i.

[111] Chakraborty N, Ghosh R, Ghosh S, Narula K, Tayal R, Datta A,

Chakraborty S (2013) Reduction of oxalate levels in tomato fruit 748 D. Paul, H. Lade and consequent metabolic remodeling following overexpression of a fungal oxalate decarboxylase1[W]. Plant Physiol 162(1):364-78. doi:10.1104/pp. 112.209197.

[112] Kim SY, Lim JH, Park MR, Kim YJ, Park TI, Se YW, Choi KG, Yun SJ (2005) Enhanced antioxidant enzymes are associated with reduced hydrogen peroxide in barley roots under saline stress. J Biochem Mol Biol 38(2):218-24. doi:10.5483/BMBRep.2005.38.2.218.

[113] Alami Y, Achouak W, Marol C, Heulin T (2000) Rhizosphere soil aggregation and plant growth promotion of sunflowers by an exopolysaccharide-producing rhizobium sp. strain isolated from sunflower roots. Appl Environ Microbiol 66(8):3393-8. doi:10.1128/AEM.66.8.3393-3398.2000.

[114] Geddie JL, Sutherland IW (1993) Uptake of metals by bacterial polysaccharides. J Appl Bacteriol 74(4):467-72. doi:10.1111/j.1365-2672.1993.tb05155.x.

[115] Dobbelaere S, Vanderleyden J, Okon Y (2003) Plant growth-promoting effects of diazotrophs in the rhizosphere. Crit Rev Plant Sci 22(2): 107-49. doi:10.1080/713610853.

[116] Bhattacharyya PN, Jha DK (2011) Plant growth-promoting rhizobacteria(PGPR): emergence in agriculture. World J Microbiol Biotechnol28(4):1327-1350. doi:10.1007/s11274-011-0979-9

[117] Khan, A.; Khan, A.L.; Muneer, S.; Kim, Y.H.; Al-Rawahi, A.; Al-Harrasi, A. Silicon and salinity: Crosstalk in crop-mediated stress tolerance mechanisms. Front. Plant Sci. 2019, 10, 1429. [CrossRef]

[118] Adhikari, A.; Khan, M.A.; Lee, K.E.; Kang, S.M.; Dhungana, S.K.; Bhusal, N.; Lee, I.J. The halotolerant rhizobacterium-pseudomonas

[119] Mahmood, S.; Daur, I.; Al-Solaimani, S.G.; Ahmad, S.; Madkour, M.H.; Yasir, M.; Hirt, H.; Ali, S.; Ali, Z. Plant growth promoting

[120] Al-Garni, S.M.S.; Khan, M.M.A.; Bahieldin, A. Plant growth-promoting bacteria and silicon fertilizer enhance plant growth and salinity tolerance in *Coriandrum sativum*. J. Plant Interact. 2019, 14, 386-396. [CrossRef]

[121] Mayak S, Tirosh T, Glick BR (2004) Plant growth-promoting bacteria confer resistance in tomato plants to salt stress. Plant Physiology and Biochemistry 42, 565-572. doi:10.1016/j.plaphy.2004.05.009

[122] Ahmad M, Zahir ZA, Khalid M, Nazli F, Arshad M (2013) Efficacy of Rhizobium and Pseudomonas strains to improve physiology, ionic balance and quality of mung bean under salt-affected conditions on farmer's fields. Plant Physiology and Biochemistry 63, 170-176. doi:10.1016/j.plaphy.2012.11.024

[123] Ashraf M, Wu L (1994) Breeding for salinity tolerance in plants. Critical Reviews in Plant Sciences 13, 17-42. doi:10.1080/07352689409701906

[124] Ashraf M, Hasnain S, Berge O, Mahmood T (2004) Inoculating wheat seedlings with exopolysaccharide-producing bacteria restricts sodium uptake and stimulates plant growth under salt stress. Biology and Fertility of Soils 40, 157-162. doi:10.1007/s00374-004-0766-y

[125] Glick BR (2012) Plant growth-promoting bacteria: mechanisms and applications. Scientifica 15, 963401.

[126] Nadeem SM, Zahir ZA, Naveed M, Arshad M (2009) Rhizobacteria containing ACC-deaminase confer salt tolerance in maize grown on salt-affected fields. Canadian Journal of Microbiology 55, 1302-1309. doi:10.1139/W09-092

[127] Nadeem S, Zahir Z, NaveedM, Nawaz S (2013)Mitigation of salinity-induced negative impact on the growth and yield of wheat by plant growth-promoting rhizobacteria in naturally saline conditions. Annals of Microbiology 63, 225-232. doi:10.1007/s13213-012-0465-0

[128] Ma Y, Prasad MNV, Rajkumar M, Freitas H (2011) Plant growth promoting rhizobacteria and endophytes accelerate phytoremediation of metalliferous soils. Biotechnology Advances 29, 248-258. doi:10.1016/j.biotechadv.2010.12.001

[129] Schulz B, Boyle C (2006) What are endophytes? In 'Microbial root endophytes'. (Eds BJE Schulz, CIC Boyle, TN Sieber) pp. 1-13. (Springer-Verlag: Berlin)

[130] Pillay VK, Nowak J (1997) Inoculum density, temperature, and genotype effects on in vitro growth promotion and epiphytic and endophytic colonization of tomato (*Lycopersicon esculentum* L.) seedlings inoculated with a pseudomonad bacterium. Canadian Journal of Microbiology 43, 354-361. doi:10.1139/m97-049

[131] Ma Y, Prasad MNV, Rajkumar M, Freitas H (2011) Plant growth promoting rhizobacteria and endophytes accelerate phytoremediation of metalliferous soils. Biotechnology Advances 29, 248-258. doi:10.1016/j.biotechadv.2010.12.001

[132] Ryan RP, Germaine K, Franks A, Ryan DJ, Dowling DN (2008) Bacterial endophytes: recent developments and applications. FEMS Microbiology Letters 278, 1-9. doi:10.1111/j.1574-6968.2007.00918.x

[133] Thomas SC, Frye S, Gale N, Garmon M, Launchbury R, Machado N, Melamed S, Murray J, Petroff A, Winsborough C (2013) Biochar mitigates negative effects of salt additions on two herbaceous plant Biochar and microbes alleviate salinity stress Functional Plant Biology K species. Journal of Environ-mentalManagement 129, 62-68. doi:10.1016/j.jenvman.2013.05.057

[134] Cantrell KB, Hunt PG, Uchimiya SM, Novak JM, Ro KS (2012) Impact of pyrolysis temperature and manure source on physicochemical characteristics of biochar. Bioresource Technology 107, 419-428.doi:10.1016/j.biortech.2011.11.084

[135] Akhtar SS, Li G, Andersen MN, Liu F (2014) Biochar enhances yield and quality of tomato under reduced irrigation. Agricultural Water Management 138, 37-44. doi:10.1016/j.agwat.2014.02.016

[136] Akhtar SS, Andersen MN, Liu F (2015b) Biochar mitigates salinity stress in potato. Journal of Agronomy and Crop Science doi:10.1111/jac.12132

[137] Xu CY, Hosseini-Bai S, Hao Y, Rachaputi RC, Wang H, Xu Z, Wallace H (2014) Effect of biochar amendment on yield and photosynthesis of peanut on two types of soils. Environmental Science and Pollution Research International 22, 6112-6125. doi:10.1007/s11356-014-3820-9

[138] Lehmann J, Rillig MC, Thies J, Masiello CA, Hockaday WC, Crowley D (2011) Biochar effects on soil biota – a review. Soil Biology & Biochemistry 43, 1812-1836. doi:10.1016/j.soilbio.2011.04.022.

[139] Akhtar SS, Andersen MN, Liu F (2015a) Residual effects of biochar on improving growth, physiology and yield of wheat under salt stress. Agricultural Water Management 158, 61-68. doi:10.1016/j.agwat.2015.04.010

Section 2

Animal Landrace

Chapter 8

Utilization and Conservation of Landrace Chickens of Nigeria: Physical and Performance Characteristics, Issues and Concerns

Cosmas Chikezie Ogbu

Abstract

The Nigerian indigenous chickens (NICs) are a critical component of the global animal genetic resources. They are distributed in rural areas, kept by the majority of the rural poor. They constitute different strains, or ecotypes local to tribes, regions, or ecological zones and are valued for their disease resistance, adaptation, and yield of valuable products and income on marginal inputs making them a low risk species. They are hence a unique and vital genetic resource and gene pool for present and long-term genetic improvement and human need for food and sustenance. The NIC is however; threatened by extinction owing to neglect, negative selection, breed substitution, and genetic introgression. There is need to draw research and policy attention to the conservation of NICs in accord with the global effort for the conservation of indigenous chickens which is probably the most neglected among farm animal species. The present review therefore, focuses on the physical and performance characteristics, genetic diversity and improvement, utilization and conservation of NIC genetic resources.

Keywords: indigenous chickens, ecotypes, chicken genetic resources, genetic diversity, ex situ conservation, in situ conservation

1. Introduction

Indigenous chicken biodiversity encompasses the genetic variants within and between native chickens distributed around the world [1]. They are domesticated but unselected and unimproved autochthonous populations characterized by tremendous variation in physical, genetic, and productive attributes [2, 3] and known by various native names; and whose attributes are best described by farmers in their home communities. Native chicken population is vital in the livelihood of resource poor house-holds and marginal rural communities in Africa, providing nutrition, cash flow reserve, recreation and cultural roles [4]. In Nigeria, ICs makeup about 80% of poultry population [5, 6]. The Nigerian indigenous chicken (NIC) is classified into three major phenotypes with regards to body weight. Dwarf, normal, and heavy types are generally distinguished [7] but more recent classifications recognize two broad body weight categories namely light body weight, and heavy body weight [8]. Plumage or feather color (pigmentation) include black, brown, gray, ash, white, red, and mottled; various color combinations [9–12] and feather patterns. Feather distribution is predominantly complete but limited in some phenotypes such as naked neck, frizzle, and short flight feather types [7]. Feather structure is predominantly normal, and seldom frizzle and silky [7]. Comb type is mostly single but rose, pea, walnut, duplex, and crest phenotypes also occur in decreasing frequencies [13–15]. Ear lobe is mostly present but absent in some dwarf ecotypes and color is mostly red or white. Wattle color is red or white while shank and skin color is often white or yellow but could be grayish, ash, blackish or bluish [9, 12, 13]. Beak color could be white, yellow, brown, or black. Some of the major physical attributes reflect environmental adaptations such as large comb, limited feather distribution, and frizzle feather structure for enhanced heat dissipation and body temperature regulation [16].

The genetic background of body size, feather structure and distribution are body size genes (DW/dw), plumage distribution genes (Na/na), feather structure genes (F/f), and numerous plumage color and color pattern genes [6, 16–18]. Morphological, physical, and performance characteristics along with biochemical, and molecular markers have been employed for genetic diversity evaluation in ICs [19–21]. Results indicate that NICs are a unique and vital genetic resource and gene pool for present and future production and breeding imperatives. The NIC is however; threatened by extinction owing to under valuation and utilization, diseases, predation, negative selection, breed substitution and genetic dilution, necessitating urgent action by research and policy makers towards the conservation of native chickens which is probably the most neglected farm animal genetic resources [16, 22–24]. The present study aims to collate information on qualitative and quantitative trait characteristics and variation, genetic diversity and improvement, and conservation issues and concerns of NICs, to draw attention to the extent of IC biodiversity and the need for action to improve the production, utilization, valuation and conservation of Nigerian's landrace chickens.

2. Physical and qualitative attributes of Nigerian indigenous chickens

Physical and qualitative trait evaluation reflects the effects of genes, and the impact of the environment; and enhances the understanding of local adaptations, which impact performance [14, 25]. Natural selection as well as mutations could throw up unique phenotypes, genes, and genotypes that have special adaptation and utility in specific environments. Characterization of the physical and qualitative traits for local adaptations facilitates selection for traits that enhance fitness and performance [13, 14]. Variation in physical and qualitative traits of NICs is expected given the diverse agro ecological climates, centuries of migration and interbreeding, natural and man-made challenges including disease, predation, and negative selection to which ICs have been subjected [6, 7, 16, 26].

2.1 Plumage type, distribution, color, and color patterns in Nigerian indigenous chickens

Plumage type (smooth plumage and frizzle plumage) and plumage distribution (complete plumage and naked neck) are genetically determined. The genetic background of plumage type is the autosomal dominant frizzle (F) and recessive smooth (f) feathering genes while the genetic basis of feather distribution is the autosomal dominant naked neck (Na) and recessive complete feathering (na) genes. Two forms of genetically determined melanin pigment (eumelanin and pheomelanin) define plumage color [27, 28] but the genetic background of plumage color in ICs is very complex. Generally, the e locus genes or alleles, secondary pattern genes, eumelanin enhancers or melanizers, eumelanin restrictors (generally called Columbian restrictors), eumelanin diluters or demelanizers, pheomelanin intensifiers or red enhancers, and pheomelanin diluters or gold diluters essentially determine plumage color and color pattern [29, 30].

The degree of expression of the e locus genes and secondary color/pattern genes determine the degree of melanization and hence deviation from ground color [28]. In the absence of pigmentation, plumage color is called silver which looks white. In wild type phenotype, the ground color is yellow to brown called gold and the presence of red enhancers boosts (intensifies) gold to a (dark) red color while presence of red diluters dilutes (tones) gold to a yellow, cream or lemon color [30]. The ground color can hence be silver, cream, yellow, gold, brown, or red, depending on the presence and dose of pheomelanin modifying genes [30].

The typical free range IC closely approximates the wild type (Red jungle fowl) plumage phenotype (most likely genotype) and is sexually dimorphic in plumage color; hens being more ground colored than roosters [30]. A significant deviation of today's ICs from the classical phenotype is expected owing to centuries of migration, natural selection (including predation mostly directed at chickens having brightly colored plumage), negative selection to fulfill cultural and ritual roles, intensification of production and artificial selection by man (increasing expression of the 'domesticated phenotype') [31, 32], and interbreeding between phenotypes (also genotypes). Consequently, plumage color phenotypes reported for adult NICs range from full ground color to full black color. Findings however, vary by source of samples (on-farm vs. market samples), system of production (intensive vs. semi-intensive vs. extensive) due to differences in flock structure (male:female; adult:grower), and environmental effects on plumage genotype and phenotype [32]. In heavy ecotype (HE) ICs, [33] reported plumage color as white, white and black, gold and black, black, barred, brown, brown and black, gold, and gold and brown. Indigenous chickens assembled from local markets within Makurdi and environ in Benue State, Northcentral Nigeria revealed black, light brown, white, spotted, and mottled phenotypes at 32.22, 12.22, 7.78, 21.11 and 26.67%, respectively [9]. In 2420 mature ICs from 100 farm families in Dekina, Kogi State, Northcentral Nigeria [10] observed plumage colors of brown, brown and black, black, black with white, white, and brown with black and white at 41.75, 35.5, 10.25, 6.50, 2.75, and 3.25%, respectively. Within Yoruba and Fulani ecotypes belonging to households in Ogbomosho, Oyo State, Southwestern Nigeria [11] observed plumage colors of white, black, brown, ash, red, and yellow in 15.07 and 20.6, 25.67 and 31.55, 9.34 and 10.69, 9.42 and 6.52, 9.11 and 12.13, and 0.00 and 2.35%, for Yoruba and Fulani ecotypes, respectively. Whitish brown and multicolor plumages were observed in 31.4 and 13.3%, and 0.00 and 2.87% of chickens, respectively. In 7091 ICs (2467 males and 4624 females reared semi-intensively) from rural households in Gwer-West, Benue State, Nigeria, [13] observed complete white, brown, and black; brown with black spotting, black with white spotting, and white with black spotting phenotypes in 15.55, 12.71, and 19.79; 12.89, 29.01, and 10.05%, respectively. In a population of Tiv and Fulani ICs reared intensively at Ekpehe in Makurdi Benue State, Nigeria, [14] reported nine plumage colors made up of three single or solid colors (brown, 11.54%, black, 3.85%, and light brown, 7.69%) and six color combinations (silver with brown, 6.25%; mottled brown, 19.20%; mottled black, 11.54%; black with brown, 23.08%; mixed gray, 3.85%, and mottled white, 3.85%) in the Tiv chickens. Within the Fulani ecotype, brown, white, and black were the solid colors at frequencies of 17.31, 3.08, and 5.77%, respectively while mixed colors included dull brown (2.69%), mixed gray (5.77%), mixed black (13.46%), black and brown (17.37%), mottled brown (13.46%), mottled white (9.46%), and mottled black (11.53%). Daikwo et al. [12] observed five single plumage color phenotypes viz. black, white, brown, ash, and red at 39.43, 23.02, 15.47, 11.13, and 9.43%, respectively and multicolored phenotypes (1.51%) in 1060 adult ICs from 208 households in Bekwara, Cross River State, Nigeria. In a recent survey of seven council wards, four villages/ward, 4 to 5 households/village, and 5 to 7 mature ICs/household, [15] observed six unicolored plumage phenotypes consisting of black, white, brown, gray, ash, and red; and a series of mixed colored plumage phenotypes viz. multicolored, black and white, gray and white, black and brown, reddish black, and ash with black. Of these single plumage colors and color mixtures, black, white and brown were predominant (36.23, 20.00, and 13.02%, respectively) while ash was the least frequent (0.57%). In a population of heavy ecotype (HE) ICs genetically improved for egg production and body weight at first egg by within ecotype selection for six generations, there was preponderance of pheomelanin-based feather colors and color combinations/patterns (white, yellow, gold, red, and brown) while in unimproved light ecotype (LE) ICs, eumelanine-based feather colors were more dominant (**Table 1, Figure 1**). The range of plumage colors and color combinations reported for the NICs indicate tremendous plumage color variation and diversity and results are similar to those of ICs from other countries in Africa [27, 34].

2.2 Distribution of comb type, beak, ear lobe, and wattle colors

Comb type in the NIC has been reported to include single, rose, pea, buttercup, walnut, and cushion varieties with single comb being the most predominant followed by rose, pea, buttercup, walnut and cushion in decreasing order. Daikwo et al. [10] observed single, pea, and rose combs with frequencies of 51.0, 28.0, and 21.0%, respectively in ICs from Dekina in Kogi State, Northcentral Nigeria while Rotimi et al. [13] observed 88.49, 7.03, 3.90, 0.32%, and 0.26% for single, rose, pea, buttercup, and cushion combs, respectively in ICs from Gwer-West, Benue State, Nigeria. From 1,700 ICs (Fulani = 900, and Tiv = 800 ecotypes), three comb types: single (99.23%), rose (0.38%) and walnut (0.39%) in Fulani ecotype and two types: single (99.62%), and walnut (0.38%) in Tiv ecotype were reported [14]. From 1,060 adult ICs (444 males and 616 females) from Bekwara in Cross River State, South–south Nigeria, [12] reported three comb types: single (88.49%), rose (7.17%), and pea (4.34%). Data presented showed male to female ratio of 95.5:83.4, 1.8:6.2, and 2.7:10.4% for single, pea, and rose combs, respectively. A similar study in the same area observed five comb types namely single (24.20%), pea (38.90%), rose (18.50%), double (13.70%), and walnut (4.70%) [15]. Four comb types (single, 43.33%; pea, 23.33%; rose, 17.78%; and walnut, 15.36%) were reported in ICs from Markudi in Benue State, Northcentral Nigeria [9] while [11] observed single comb (94.27 and 80.44), rose (2.75 and 11.34), and pea (2.98 and 8.21%) for Yoruba and Fulani ecotypes, respectively in ICs from Ogbomosho in Oyo State, Southwestern Nigeria. Comb color did not vary within and between populations and sexes being 100% red [11, 14]. Elsewhere in Africa [27] observed a preponderance of pea comb (range, 49–56%; overall, 53%) followed by rose comb (range, 12–22%, overall, 16%) out of the five comb types (single, rose, pea, walnut, and duplex) present in five IC ecotypes or strains of Ethiopia. The authors reported higher percentage rose, and pea combs in females compared to males, while more males had duplex, single, and walnut combs compared to females.

Beak color in NICs was reported to include white, yellow, brown, black, ash, pink, and orange. White beak (41.16%), black (31.61%), and yellow (27.23%) were observed in IC population from Gwer-West in Benue State, Northcentral

Trait		[1]LE (No.)	Freq. (%)	[2]HE (No.)	Freq. (%)
Comb type	Single	61	100.00	30	93.75
	Rose			2	6.25
Comb color	Red	46	75.41	32	100.00
	Red and black	11	18.03		
	Black	4	6.56		
Wattle color	Red	39	63.93	32	100.00
	Red-black	15	24.59		
	Pink	7	11.48		
Beak color	Black	25	40.98	1	3.13
	Black and slate	14	22.95		
	Black and brown	1	1.64		
	Yellow	5	8.20	4	12.50
	Yellowish brown	1	1.64		
	Brown	15	24.59	10	31.25
Ear lobe	Present	51	83.61	32	100.00
	Absent	10	16.39		
Ear lobe color	Red	32	62.75	16	50.00
	Pink	19	37.25		
	White			8	25.00
	White and red			8	25.00
Skin color	Yellow				
	Yellowish white	34	55.74	4	12.50
	Blackish (slate)				
	Ash-gray	25	40.98		
	White	2	3.28	28	87.50
Shank color	Black/ash/gray	31	50.82	1	3.13
	Yellow	16	26.23	10	31.25
	White/slate	3	4.92		
	Yellowish-white (cream)	10	16.39	21	65.63
Plumage color (sex)	Black	16 (F)	26.23	1 (F)	3.13
	Black and gold/yellow	2 (F)	3.28		
	Black and white	10 (F)	16.39	1 (F)	3.13
	Gold, ash, black	1 (F)	1.64		
	Brown and gold	1 (F)	1.64	1 (F)	3.13
	Gold, brown, black, ash	3 (F)	4.92	1 (F)	3.13
	Red, gold, black	2 (M)	3.28	2 (M)	6.25
	Red and white			1 (M)	3.13
	Red, gold, white			8 (M)	25.00
	White, gold, brown			1 (M)	3.13
	White			3 (M)	9.38

Trait		[1]LE (No.)	Freq. (%)	[2]HE (No.)	Freq. (%)
	White and gold			3 (M)	9.38
	Brown and black	14 (F)	22.95	1 (F)	3.13
	Brown	6 (F)	9.84	1 (F)	3.13
	Brown and white spots	6 (F)	9.84	3 (F)	9.38
	Gold, white, black			1 (M)	3.13
	Mottled (brown, black, ash, white, gold)			1 (F)	3.13
	White and brown			3 (F)	9.38

[1]*LE: unselected light ecotype (bantam or dwarf) chickens;*
[2]*HE: sixth generation heavy ecotype ICs*

Table 1.
Qualitative traits of light (LE) and heavy (HE) ecotype NICs.

Nigeria [13] while [14] observed mostly brown (44.23 and 43.90%), black (35.00 and 36.50%), and white beak (20.74 and 19.60%) for Tiv and Fulani IC ecotypes, respectively. Yellow, brown, ash, white, pink, and orange colored beak were observed in ICs from Bekwara, Cross River State, Nigeria at 21.10, 11.10, 9.10, 6.80, 1.90 and 1.10%, respectively [15]. In the study by [34], beak color was yellow (32.48%), white (33.73%), brown (26.30%), and black (7.75%).

White, red, brown, yellow, ash, and black were the range of ear lobe colors reported for NIC by several studies [9, 11, 13, 14] with white followed by red ear lobes being predominant (range: 52.00–79.37 and 20.63–45.00%, for white and red ear lobes, respectively) in the populations studied by [9, 11, 13] while [14] reported preponderance of brown and black ear lobes.

Wattle color was reported to be white and red by [13] with white wattle being the more frequent (68.02 vs. 31.98%). Reports describing wattle color are very scanty. **Table 1** shows the comb, beak and wattle colors observed in improved HE and unimproved LE NICs while **Figure 1** shows birds with some plumage color phenotypes.

2.3 Body, eye, and shank colors of Nigerian indigenous chickens

Three shank colors: yellow (18.89 and 27.23%), white (38.89 and 41.16%) and black (42.22 and 31.61%) were reported by [9, 13], respectively in ICs from Benue State (Northcentral Nigeria) while three shank colors: white (36.15%), green (12.69%), and black (51.15%), and two shank colors (white, 70.00% and black, 30.00%) were reported in Tiv and Fulani ICs, respectively by [14]. The authors observed three eye colors namely yellow (50.77%), white (20.00%), and brown (29.23%) in Tiv ecotype and two colors in Fulani ecotype namely yellow and brown at 76.90 and 23.10%, respectively. All the birds (100%) had white skin. From ICs of Cross River State, [12] reported five eye colors (black, brown, dark red, orange, and pink), and two skin colors of white and yellow. Of the five eye colors, black was most frequent at 44.72% followed by brown (27.74%) while pink was the least frequent at 5.09%. White skin predominated over yellow skin in the sampled population (75.85 vs. 24.15%). Odah et al. [15] reported 10 shank colors in IC population from the same state. These were yellow, black, white, greenish, milky, ash, dark-ash, pink, red, and light brown with yellow being the most frequent (31.90%) followed by black (19.60%), and white (18.50%) while red and light brown were the least frequent (0.40%, respectively). In this same population eye color were six namely black, light brown, dark brown, dark red, orange, and pink with black, light brown and dark brown predominating (44.72, 14.91 and 12.83%, respectively)

Figure 1.
Plumage colors of some landrace chickens of Nigeria (unimproved and improved ecotypes). (A): Unimproved light ecotype (dwarf) ICs of various plumage and shank colors (females). (B): Unimproved light ecotype (dwarf) ICs of black plumage and mostly black shank color (females). (C): Heavy ecotype (HE) ICs (G_0 generation) subjected to selection for growth and egg production traits. (D): Predominantly brown 2nd generation HE ICs (females). (E): Predominantly black plumage 2nd generation HE ICs (females). (F): Black, browm, and white plumage female progeny of 6th generation HE ICs. (G): Predominantly white plumage male progeny of 6th generation HE ICs. (H): Red/gold and white plumage male progeny of 6th generation HE ICs.

and the least frequent being pink (5.09%). In Dekina in Kogi State, Northcentral Nigeria, four shank colors were reported namely, black (13.75%), yellow (40.50%), black and yellow (37.25%), and white (8.50%) [10].

3. Productive characteristics of Nigerian indigenous chickens

Body weight, and morphometric traits; egg production traits, semen quality, fertility, hatchability, and chick survival; feed intake and feed efficiency have been used to characterize NICs for productive potentials. Body weight, and morphometric evaluations quantify growth potential, and meat yield while Egg production, semen traits, fertility, hatchability, and chick survival evaluate laying and reproductive capacities. These variables determine commercial value which, in addition to cultural and social utility, constitute conservation value.

3.1 Body weight, and linear body traits of Nigerian indigenous chickens

Early reports classified NICs based on location/tribal ecotypes. In southeastern Nigeria, [35] identified three location ecotypes (Nsukka, Owerri, and Awgu ecotypes) while [36] identified two tribal ecotypes (Yoruba and Fulani ecotypes). Odubote [16] observed that none of the location or tribal ecotypes was unique in any of the attributes and that the only striking phenotypic difference between these ecotypes is in their mature body size. The author hence distinguished three types namely dwarf, normal size and heavy body (Fulani) types. A more recent classification advocated only two categories based on mature body weight namely, light body weight ecotype (light ecotype, LE) and heavy body weight ecotype (heavy ecotype, HE) [8]. The LE represents the chicken type from rainforest and derived savannah agro-ecological zones, whose mature body weight ranges from 0.68–1.5 kg and includes dwarf and normal size types referred to as rain forest, swamp, or Yoruba chickens by some authors [6, 11, 17, 22] while the HE are those of the guinea savannah, Sahel savannah and some montane regions, whose mature body weight ranges from 0.90–2.5 kg referred to as Fulani, and Tiv chickens by some authors [11, 21].

The wide range of body weight within ecotypes reflect genotypic differences, differences in husbandry system, and level of input; body weight being generally lower in free range, scavenging system. Early studies reported mature (> 20 weeks) body weight range of 1.0 to 1.76 kg for extensive system [37, 38] and 768 to 1096 g at 20 weeks of age for on-station populations [39–42]. More recently, a mean range of 1.32 to 2.0 kg was reported for extensive system [10, 13, 15, 43] while for intensive system, 20 week body weight was reported as 771.11 and 765.94 g for LE parent and inbred projeny, respectively [44]; and 1.42, 1.39, and 1.30 kg for normal feather, naked neck, and frizzle ICs, respectively [45]. Momoh et al. [46] showed that HE and LE ICs differed significantly in body weight from hatch to 20 weeks of age (**Table 2**). Oleforuh-Okoleh et al. [52] reported higher body weight in normal feathered ICs compared to naked neck chickens at 4 and 8 weeks of age (312.06 ± 7.71 vs. 287.13 ± 6.17 g, and 931.72 ± 23.85 vs. 844.30 ± 21.84 g, respectively) but similar values at 12 and 16 weeks of age. Body length, chest girth, leg length, and shank circumference were also higher in normal feathered chickens at 4 weeks of age while chest girth, leg length, and shank length were higher at 8 weeks of age while [15] reported linear body values of 42.7 ± 0.03, 55.8 ± 0.21, 12.2 ± 0.04, 9.8 ± 0.02, 25.0 ± 0.70, and 8.9 ± 0.50 cm for body circumference, body length, shank length, keel length, wing length, and neck length, respectively in ICs from Bekwara in Cross River State, South–south Nigeria. Sanusi and Oseni [53] evaluated Fulani ecotype chickens under intensive and pasture production systems and reported significant effect of sex of chicken on body weight from 10 to 20 weeks of age as well as significant interaction effect of sex and production system on body weight. Males averaged 1343.43 ± 55.2 vs. 1295.57 ± 59.12 g while females averaged 938.66 ± 60.3 vs. 1061.805 ± 59.9 g for intensive vs. pasture systems, respectively.

DOI: http://dx.doi.org/10.5772/intechopen.96580

Genetic resource	Hatch	4	8	12	16	20	Reference
DSIC (male)	29 ± 1.0	124 ± 9.2	311 ± 26.4	702 ± 55.3		1096 ± 84.1	[44]
DSIC (female)	23 ± 1.6	104 ± 14.5	262 ± 4.8	605 ± 67.5		948 ± 130.6	
RFIC (male)	24 ± 0.8	99 ± 6.6	255 ± 19.7	615 ± 41.3		810 ± 46.7	
RFIC (female)	25.6 ± 0.7	104 ± 5.9	242 ± 17.1	533 ± 35.7		768 ± 36.6	
[1]HE	30.30 ± 0.17	151.41 ± 1.74	344.19 ± 4.14	667.98 ± 6.30	791.52 ± 6.24	911.59 ± 6.33	[47]
HE	30.2 ± 0.06	157 ± 0.45	350 ± 3.01	720 ± 9.47	840 ± 9.35	976 ± 11.2	[48]
LE	24.2 ± 0.05	139 ± 2.24	299 ± 3.01	560 ± 4.31	707 ± 4.89	831 ± 5.52	
HE x LE	25.1 ± 0.04	147 ± 2.13	335 ± 2.81	700 ± 4.21	819 ± 4.86	937 ± 7.32	
LE x HE	28.6 ± 0.07	148 ± 2.03	331 ± 2.43	693 ± 3.51	806 ± 4.18	934 ± 4.54	
NF		312.06 ± 7.71	931.72 ± 23.85	1180.59 ± 32.45	1635.08 ± 43.62		[49]
Na		287.13 ± 6.17	844.30 ± 21.84	1158.15 ± 25.71	1587.98 ± 40.00		
NF x FF	25.48 ± 0.40	86.3 ± 0.54	267.78 ± 3.68	620.22 ± 9.99	819.14 ± 9.30	1040.52 ± 12.34	[50]
FF x NF	26.51 ± 0.38	87.18 ± 0.51	268.57 ± 3.52	608.15 ± 10.13	821.59 ± 8.83	1047.45 ± 13.47	
NF x Na	26.10 ± 0.19	83.57 ± 0.69	264.11 ± 3.19	639.49 ± 7.94	842.29 ± 5.88	1088.20 ± 12.21	
Na x NF	25.76 ± 0.43	80.91 ± 0.87	257.16 ± 3.01	500.53 ± 7.11	793.95 ± 5.84	1017.63 ± 10.79	
FF x Na	28.61 ± 0.34	91.87 ± 0.78	283.50 ± 2.41	526.81 ± 7.84	734.41 ± 7.38	1040.49 ± 13.06	
Na x FF	28.95 ± 0.45	88.29 ± 0.91	270.13 ± 1.92	623.18 ± 7.10	817.42 ± 6.71	1121.78 ± 9.94	
E x NF	26.45 ± 0.35			508.60 ± 29.85			[51]
E x Na	26.00 ± 0.26			519.43 ± 35.46			
E x F	26.50 ± 0.73			609.58 ± 16.86			
NF x E	29.84 ± 0.32			1039.15 ± 52.18			
Na x E	30.83 ± 0.59			910.88 ± 67.15			

Genetic resource	Hatch	4	8	12	16	20	Reference
F x E	30.22 ± 0.30			1141.88 ± 42.28			
E x (E x NF)	24.19 ± 0.50			1014.38 ± 71.90			
E x (E x Na)	26.42 ± 0.54			956.11 ± 69.54			
E x (E x F)	23.71 ± 0.40			645.00 ± 34.51			
E x (NF x E)	30.48 ± 0.38			1752.23 ± 42.49			
E x (Na x E)	30.00 ± 0.43			1223.13 ± 74.60			
E x (F x E)	31.33 ± 0.47			1976.67 ± 97.60			

DSIC: derived savannah IC; RFIC: rainforest IC; HE, LE: heavy, light, ecotype; NF, Na, FF: normal feather, naked neck, frizzle, E: exotic broiler;
[1]HE: 0–8 wk. (sexes combined), 12–20 wk. (female).

Table 2.
Body weight of various genetic groups involving IC eco- or geno-types as reported in different ecological zones.

3.2 Egg production, semen quality, fertility and hatchability potentials

Egg production of NICs is reported to be very low especially in traditional scavenging system. Egg production ranged from 22 to 80 eggs/hen/year [10, 15, 47, 50, 51, 54], laid in 2 to 3 clutches of size 4 to 14 eggs [10, 38, 55–57] and of weight, 25 to 35 g [10, 15, 58]. Reasons for poor egg production include poor genetic potential, disease, poor nutrition, broodiness and rearing of chicks, and social behavior [6, 10, 15]. Hatchability values reported across zones, ecotypes, and populations ranged from 60 to 100% [10, 15, 51, 55–57]. For on-station populations, a range of 35 to 175 eggs laid in 90 to 500 days, and of mean weight 28.78 to 43.9 g, was reported by various studies using deep litter or battery cage systems [35, 39, 59–64]. Age at first egg (AFE) ranged between 148.4 and 176.9 days (d) [60, 65]. More recently, [63] reported AFE of 156 to 159 d, and egg number and egg weight to 90 d of 34.04 ± 1.15 to 37.38 ± 2.21 eggs and 35.27 ± 0.31 to 35.73 ± 0.59 g, respectively over three generations while [66] reported AFE, body weight at first egg (BWFE), egg production (EN), egg weight (EW), clutch size (CS), and pause length (PL) to range between 22 and 31 and 20 and 23 d, 1350 and 1650 and 1300 and 1440 g, 78 and 174 and 58 and 128 eggs, 35.72 and 52.50 and 35.36 and 50.61 g, 3 and 9 and 2 and 6 eggs, and 1 and 3 and 1 and 6 d, for Fulani and Yoruba ecotypes, respectively in Southwest Nigeria. Gwaza et al. [67] reported AFE of 199.72 ± 0.089 and 195.30 ± 0.104 d and BWFE of 1.486 ± 0.104 and 1.186 ± 0.022 kg for Tiv and Fulani ICs, respectively. In the high rainforest zone of Nigeria, [17] observed no effect of genotype on fertility and embryo mortality between normal feathered, frizzle and naked neck ICs but percent hatchability was highest in normal feathered (86.36%). Fertility (range: 76.67–90.53), hatchability (range: 83.50–91.36), dead in shell (range: 8.23–9.46), and weak in shell (range: 0.32–1.32) did not differ significantly between naked neck, normal feathered, and frizzled ICs and an exotic broiler strain. For semen quality [68] reported higher mean ejaculate volume, sperm concentration, sperm motility, and vigor in local compared to exotic cocks at different collection frequencies and intervals. Omeje and Udeh [69] had shown that feed restriction adversely affected semen production in exotic than local cocks and that only the local cock yielded semen at once in four days feeding. Naked neck and normal feathered ICs had significantly higher sperm concentration and motility compared to Nera Black, White Leghorn, Giriraja, and an indigenous breed (FUNAAB Alpha) [70]. Ajayi et al. [71] showed that naked neck cocks had higher semen concentration than frizzle and normal feathered cocks (4.85 x 10^9 ± 0.03/ml vs. 3.26 x 10^9 ± 0.94 and 3.33 x 10^9 ± 0.57/ml, respectively) and there was higher sperm motility in naked neck and frizzled chickens compared to normal feathered while normal feathered and frizzled had higher semen volume compared to naked neck. Udeh et al. [72] studied the value of linear body measures to predict semen traits of local and exotic cocks and reported higher semen volume in local compared to exotic cocks, and positive correlation between wing length and percent live sperm in local chickens while beak length, sperm concentration, and motility; comb length and sperm concentration; and shank length and sperm motility were positively correlated in exotic cocks. Oke and Ihemeson [45] had observed no effect of genotype on total reproductive organ weight in normal feathered, frizzle and naked neck ICs (14.1, 11.2, and 11.6 g, respectively) but higher ($p \leq 0.05$) testis weight, semen volume, sperm concentration and motility in normal feathered (11.7 g, 0.25 ± 0.02 ml, 270 x 10^9 ± 5.99/ml, and 77%, respectively) and naked neck (10.1 g, 0.24 ± 0.02 ml, 250 x10^9 ± 6.00/ml and 65.8%, respectively) chickens compared to the frizzled genotype (7.07 g, 0.15 ± 0.03 ml, 198 x 10^9 ± 11.5/ml and 52.5%, respectively).

4. Genetic improvement of productivity of Nigerian indigenous chickens

Genetic improvement of NICs in growth traits, egg production, fertility and hatchability has been the objective of numerous studies. These studies involve crossbreeding between ecotypes, and genotypes; ICs with exotic breeds/strains; and selection within ecotypes.

4.1 Crossbreeding between Nigerian indigenous chicken ecotypes, and genotypes

The extensive genetic diversity between IC ecotypes as well as within and between population variations in productive traits provide opportunity for improvement of performance through within and between population selective breeding. Ogbu and Omeje [26] reported high within population variation in growth traits in NICs which could be exploited for genetic improvement while [73] recorded improved growth performance following positive assortative mating in NIC populations. Egahi et al. [48] evaluated the effect of crossbreeding between NIC genotypes and reported the body weight of progenies of crosses between normal feathered, frizzle, and naked neck ICs to range from 25.48 ± 0.40 to 28.95 ± 0.45 g for hatch weight, 80.91 ± 0.87 to 91.87 ± 0.78 g, 257.16 ± 3.01 to 283.50 ± 2.41 g, 500.53 ± 7.11 to 639.49 ± 7.94 g, 734.41 ± 7.38 to 842.29 ± 5.88 g, and 1017.63 ± 10.79 to 1121.78 ± 9.94 g, for 4, 8, 12, 16, and 20 weeks of age, respectively (**Table 2**) while [67] observed significant effect of sire, dam, and ecotype on AFE and BWFE of Fulani and Tiv ICs and positive genetic correlations between AFE, BWFE, and EW; and EW, egg length (EL), and egg diameter (ED). Additive genetic heritability (h^2) of AFE, BWFE, EW, EL, and ED for Fulani and Tiv ICs were 0.358 and 0.438, 0.420 and 0.398, 0.482 and 0.642, 0.182 and 0.000, and 0.051 and 0.309, respectively. For egg production pattern (clutch size, clutch number, pause number, and pause length), h^2 values were 0.358, 0.412, 0.045, and 0.036, respectively in Fulani chickens and 0.428, 0.391, 0.063, and 0.048, respectively in Tiv chickens [74]. High positive genetic correlations (range: 0.78 to 0.88) were reported between BWFE and AFE, BWFE and EW, EW and EL, and EW and ED. Agu et al. [75] reported significant effect of sire on AFE, weight of first egg (WFE), egg production (EN), egg mass (EM), egg weight (EW), thigh length, back width, and neck length in HE ICs of Southeastern Nigeria. Heritability values for EW, EN, and EM was 0.31 ± 0.30, 0.16 ± 0.13, and 0.28 ± 0.24, respectively and ranged from 0.13 ± 0.23 to 0.52 ± 0.24 from 4 to 20 weeks of age for thigh length, 0.23 ± 0.23 to 0.41 ± 0.29 for back width, and 0.10 ± 0.18 to 0.52 ± 0.44 for neck length. Momoh et al. [46] showed that main (HE x LE) and reciprocal (LE x HE) crossbred progenies were similar in body weight to the HE chickens but superior to the LE chickens. Momoh and Nwosu [76] evaluated the genetic parameters of body weight (BW), body weight gain (BWG), and feed conversion ratio (FCR) in HE, LE, HE x LE, and LE x HE populations and reported h^2 values of 0.17 ± 0.19, 0.08 ± 0.10, and 0.19 ± 0.22 for BW at hatch, respectively. The corresponding values for BW from week 4 to week 20 of age ranged from 0.16 ± 0.18 to 0.43 ± 0.26, 0.16 ± 0.13 to 0.25 ± 0.17, and 0.20 ± 0.21 to 0.36 ± 0.28, respectively. For daily gain from 4 to 20 weeks, 0.03 ± 0.11 to 0.12 ± 0.14, 0.21 ± 0.15 to 0.89 ± 0.50, and 0.10 ± 0.16 to 0.80 ± 0.14, respectively were reported while 0.13 ± 0.16 to 0.41 ± 0.25, 0.10 ± 0.10 to 0.46 ± 0.24, and 0.11 ± 0.16 to 0.24 ± 0.23, respectively were reported for FCR.

4.2 Crossbreeding of Nigerian indigenous chicken ecotypes, and genotypes with exotic breeds

Crossbreeding NICs with exotic breeds/strains is advocated to exploit the high genetic distance and variation between ICs and exotic strains believed to enhance hybrid vigor, heterosis, and breed complementarity. Omeje and Nwosu [58] evaluated progenies of crosses between NIC (LC) and Gold link (GL, an exotic breed) and reported reduced age at first egg (AFE) in LC x GL progenies compared to LC, GL, and GL x LC (155.4 ± 1.49 vs. 157.8 ± 3.21, 169.2 ± 1.65, and 169.7 ± 3.74 d, respectively). Authors also reported superior egg weight for GL, GL x LC, and LC x GL compared to LC (53.44, 47.74, and 47.02 vs. 38.63 g, respectively). The corresponding values for egg mass was 12.12, 10.18, and 8.89 vs. 5.64 kg, respectively. An improvement in annual egg production from 146 eggs/hen for LC to 213 eggs/hen for GL x LC was reported by [77]. Fewer but longer pauses and shorter but more pauses were observed in LC and LC x GL; and GL and GL x LC, respectively. It was also observed that hybrids of crosses involving LC, Yaffa (Y) and GL exotic chickens [LC(Y x GL), GL (Y x LC), and Y x GL] were superior to LC in egg weight (51.91, 52.07, and 54.22 vs. 40.36 g, respectively), and egg mass (5.40, 5.37, and 6.10 vs. 3.32 kg, respectively) [78] while [79] reported superior body weights for GL, GL x LC, and [GL(GL x LC)] in growth and egg production compared to LC attributed to dominance, epistasis, and/or maternal effects. Oluyemi [80] reported heterosis of 12 week body weight in progenies of LC x White Rock and LC x Rhode Island Red (RIR) to range from 4.0 to 12.4% while significant improvement in BWFE, WFE, ASM, egg production (EN_{90}) and egg weight (EW_{90}) to 90 d was observed in LC x RIR males backcrossed to RIR dams [81]. Ukpong [82] observed improved meat yield in crosses of LC x Abor acre (AA) broiler chickens relative to LC while [49] reported improved growth performance and feed conversion in F_2 (main and reciprocal backcross groups) compared to F_1 counterparts in crosses of Abor Acre broiler breeder and native chicken genotypes (**Table 2**). Nwachukwu et al. [18] had shown that main crossbred progenies of AA x LC genotypes were inferior in body weight at first egg to their reciprocal crossbred counterparts (960.00, 812.50 and 1030.00 vs. 1891.67, 1576.50 and 2072.00 g, respectively). The latter group also had higher values for WFE, EN_{90}, egg length, yolk index, albumen weight, and Haugh unit and crosses involving the frizzle genotype were superior to crosses involving other IC genotypes. Adeleke et al. [83] crossed complete feathered, frizzle and naked neck ICs to Anak titan (AT) broilers and reported significant effect of sire, dam, and progeny genotype on growth traits. Anak titan sire significantly improved 8 to 20 week body weight compared to IC sires. Significant sire genotype effect on fertility and percent dead in shell was also reported in IC genotypes crossed to AT [84]. Frizzled sire had highest fertility (90.5%) and produced eggs with highest hatchability (91.4%) and least embryo mortality (7.5%) while AT dams produced eggs with highest fertility and hatchability (88.2 and 94.6%, respectively). Main and reciprocal crosses involving the frizzle genotype were also better in the traits studied [84]. Ayorinde et al. [85] observed superior body weight in Fulani ecotype X Dominant black (FE x DB) progenies compared to FE, DB, and DB x FE at 21 weeks of age (1408.50 ± 3.5 vs. 1350.60 ± 4.5, 1388.60 ± 3.2, and 1375.00 ± 3.2 g, respectively). All crossbred genotypes were superior in early (0 to 13 weeks) body weight to FE. Udeh and Omeje [86] reported heterosis of body weight in native and exotic inbred chicken crosses with native X exotic being higher than exotic X native, and native backcrosses being higher than exotic backcrosses. The authors concluded that body weight heterosis resulted from complete dominance in native backcrosses while 2–3 locus parental epistasis involving complementary genes were responsible for heterosis observed in exotic backcrosses. Udeh [87] reported significant differences in age at first egg (AFE), BWFE and WFE among native X exotic inbred chicken groups. Inheritance of AFE and WFE was attributed to additive (e.g., sire) and non-additive (e.g., dam) genetic effects while dominance effect was responsible for inheritance of BWFE. Udeh [88] showed that crossing IC with inbred progenies of H and N brown nick, and Black Olympia, improved BW and BWG from hatch to 20 weeks of age relative to IC due to significant direct additive, maternal additive and direct heterotic effects. Significant genotype effect on fertility, and hatchability and improved BW and EW in LE x Isa Brown progenies compared to LE was reported by [69].

Reported heritability (h^2) estimates of production traits in crosses between local chickens and exotic breeds vary widely being specific for populations, point of estimation, and age of birds. Akinokun and Dettmers [89] reported values of 0.15, 0.02 and 0.25 for age at sexual maturity (ASM), 4 months, and 8 months egg production, respectively; 0.20 to 0.54 for egg weight to 7 months of lay; 0.51, 0.41 and 0.27 for 4, 12, and 20 week body weight, respectively; and realized heritability of 0.27 and 0.24 for 4 months egg production in 2nd and 3rd generation, respectively. Oluyemi [90] had reported h^2 value of 0.31 for 12 week body weight while [91] reported values of 0.35 to 0.74, 0.31 to 0.89, and 0.27 to 0.49 for body weight from sire, dam, and sire + dam variance components in progenies of crosses involving ICs, Yaffa and Goldlink. The same authors reported heritability of 0.46 ± 0.24 for egg weight and 0.36 ± 0.18 for shell thickness. Udeh [92] reported h^2 values of 0.08 to 0.80, 0.03 to 0.69, and 0.22 to 0.47 for BW, shank length, and wing length, respectively, and positive genetic correlation (except for SL and WL) and phenotypic correlation coefficients that ranged from 0.18 to 0.96 and 0.10 to 0.91, respectively among BW, SL, and WL at different ages in NICs.

4.3 Genetic improvement of Nigerian indigenous chickens through selection

Relatively few studies that are far in between have been undertaken to evaluate selection response in NICs. The earliest report on genetic selection [80] observed poor selection response in body weight in NICs over 7 generations while [93] reported genetic gain of 2.20 and 2.48 eggs for first and second generations, respectively. Recently, a number of studies demonstrated significant improvement of growth and egg production traits. In light ecotype (LE) IC, [63] reported improvement in BWFE, EN, EW, and WFE but increased AFE following three generations of index selection (G_0 to G_2). Values reported for selected vs. control groups ranged from 962.50 ± 23.33 to 1062.90 ± 18.06 vs. 880.14 ± 16.72 to 892.10 ± 18.85 for BWFE, 33.40 ± 1.23 to 47.18 ± 2.36 vs. 34.04 ± 1.15 to 37.38 ± 2.21 eggs for EN, 36.51 ± 0.55 to 38.64 ± 0.49 vs. 35.27 ± 0.31 to 35.73 ± 0.59 g for EW, 30.62 ± 0.92 to 31.92 ± 0.63 vs. 29.44 ± 0.37 to 29.99 ± 0.66 g for WFE, and 159.47 ± 1.97 to 164.78 ± 2.40 vs. 158.40 ± 1.13 to 159.48 ± 1.47 d for AFE. From the same population cumulative selection differential (CumΔs) of 269.38 g, 1.58 g, and 3.88 eggs and realized genetic gain per generation of 94.22 g, 0.84 g, and 4.85 eggs, for BWFE, EW, and EN, respectively were reported [94]. Pooled heritability estimates over the three generations was 0.56, 0.44, and 0.28 for BWFE, EN, and EW, respectively while genetic correlation values were 0.41 for BWTE and EW, −0.18 for BWFE and EN, and − 0.23 for EN and EW [95]. Ogbu et al. [96] estimated the economic, and relative economic weights of BW, EW and EN to 16 weeks of lay in heavy ecotype IC (HE) over three generations (G_0 to G_2) for use in construction of selection indices and reported values of 7.47 and 3.15, 13.67

and 5.77, and – 2.37 and – 1.00, respectively in G_0,; 13.07 and 3.82, 23.69 and 6.93, and – 3.42 and – 1.00, respectively in G_1; and 16.80 and 2.89, 30.75 and 5.28, and – 5.82 and – 1.00, respectively in G_2 generation. Using an index of weighted breeding values that considered the heritability, relative economic weight, and standardized trait values, [97] reported expected average direct genetic gain per generation for short term (16 weeks) egg production of 12.58 eggs, 1.98 g, and 25.04 g for EN, EW, and BWFE, respectively; realized genetic gain of 2.19 and 1.59 eggs for EN, 1.65 and 0.26 g for EW, and – 25.60 and 123.64 g for BWFE for G_0 and G_1, respectively; and corresponding values for ratio of realized to expected genetic gain of 2.27 and 1.22, 3.15 and 0.24, and 0.95 and 2.21, respectively. The authors reported h^2 estimates that ranged from 0.12 to 0.24 for EN, 0.34 to 0.43 for EW and 0.57 to 0.69 for BWFE. For males, improvement in 39 week body weight was observed with realized genetic gain of 284.22 and 111.87 g for G_0 and G_1, respectively and average expected gain of 508.50 g per generation following mass selection. Ogbu [98] had reported improvement in BW from hatch to 39 weeks following mass selection in male HE IC with final body weight increased from 1372.66 ± 16.46 g in G_0 to 1768.75 ± 33.15 g in G_2 implying a cumulative gain of 925.76 g over three generations. The author reported h^2 estimate of 0.24 ± 0.27 to 0.59 ± 0.45 and 0.13 ± 0.49 to 0.25 ± 0.31 across the three generations for BW from 12 to 20 and 39 weeks of age, respectively. Agbo [99] furthered the selection for improved growth and egg production in HE ICs from 4th to 6th generation and reported improvement in short term (16 weeks) EN and EW from 89.98 ± 0.81 eggs and 43.52 ± 0.08 g, respectively in G_4 to 94.98 ± 0.51 eggs and 45.06 ± 0.12 g, respectively in G_6, and mean realized genetic gain of 119.18 g for 39 week BW in males. The author reported h^2 values of range 0.28 to 0.52 for EW, 0.14 to 0.45 for EN, and 0.23 to 0.69 for BWFE and relative economic weight of 2.02 to 2.24, 2.45 to 2.78, and – 1.00 for EW, EN, and BWFE, respectively. These studies indicate that NICs can be improved for commercial utility as layer or dual purpose bird (meat and egg production) using within ecotype selection.

5. Genetic diversity and distance within and between ecotypes, and genotypes

Studies to evaluate genetic diversity and distance within and between NICs involved phenotypic and molecular evaluation of different ecotypes, genotypes and populations [19–21]. Ige [100] using correlation and regression models estimated genetic parameters of BW and linear body traits to evaluate genetic distance between Yoruba (YE) (light) and Fulani (FE) (heavy) IC ecotypes. Correlation coefficients ranged from 0.30 to 0.89 and 0.40 to 0.99 in male and female FE, respectively and from 0.20 to 0.88 and 0.15 to 0.85 in female and male YE, respectively. Coefficient of determination (R^2) ranged from 0.20 to 0.91, 0.10 to 0.76, and 0.22 to 0.94 for linear, quadratic and cubic functions, respectively in YE and 0.55 to 0.94, 0.64 to 0.81, and 0.55 to 0.86, respectively in FE. The IC ecotypes showed strong discriminatory power (98.29%) but low genetic distance (Euclidean genetic distance = 11.2) indicating close relationship. Using canonical discriminant analysis [19] evaluated the diversity among NIC genotypes and reported highest discriminatory power in Body weight, thigh length, and body width. Mahalanobis distance measure indicated closer relationship between normal feathered and naked neck (3.371) compared to normal feathered and frizzle genotype (4.626). Gwaza et al. [101] however reported wide genetic diversity in body dimensions among isolated populations of Tiv chickens. Ukwu et al. [102] evaluated within ecotype genetic diversity at the hemoglobin (Hb) locus in Tiv chickens and observed three Hb genotypes (Hb^{AA}, Hb^{AB}, and Hb^{BB}) at frequencies 0.40, 0.32, and 0.24, respectively resulting from two Hb alleles at frequencies 0.60 for Hb^{A} and 0.40 for Hb^{B}. Tiv chickens showed moderate Hb heterozygosity of 0.48. Adenaike et al. [20] investigated genetic diversity in NIC genotypes and Nera Black chickens based on variation in *zyxin* and TNFRSF1A genes. Highest nucleotide substitution per site (D_{xy} = 0.081) was reported for TNFRSF1A gene sequences in normal feathered and naked neck chickens while frizzle and Nera Black chickens had the lowest value of D_{xy} = 0.065. For *zyxin* gene sequences, normal and frizzle feathered chickens had highest D_{xy} value of 0.6551 vs. 0.0739 for Nera Black and naked neck. Mean haplotype diversity and average number of nucleotide difference in TNFRSF1A gene sequences was highest in Nera Black (0.923 and 3.967, respectively) while frizzle chickens had the corresponding lowest values (0.00489 and 3.143, respectively). The authors inferred high nucleotide divergence, haplotype diversity and restricted gene flow among the chicken genotypes. Gambo et al. [21] studied diversity and genetic distance within and between Tiv (TE) and Fulani (FE) ecotypes based on blood proteins (Hb, albumin, transferrin and carbonic anhydrase) electrophoresis. Two Hb genotypes (Hb^{AA} and Hb^{AB} at frequencies 0.125 and 0.875, respectively in TE, and 0.538 and 0.462, respectively in FE), three albumin genotypes (AB, and AC at frequencies 0.026 and 0.974, respectively in TE, and AA and AC at frequencies 0.077 and 0.923, respectively in FE), six transferrin genotypes (AA, AB, AD, BB, BD, and DD at frequencies 0.054, 0.027, 0.297, 0.162, 0.378, and 0.082, respectively in TE and AA, AD, and DD at frequencies 0.568, 0.243, and 0.189, respectively in FE), and four carbonic anhydrase genotypes (AA, AB, AC, and BB at frequencies 0.20, 0.175, 0.525, and 0.100, respectively in TE and AA, AB, and AC at frequencies 0.263, 0.026, and 0.711, respectively in FE) were reported. Thus Hb^{AA} and Hb^{AB} were most abundant in TE and FE (0.875 and 0.538, respectively). Genotype AC for albumin, BD and AA for transferin and AC for carbonic anhydrase were the most frequent in the two ecotypes while albumin genotypes AA and AB were absent in TE and FE, respectively. The authors inferred a common origin for the two ecotypes but positive genetic distance between them attributable to divergence from one locality to another. Ige and Salako [103] employed direct gene counting and dendogram following cellulose acetate electrophoresis to evaluate genetic variation at the transferrin locus and established genetic relationships within and between FE and YE chickens. The authors reported six phenotypes (AA, AB, AC, BB, BC, and CC at genotypic frequencies 12.5, 10.0, 7.5, 35.0, 17.5, and 15.0%, respectively in YE, and 11.19, 16.6, 2.8, 22.2, and 27.7%, respectively in FE) controlled by three co-dominant alleles (Tf^{A}, Tf^{B}, and Tf^{C} at frequencies 0.35, 0.20, and 0.43, respectively in YE and 0.21, 0.32, and 0.44, respectively in FE). Dendogram clustering analysis indicated 72% genetic similarity within FE, 58% within YE, 70% between the two ecotypes, and no genetic relationship between transferrin locus and phenotypic traits such as sex, plumage color, and comb type of chicken. Ige et al. [22] considered the variation at globulin (95SKDa), transferrin (66KDa), albumin (36KDa), and post albumin (29KDa) loci using sodiumdodecylsulphate polyacrylamide gel electrophoresis to evaluate the genetic similarity of YE ICs. Similarity indices for transferrin, albumin, globulin, and post albumin were 58, 19, 18, and 40%, respectively indicating genetic similarity at the transferrin locus but wide variation at the other blood protein loci. The authors also inferred that the YE IC populations were still under natural selection. Adeleke et al. [104] had reported mean genetic similarity index of 55% between normal feathered, frizzle feathered and naked neck IC genotypes using blood protein polymorphisms and inferred clearly separated genotypes with naked neck genotype being the most diverged.

6. Conservation of Nigerian indigenous chicken genetic resources: issues and concerns

Nigerian ICs have evolved as homeostatic populations with adaptive and neutral diversity and capacity to respond to changing environmental conditions in diverse ways [6, 7, 16]. Experts believe that diverse unselected indigenous animal resources that harbor high proportions of neutral and adaptive diversity represent equivalent genetic resources in the absence of wild ancestors, and should be conserved with high national priority [105]. Emerging diseases, climate change, and changes in nutritional needs of humanity are unforeseeable. Consequently, overall genetic resources defined by adaptive and neutral diversity must be maintained in order to conserve the potential to react to future challenges [105–107].

There is dearth of data on the extinction risk status of NICs but commentators agree that IC genetic resources are the most endangered and under conserved animal genetic resources [6, 108, 109] with extinction risk of 33% [110] and about 40% of breeds with unknown extinction risk status [111]. Studies have shown the dwindling frequency or apparent loss of rare NIC phenotypes such as the crested head, feathered shank or ptylopody, polydactyl or 5 toed, short flight feathered, and dwarf types, naked neck, frizzle, and silky, and major genes such as naked neck (Na), frizzle (F), dwarf (Dw), ptylopody (Fsh), and polydactyl (Po) believed to enhance survival and performance in tropical environments [6, 16, 112].

6.1 Drivers of erosion and loss of indigenous chicken genetic resources

The declining animal genetic resources in developing countries has been blamed on a number of factors defined by scholars to threaten production, utilization, and conservation of native animal populations including ICs [24, 113, 114]. A brief overview of these factors will provide the background for suggested mitigation strategies.

a. Pressure to substitute indigenous types with exotic breeds

The notion within professionals, and policy makers that husbandry of landrace chickens is an economic waste put pressure on farmers to cull local strains in favor of exotic breeds [23] leading to loss, sometimes irretrievably, rare IC genetic resources [23, 24].

b. Introgression of exotic genes into native animal genetic base

To meet growing animal food demands due to increasing human population and rise in income, policy makers, researchers, and farmers advocate crossbreeding of local strains and exotic breeds without long-term breeding objectives [23, 98, 115]. These activities dilute and narrow indigenous animal genetic base and lead to loss of important traits for survival and production in scavenging husbandry system and harsh (disease endemic and high temperature) environments [23, 24, 115, 116]. Reduced fitness of resulting hybrids have been reported in chickens and other species [24, 117], and no commercial breed has resulted from decades of crossbreeding involving NICs and exotic breeds [98]. Similar scenario has been reported in other African countries [23, 24, 115, 118, 119].

c. Radical shift in production system and poor economic valuation of ICs

The shift from small scale, subsistence production to large scale, intensive holdings alienates indigenous strains [24, 120, 121], resulting in loss of IC genetic resources [111]. Following the adoption of backyard or family poultry that employs exotic breeds, the family local chicken was substantially eliminated [121]. In addition, native chickens have attracted poor economic appeal because only direct use commercial products (without adaptive potentials) have been used in economic assessment of chicken genetic resources [23, 24].

d. Globalization and livestock revolution

The expanding industrial poultry production coupled with increased farm input costs necessitate that breeds that produce more efficiently are adopted in place of indigenous strains [122–124]. Furthermore, increased globalization of animal production and worldwide movement of germplasm fuel breed replacement against ICs leading to decline and loss in IC genetic resources [23, 117, 122].

e. Disease, Predators, negative selection and cultural practices

Endemic diseases and predators cause the loss of ICs [23, 48] and force many marginal families to discontinue investment in IC production [23]. Culling the best chickens for income and other functions (food, festivals, cultural, religious or ritual practices, i.e., negative selection) [16, 125], without a breeding programme for their replacement, depletes IC genetic resources.

6.2 Key concerns on loss of indigenous chicken genetic resources

The concern about loss of IC genetic resources stems from the multifaceted economic, environmental and socio-cultural consequences highlighted by numerous studies [6, 7, 16, 24, 109, 113].

a. Extinction of native strains and loss of overall genetic diversity and heritable variations with co-adapted genes and gene complexes that may be useful to meet breeding goals now and in the long-term [24, 105, 126].

b. Depletion of adaptive genes and traits, reduced fitness and evolutionary potentials, resulting in reduced flexibility of future breeding options [6, 108, 109, 111].

c. Limited gene flow between populations, genetic bottleneck, low effective population size and heightened inbreeding, loss of reproductive capacity, fecundity, and survivorship [105, 122, 126].

d. Loss of key genetic attributes for chicken production in marginal environments, and in disease and parasite endemic regions [16, 24, 105], resulting in loss of livelihood [24, 109], irreversible social disintegration, and human migration [24, 127, 128].

e. Loss of cultural identity and pride since ICs are associated with peoples, and tribes, and are integral part of their culture, tradition and wellbeing [23, 109, 129].

f. Loss of valuable animal models for biomedical research and training [24, 130] and loss in country gross domestic product (GDP) accounted for by indigenous chicken production subsector [24].

6.3 Schemes to conserve Nigerian indigenous chicken genetic resources

Conservation of IC genetic resources can be ex situ or in situ or both. Ex situ conservation involves ex situ in vivo which is conservation of genetic resources away from their traditional breeding and production environments or locality as in live animal facilities (zoological gardens, parks, on-station farms) [131, 132], and ex situ cryopreservation which is conservation of genetic materials such as semen, oocytes, embryos, and deoxyribonucleic acids (DNAs) in gene banks [131, 132]. In situ conservation on the other hand is conservation of living genetic resources in their natural habitat or production environment such as on-farm populations or community based production facilities [24, 131].

Animal research stations, parks, and zoological gardens exist in few states in Nigeria but ICs are not currently included in these facilities. Ex situ in vivo facilities face challenges such as inadequate infrastructure, low technical capacity, obsolete legislation, and epileptic funding. Apart from these, ex situ in vivo conservation implies that animals are kept in managed environments such that natural selection may no longer be solely responsible in shaping attributes [105, 132]. Being isolated and with limited gene flow, ex situ in vivo populations express limited genetic diversity with time [105, 132]. Coupled with decreasing census number, they become genetically distinct from the source population [24, 105]. The resulting small effective population, high rate of inbreeding, loss of fitness, reproductive failure, poor fecundity and survivorship lead to high extinction risk [105, 109].

Ex situ conservation in gene banks is an effective means of conserving critical genetic resources from rare and extremely threatened animal species. Large biological materials representing tremendous genetic diversity can be stored in gene banks. These materials are however, none homeostatic entities that do not respond to environmental challenges and so do not undergo natural selection and evolution. Consequently, they reflect only the genetic diversity of the source population at point of collection. Semen, oocytes, DNA, etc. do not reflect the historical, ecological, cultural, traditional, and economic values of the source population and do not fulfill these roles. Furthermore, ex situ conservation runs on sophisticated technology, and infrastructure, high level technical and human resources, and huge capital investments which many third world countries including Nigeria may not afford at present [24, 133]. Despite the many treaties on conservation of biological diversity, the topic of conservation is not a priority in many third world countries battling to feed increasing human populations [24, 133].

In situ conservation aims to maintain animal populations in original habitats under the management of traditional keepers. In addition to minimizing ecological disruption, it is dynamic, allowing genes to evolve subject to the environment [133]; encourages native production systems, historical, and socio-cultural roles and values [24], livelihood, food security, panmixis, gene flow and biodiversity, natural selection and evolution, and development of adaptive capacity [24, 122, 133].

To be successful, in situ conservation requires the participation of farmers who become owners and managers of the conservation [122, 133]. It must be implemented using community-based management initiatives aimed to enhance IC production and returns, and promote food security. It has been emphasized that substitution of ICs with exotic breeds and adoption of modern technology and breeding practices that enhance productivity of chickens cannot be halted because providing adequate animal food for the increasing human population from finite resources is of primary concern to policy makers and farmers [122, 133]. Furthermore, profit motives drive economic endeavors. Consequently, a convincing instrument that could sway farmers to rear ICs is to provide economic incentives as compensation for roles in IC conservation.

6.4 Optimum conservation strategy for indigenous chicken genetic resources

The combination of ex situ and in situ conservation could be optimal for IC genetic resources [24, 109, 133]. Ex situ in vivo conservation could maintain diverse NICs in production and genetic improvement facilities with well-defined improvement programme [23, 109] that aim to preserve heritable variations, prevent fixation of deleterious alleles, and retain high reproductive fitness and adaptive potential [109] while in situ conservation would maintain ICs as on-farm populations. The segregation of conserved and genetic improvement populations will enable a separated but connected programme whereby conserved populations feed genetic improvement while maintaining high genetic diversity to ensure resilience and adaptability [109]. On-farm conservation implemented with well defined, and appropriately valued productive and adaptive traits, with market niches for all potentials, and with well informed, and adequately motivated IC farmers, will serve as reservoir of raw genetic diversity and together with ex situ populations, supply critical materials for conservation in gene banks. To determine appropriate compensation for rearing ICs, the model proposed by [133] for determining compensation for on-farm conservation of landrace crop varieties could be adopted. Using this model, economic incentive or compensation is the difference in net revenue or net profit between IC and exotic chicken production by each participating farmer. A number of national and international agencies are deploying this economic incentive strategy to encourage farmers to conserve landrace crops on-farm [133]. To secure effective cooperation, communities must be made aware of the costs and benefits of interventions, the capacity of proposed actions to achieve set objectives, and the economic benefits accruable to the farmer and community [133]. A rational farmer will usually shift from IC or exotic chicken production after comparing the expected annual profit of each enterprise. The proposed compensation framework will help to determine the critical diversity to conserve, the current diversity that maximizes total productivity, the risk of diversity loss due to changing economic, production and technological constraints and the optimal cost of conservation [133].

Critical to NIC conservation is proper characterization and economic valuation to enhance conservation value [6, 16, 108, 109]. Characterization should aim at comprehensive knowledge of the ecotype phenotypic and genotypic characteristics including data on population size and structure, geographical distribution, the production environment, and within and between breed genetic diversity [109, 122]. Conservation value is enhanced by appraisal of the relative importance of NICs from the farmer's perspective and the value placed on the characteristics, and maintenance of IC diversity [109, 122]; creation of appropriate breeding objectives to maximize the value and contribution of ICs to livelihood, and food security; and provide incentives including knowledge and infrastructure for local communities to keep and maintain IC genetic resources in their respective ecological context, thereby achieving conservation as well as maintaining rural livelihood and food security. Conservation priority should be given to ICs proven to be free of admixes of foreign genetic elements, and should be focused on the zones that contain maximum IC diversity to minimize the cost of conservation [108, 133].

7. Conclusion

Nigerian ICs are a genetically complex and critical animal genetic resource characterized by unique genetic attributes, diversity, and heritable variations. Conservation of the total diversity of NICs is very crucial because genetic complexity is requisite for evolutionary adaptation and adaptation is key to long-term survival. There is need to prioritize conservation of NICs (an invaluable national heritage) to stem the loss of IC genetic resources. Adoption of suggested conservation strategies would lead to the realization of this goal.

Author details

Cosmas Chikezie Ogbu
Department of Veterinary Biochemistry and Animal Production, College of Veterinary Medicine, Michael Okpara University of Agriculture Umudike, Abia State, Nigeria

*Address all correspondence to: ogbu.cosmas@mouau.edu.ng; ccogbu07@gmail.com

References

[1] Khobondo JO, Okeno TO, Lihare GO, Wasike CB, Kahi AK. (2014). The past, present and future genetic improvement of indigenous chicken of Kenya. Animal Genetic Resources, 125.

[2] Kingori AM, Tuitoek JK, Muiruri HK, Wachira AM, Birech EK. (2007). Protein intake of growing chicken on free-range and their response to supplementation. International Journal of Poultry Science 6: 617-621.

[3] Khobondo JO, Muasya TK, Miyumo S, Okeno TO, Wasike CB, Mwakubambanya R, Kingori AM, Kahi AK. (2015). Genetic and nutrition development of indigenous chicken in Africa. Livestock Research for Rural Development 27 (7). HTML.

[4] Emuron N, Magala H, Kuginza DR, Kyazxe FB, Kyyarisuma CC. (2010). Factor influencing the trade of local chicken. Livestock Research for Rural Development 22 (4): http//www.Irrd.org/irrd22/4/.

[5] Fayeye TR, Ayorinde KL, Ojo V, Adesina OM. (2006). Frequency and influence of some major genes on body weight and body size parameters of Nigerian local chickens. Livestock Research for Rural Development 18 (3): article 37. Retrieved from http://www.Irrd.org/Irrd18/3/fayeye18037.htm.

[6] Ajayi FO. (2010). Nigerian indigenous chicken: A valuable genetic resource for meat and egg production. Asian Journal of Poultry Science 4 (4): 164-172.

[7] Odubote K. (2015a). The local chicken of Nigeria – A review. Researchgate publications. Doi 10.13140RG2.2.2193.5520.

[8] Momoh OM, Ehiogu NG. Nwosu CC. (2007). Egg production of two Nigerian local chicken ecotypes under improved management. Proceedings 32nd Annual conference of Nigerian Society for Animal Production, March 18-22, University of Calabar, Nigeria, pp. 278-281.

[9] Eganhi JO, Dim NI, Momoh OM, Gwaza DS. (2010). Variation in qualitative traits in the Nigerian local chicken. International Journal of Poultry Science 9 (10): 978-979.

[10] Daikwo IS, Okpe AA, OchejaJO. (2011). Phenotypic characterization of local chicken in Dekina. International Journal of Poultry Science 10 (6): 444-447.

[11] Ige AO, Salako AE, Yakubu A, Adeyemi SA. (2012). Qualitative traits characterization of Yoruba and Fulani ecotype indigenous chickens in derived savanna zone of Nigeria. International Journal of Poultry Science 11 (10): 616-620.

[12] Daikwo SI, Odah EO, Ogah DM and Baba-Onoja EBT. (2018). Qualitative trait variation in indigenous chickens of Bekwara, Nigeria. Asian Research Journal of Agriculture 9 (1): 1-6.

[13] Rotimi EA, Egahi JO, Adeoye AA. (2013). Phenotypic characterization of indigenous chicken population in Gwer-West, Benue State, Nigeria. World Scientific News 53 (3): 343-353.

[14] Gwaza DS, Dim NI, Momoh OM. (2018a). Distribution of qualitative traits within and between two populations of Nigerian local chicken ecotypes. Journal of Research Reports Genetics 2 (1): 6-14.

[15] Odah EO, Daikwo SI, Mbap ST, Okpanachi U. (2019). Phenotypic characterization of local chickens (*Gallus Gallus domesticus*) in Bekwarra

Cross River State, Nigeria. JSM Veterinary Medicine and Research 2: 7.

[16] Odubote K. (2015b). Genetic diversity of the Nigerian local chickens. Researchgate Publications. Retrieved from https://www.researchgate.net/publication/277014164. Doi: 10.13140/RG.2.1.2837.9044.

[17] Ajayi FO, Agaviezor BO. (2016). Fertility and hatchability performance of pure and crossbred indigenous chicken strains in the high rainforest zone of Nigeria. International Journal of Livestock Production 7 (12): 141-144.

[18] Nwachukwu EN, Ibe SN, Ejekwu K. (2006). Short term egg production and egg quality characteristics of main and reciprocal crossbred normal local, naked neck and frizzle chicken x exotic broiler breeder stock in a humid tropical environment. Journal of Animal and Veterinary Advances 5 (7): 547-551.

[19] Ogah OM. (2013).Canonical discriminant analysis of morphometric traits in indigenous chicken genotypes. Trakia Journal of Sciences 2: 170-174.

[20] Adenaike AS, Peters SO, Fafiolu AO, Adeleke MA, Takeet MI, Wheto M, Adebambo OA, Ikeobi CON. (2019). Genetic diversity of *zyxin* and TNFRSF1A genes in Nigerian local chickens and Nera Black chicken. Agric. Conspec. Sci. 84 (3): 305-311.

[21] Gambo D, Momoh OM, Gwaza DS, Osaiyuwu OH, Addas PA. (2020). Biochemical diversity and genetic distance within and between Tiv and Fulani local chicken of Nigeria. IOSR Journal of Agriculture and Veterinary Science 13 (3): 19-26.

[22] Ige AO, Salako AE, Akinyemi MO, Adedeji TA, Ojedapo LO, Oyelade RT. (2014). Genetic similarity of Yoruba ecotype indigenous chicken using polyacrylamide gel electrophoresis. Journal of Biology, Agriculture and Healthcare 4 (6): 55-59.

[23] Mtileni BJ, Muchadeyi FC, Maiwashe A, Chimonyo M, Dzama K. (2012). Conservation and utilization of indigenous chicken genetic resources in Southern Africa. World Poultry Science Journal 68: 727-748.

[24] Adebabay KB, Tesfaye K, Belay G, Assefa G. (2016). The state of conservation of Animal genetic resources in developing countries: A review. International Journal of Pharma Medicine and Biological Sciences 5 (10): 58-66.

[25] Musa AA, Abdulmalik SE, Shoyombo AJ, Akinsola OM, Usman T. (2018). Morphological characterization of Nigerian chicken genotypes using multivariate analyses. International Journal of Poultry Science 17: 560-567.

[26] Ogbu CC, Omeje SI (2011). Within population variation in performance traits in the Nigerian indigenous chicken (NIC). International Journal of Science and Nature 2 (2): 192-197.

[27] Dana N, Tadelle D, Liesbeth H, van der W, Arendonk JAM. (2010). Morphological features of indigenous chicken populations of Ethiopia. Animal Genetic Resources 46: 11-23.

[28] McGraw KJ, Safran RJ, Wakamatsu K. (2005). How feather colour reflects its melanin content. Functional Ecology 19: 816-821.

[29] Gunnawsson U, Hellstrom AR, Tixier-Boichard M, Minvielle F, Bed'homB, Ito S, Jensen P, Rattink A, Vereijken A, Andersson L. (2007). MutationinSLC45A2 cause plumage colour variation in chicken and Japanese quail. Genetics 175: 867-877.

[30] Poultry Genetics-Mutations sub Menu 1 (2007). Retrieved from

http://home.ez web.com.au/_kazballea/genetics/mutations1_print html.

[31] Jensen P. (2006). Domestication -from behaviour to genes and back again. Applied animal behaviour science 97:3-15.

[32] Karlsson A-C. (2014). Effects of domestication related genes on behavior, physiology and gene expression in chickens. Linkoping Studies in Science and Technology. Dissertation No. 1633, pp. 1-2.

[33] Ohagenyi IJ. (2009). Some biometric and allometric growth traits of purebred heavy ecotype of the Nigerian local chicken. MSC Dissertation. Department of Animal Science University of Nigeria Nsukka, pp. 56-57.

[34] Emebet MB. (2015). Phenotypic and genetic characterization of indigenous chickens in Southwest Showa and Gurage zones of Ethiopia. PhD Dissertation College of Veterinary Medicine and Agriculture, Addis Ababa University. pp. 48-54.

[35] Nwosu CC. (1979). Characterization of the local chicken of Nigeria and its potential for egg and meat production. Proc. 1st National Seminar on poultry production NAPRI, Zaria, Vol. 1: 187-210.

[36] Olori VO. (1992). An evaluation of two ecotypes of the Nigerian Indigenous Chicken. M.Sc. Thesis, Obafemi Awolowo University, Ile- Ife, Nigeria. 108p.

[37] Nwosu CC, Gowen F, Obioha FC, Okpan IA, Onura GI. (1985). A biometrical study of the conformation of the native chickens. Nig. J. Anim. Prod. 12: 141-146.

[38] Atteh JP. (1990). Rural Poultry production in western Middle belt region of Nigeria. In Sonaiya E.B. (ed.) Rural poultry in Africa. Proceedings of an International Workshop held at the Obafemi Awolowo University, Ile- Ife, Nigeria. 13-16 November, 1989. Thelia Publishers, Nigeria. pp. 211-220.

[39] Hill DH, Mobede ANA. (1961). Poultry production at University College, Ibadan, 1950-58. Tech. Bull. No2. Fac. Agric., University College, Ibadan. 53p.

[40] Nwosu CC, Asuquo BO. (1984). Heritability estimates of body weight in the local chickens. Proceeding of the 9th Ann. Conf. Nig. Soc. Anim. Prod. pp. 41-48.

[41] Nwosu, CC. Asuquo BO. (1985). Heritability and correlation estimates of body weight in the local chicken. Proceedings of the 9th Annual conference of Nigerian Society for Animal Production, March 25-29, 1984, University of Nigeria, Nsukka, Pp: 49-56.

[42] Adedokun TA, Sonaiya BB. (2001). Comparison of the performance of Nigerian indigenous chicken from the three agro-ecological zones. Livestock Research for Rural Development 13: 1-6.

[43] Egena SSA, Ijaiya AT, Ogah DM, Aya VE. (2014). Principal component analysis of body measurements in a population of indigenous Nigerian chickens raised under extensive management system. Slovak Journal of Animal Science 47 (2): 77-82.

[44] Udeh I, Omeje SI. (2011a). Growth and short term egg production of two exotic (layer type) and the local chicken compared with their F_1 inbred progenies. International Journal of Poultry Science 10 (3): 221-224.

[45] Oke UK, Ihemeson C. (2010). Effect of genotype on the morphometric differentiation of the reproductive organs and sperm reserves in the Nigerian local chicken. Livestock

research for Rural Development 22 (3) html.

[46] Momoh OM, Nwosu CC, Adeyinka IA (2010). Comparative evaluation of two Nigerian local chicken ecotypes and their crosses for growth traits. International Journal of Poultry Science 9 (8): 738-743.

[47] Ikeobi CON, Ozoje MO, Adebambo OA, Adenowo JA, Osonowo OA. (1996). Genetic differences in the performance of local chicken in south-western Nigeria. Nigerian Journal of Genetics XI: 1996; 30-39.

[48] Egahi JO, Dim NI, Momoh OM. (2013). Crossbreeding and reciprocal effect on egg weight, hatch weight, and growth pattern and the interrelationship between these traits in three genetic groups of native chickens of Nigeria. Asian Journal of Biological Sciences 6 (3): 187-191. Doi: 10.3923/ajbs.2013.187.191.

[49] Nwachukwu EN, Ogbu CC. (2014). Effect of feathering genes on growth performance of F_2 backcrosses and comparison of F_1 and F_2 crosses of *Abor acre* broiler breeder x native chickens in a humid tropical environment. Nigerian Journal of Animal Production 41 (2): 19-33.

[50] Kekeocha C.C.(1984) Pfizer poultry production handbook. Macmillan London. 166p.

[51] Dafwang II. (1990). A Survey of rural poultry production in Lafia Area in the middle belt region of Nigeria. In Sonaiya E.B. (ed.) Rural poultry in Africa. Proceedings of an International Workshop held at the Obafemi Awolowo University, Ile- Ife, Nigeria. 13-16 November, 1989. Thelia PublisheRs, Nigeria. pp 221-235.

[52] Oleforuh-Okoleh VU, Kurutsi RF, Ideozu HM. (2017). Phenotypic evaluation of growth traits in two Nigerian local chicken genotypes. Animal Research International 14 (1): 2611-2618.

[53] Sanusi AR, Oseni SO. (2020). Nigerian Fulani ecotype chickens: growth performance under two production systems. Genetics and Biodiversity Journal 4 (1): 14-21.

[54] Sonaiya EB. (2000). ANRPD progress reports Nov. 1989- June 1995. In: sustainable rural poultry production in Africa. Ed. Proceeding of an international workshop. Addis-Ababa, Ethiopia ANRPD. 2000; 134-143.

[55] Hassan WA, Pemida OF, Fajinmi AO. (1990). Some aspects of poultry production at village level in North-west, Nigeria. In Sonaiya E.B. (ed.) Rural poultry in Africa. Proceedings of an International Workshop held at the Obafemi Awolowo University, Ile- Ife, Nigeria. 13-16 November, 1989. Thelia Publishers, Nigeria. pp 205-210.

[56] Otchere EO, Adeoye AT, Gefu JO, Adewuyi AA. (1990). Preliminary observations on village poultry production, North-Central Nigeria. In Sonaiya E.B. (ed.) Rural Poultry in Africa. Proceedings of an International Workshop held at the Obafemi Awolowo University, Ile- Ife, Nigeria. 13-16 November, 1989. Thelia Publishers, Nigeria. pp196-200.

[57] Sonaiya EB, Olori VO. (1990). Village poultry production in South Western Nigeria. In Sonaiya E.B. (ed.) Rural Poultry in Africa. Proceedings of an International Workshop held at the Obafemi Awolowo University, Ile- Ife, Nigeria. 13-16 November, 1989. Thelia Publishers, Nigeria. 266p.

[58] Omeje SSI, Nwosu CC.(1983) Egg production patterns in local chickens and their crosses in the short- term. Nig. J. Anim. Prod. 10(2) 91-96.

[59] Akinokun O. (1974). The Ife foundation stock of Nigeria chicken (2). Frequency of crests and mode of inheritance. Proc. 2nd Annual Conference of Genetics Society of Nigeria pp. 25-29.

[60] Akinokun O, Dettmers A. (1979). Genotype environment interaction in an exotic commercial egg strain and the local chickens of Nigeria. Ife J. of Agric. 1: 57-62.

[61] Akpan GN. (1987). Analysis of persistency in egg production of local chickens kept in battery cages for four years. Final year project Report, Department of Animal science University of Nigeria Nsukka, Nigeria.

[62] Nwosu CC. (1990). Review of Indigenous poultry research in South-Eastern Nigeria. In Sonaiya E.B. (ed.) Rural poultry in Africa. Proceedings of an International Workshop held at the Obafemi Awolowo University, Ile- Ife, Nigeria. 13-16 November, 1989. Thelia Publishers, Nigeria. pp 62-77.

[63] Oleforuh-Okoleh V, Nwosu CC, Adeolu AI, Udeh I, Uberu CPN, Ndofor-Foleng HM. (2012). Egg production performance in a Nigerian local chicken ecotype subjected to selection. Journal of Agricultural Science 4 (6): 180-186.

[64] Ndofor-Foleng HM, Oleforuh-Okoleh V, Musongong GA, Ohagenyi J, Duru UE. (2015). Evaluation of growth and reproductive traits of Nigerian local chicken and exotic chicken. Indian Journal of Animal Research 49 (2): 155-160.

[65] Akinokun O. (1975). The problem of improvement of poultry production in Nigeria. Nig. Agric. J. Vol 11 (2): 61-71.

[66] Sola-Ojo FE, Ayorinde KL, Jatto OM, Toye AA. (2013). Comparative studies oftwo Nigerian ecotypes chicken kept in battery cages for laying performance and egg quality traits. Asian Journal of Agriculture and Rural Development 3 (2): 30-45.

[67] Gwaza DS, Dim NI, Momoh OM. (2016a). Genetic improvement of egg production traits by direct and indirect selection of egg traits in Nigerian local chicken. Advances in Genetic Engineering 5: 148. Doi: 10.4172/2169-0111.1000148.

[68] Udeh I, Ugwu SO, Ojeh AI (2000). Comparison of the physical characteristics of semen under different collection frequencies in the native and exotic cocks. Tropical Journal of Animal Science 3 (2): 131-136.

[69] Omeje SI, Udeh I. (1998). Effect of feed restriction on body weight and semen characteristics of native and exotic (broiler) cocks. Journal of Applied Animal Research 14 (1): 81-86. Doi: 10.1080/09712119.9706219.

[70] Peters SO, Shoyebo OD, Ilori BM, Ozoje MO, Ikeobi CON, Adebambo OA. (2008). Semen quality traits of seven strains of chickens raised in the humid tropics. International Journal of Poultry Science 7 (10): 949-953.

[71] Ajayi FO, Agaviezor BO, Ajuogu PK. (2011). Semen characteristics of three strains of local cocks in the humid tropical environment of Nigeria. International Journal of Animal and Veterinary Advances 3 (3): 125-127.

[72] Udeh I, Ugwu SOC, Ogagifo NL. (2011). Predicting semen traits of local and exotic cocks using linear body measurements. Asian Journal of Animal Sciences 5 (4): 268-276.

[73] Ogbu CC. (2012a). Effect of positive assortative mating on between and within line variation in performance traits of the Nigerian indigenous chickens (NIC). International Journal of Science and Nature 3 (1): 20-23.

[74] Gwaza DS, Dim NI, Momoh OM. (2016b). Estimation of genetic parameters for traits of egg production patterns in local laying hens. Academia Journal of Biotechnology 4 (4): 111-114.

[75] Agu CI, Ndofor-Foleng HM, Nwosu CC. (2012). Evaluation of economic traits in progenies of Nigerian heavy ecotype chicken as genetic material for development of rural poultry production. African Journal of Biotechnology 11 (39): 9501-9507.

[76] Momoh OM, Nwosu CC (2008). Genetic evaluation of growth traits in crosses between two ecotype of Nigerian local chicken. Livestock Research for Rural Development 20 (10) html.

[77] Omeje SSI, Nwosu CC. (1985). Effect of three way crossing on the egg production of the local chicken. East. Afr. Agric. For. J. 51 (1) 17-21.

[78] Omeje SSI, Nwosu CC. (1987). Further explanation of the genetic basis of body weight heterosis in local X Gold Link chicken crosses. Proc. 11th Ann. Conf. Nigeria Soc. Anim. Prod. pp. 41-48.

[79] Omeje SSI, Nwosu CC. (1988). Utilization of the Nigerian chicken in poultry breeding: assessment of crossbred heterosis in growth and egg production. J. Anim. Breeding and Genetics. 105:417-425.

[80] Oluyemi JA. (1979). Potentials of the indigenous species of poultry for meat and egg production in Nigeria. In: Poultry production in Nigeria. Proc. 1st national seminar on poultry production, Zaria. 163-186.

[81] Amao SR. (2019). Production potential of backcrossed Nigerian indigenous chickens with exotic birds under Southern guinea savannah zone of Nigeria. 1. Egg production performance. Journal of Animal and Veterinary Sciences 6 (1): 1-7.

[82] Ukpong, UJ. (1987). Estimation of heterosis in body weight of F1 crosses between the local and Arbor Acre Broiler chickens. M.Sc. Thesis, University of Nigeria, Nsukka.

[83] Adeleke MA, Peters SO, and Ozoje MO. (2011a). Growth performance of Nigerian local chickens in crosses involving an exotic broiler breeder. Tropical Animal Health and Production 43: 643-650.

[84] Adeleke MA, Peters SO, Ozoje MO, Ikeobi CON, Bamgbose AM, Adebambo OA. (2015). Effect of crossbreeding on fertility, hatchability and embryonic mortality of Nigerian local chickens. Tropical Animal Health and Production 44: 505-510.

[85] Ayorinde, KL, Sola-Ojo FE, Toye AA. (2012). A comparative study of growth performance and feed efficiency in dominant Black strain, Fulani ecotype chicken and progeny from their reciprocal crosses. Asian Journal of Agricultural and Rural Development 2 (2): 120-125.

[86] Udeh I, Omeje SI. (2001). Heterosis of body weight in native by exotic inbred crosses. Tropical Journal of Animal Science 4 (1): 1-14.

[87] Udeh I. (2010). Mode of inheritance and interrelationship among age at first egg, body weight at first egg and weight of first egg in local by exotic inbred chicken crosses. International Journal of Poultry Science 9 (10): 948-953.

[88] Udeh I. (2015). Estimation of crossbreeding parameters for growth traits in crosses between Nigerian indigenous and exotic chickens. Global Journal of Animal Scientific Research 3 (2): 435-440.

[89] Akinokun O. Dettmers A. (1978a). comparison of an exotic commercial strain with the local chicken of Nigeria

(2) Heritabilities, genetic correlations and short term response to selection. Nig. J. Genetics 2: 64-70.

[90] Oluyemi JA. (1974). Evaluation and improvement of the indigenous fowl of Nigeria. Ph.D. Thesis. University of Ibadan.

[91] Nwosu CC, Omeje SSI, Ikeme I. (1987). Effects of genotype, age and egg size on measures of shell quality of local and crossbred hens. J. Anim. Prod. Res. Vol. 7 (1) 19-27.

[92] Udeh I. (2017). Genetic parameters for some growth traits of Nigerian local chicken. Biotechnology in Animal Husbandry 33 (1): 65-71.

[93] Akinokun O, Dettmers A. (1978b). Repeatability of egg production and egg weight of an exotic and local breeds and strains of chicken. Nig. J. Genetics 2:96-100.

[94] Oleforuh-Okoleh VU. (2011). Estimation of genetic parameters and selection response for egg production traits in a Nigerian local chicken ecotype. ARPN Journal of Agriculture and Biological Science 6 (12): 54-57.

[95] Oleforuh-Okoleh VU. (2013). Genetic gains from within breed selection for egg production traits in a Nigerian local chicken. ARPN Journal of Agriculture and Biological Sciences 8 (12): 788-792.

[96] Ogbu, CC, Nwachukwu EN, Nwosu CC. (2014). Determination of relative economic weights of growth and egg production traits in Nigerian indigenous chickens. Nigerian Journal of Animal Production 41 (2): 1-11.

[97] Ogbu CC, Nwosu CC. (2017). Genetic response to short-term index selection in females and mass selection in males of Nigerian heavy local chicken ecotype. Nigerian Journal of Animal Production 44 (2): 1-17.

[98] Ogbu CC. (2012b). Phenotypic response to mass selection in the Nigerian indigenous chickens. Asian Journal of Poultry Science 6 (3): 89-96.

[99] Agbo MC. (2016). Genetic response of productive traits of the Nigerian heavy local chicken ecotype using selection index from fourth to sixth generation. PhD thesis, Department of Animal Science, University of Nigeria, Nsukka.

[100] Ige AO. (2013). Estimation of genetic parameters in Yoruba and Fulani ecotype indigenous chickens of Nigeria. Transnational Journal of Science and Technology 3 (10): 1-25.

[101] Gwaza DS, Igbayima WM, Chia SS. (2015). Genetic distance between populations of Tiv local chickens in the derived guinea savannah zone of Nigeria. IOSR Journal of Agriculture and Veterinary Science 8 (2) ver. 2: 103-106.

[102] Ukwu H, Ezechukwu D, Odache F, Okopi V, Egbere M, Sunday V. (2018). Preliminary assessment of within ecotype genetic diversity at the haemoglobin locus in the Tiv local chickens in Makurdi, Nigeria. International Journal of Livestock Research 8 (4): 22-29.

[103] Ige AO, Salako AE (2014). Transferrin genetic types in Fulani and Yoruba ecotype of Nigerian indigenous chickens. Iranian Journal of Applied Animal Science 4 (1): 1919-196.

[104] Adeleke MA, Peters SO, Ozoje MO, Ikeobi CON, Adebambo AO, Olowofeso O, Bamgbose AM, Adebambo OA. (2011c). A preliminary screening of genetic lineage of Nigerian local chickens based on blood protein polymorphisms. Animal Genetics Research 48: 23-28.

[105] Medugorac I, Veit-Kensch CE, Ramljak J, Brka M, Markovic B,

Stajanovic S, Bytyqi H, Kochoski L, Kume K, Grunenfelder H-P, Bennewitz J, Forster M. (2011). Conservation priorities of genetic diversity in domesticated meta populations: a study in taurine cattle breeds. Ecology and Evolution 408-420. Doi: 10.1002/ece 3.39.

[106] Taberlet P, Valentini A, Rezaei HR, Naderi S, Pompanon F, Negrini R, Ajmone-Marsan P. (2008). Are cattle, sheep, and goats endangered species? Mol. Ecol. 17:275-284.

[107] Medugorac, I, Medugorac A, Russ I, Veit-Kensch CE, Taberlet P, Luntz B, Mix HM, F¨orster M. (2009). Genetic diversity of European cattle breeds highlights the conservation value of traditional unselected breeds with high effective population size. Mol. Ecol. 18:3394-3410.

[108] Ramadan S, Kayang BB, Inoue E, Nirasawa K, Hayakawa H, Ito S, Inoue-Murayama M. (2012). Evaluation of genetic diversity and conservation priorities for Egyptian chickens. Open Journal of Animal Sciences 2 (3): 183-190.

[109] Mtileni B, Dzama K, Nephawe K, Rhde C. (2016). Estimate of effective population size and inbreeding in South African indigenous chicken population: implication for conservation of unique genetic resources. Tropical Animal Health and Production 48: 943-950.

[110] Hoffmann I. (2009). The global plan of action for animal genetic resources and the conservation of poultry genetic resources. World's Poultry Science Journal, 65, 286-297.

[111] FAO (2007). The state of the world's animal genetic re-sources for food and agriculture. FAO, Rome. http://www.fao.org/docrep/010/a1250e/a1250e00.htm

[112] Fayeye TR, Ayorinde KL, Ojo V, Adesina OM. (2006). Frequency and influence of some major genes on body weight and body size parameters of Nigerian local chickens. Livestock Research for Rural Development 18 (3) html.

[113] Rege JEO, Gibson JP. (2003). Animal genetic resources and economic development issues in relation to economic valuation. Ecological Economics 45 (3): 319-330.

[114] Tisdell C. (2003). Socioeconomic causes of loss of animal genetic diversity: Analysis and assessment," Ecological Economics 45: 365-376.

[115] Safaloah ACI. (2001). Village chicken upgrading programme in Malawi. World's Poultry Science Journal 57: 179-188.

[116] Taberlet P, Pansu J, Pompanon F. (2011). Conservation genetics of cattle, sheep, and goats. Comptes Rendus Biologies 334: 247-254.

[117] Philipsson, JE, Rege O, Zonabend E, Okeyo AM. (2011). Sustainable breeding programmes for tropical farming systems. In: *Animal Genetics Training Resource*, J. M. Ojango, B. Malmfors, and A. M. Okeyo, Eds. International Livestock Research Institute, Nairobi, Kenya, and Swedish University of Agricultural Sciences, Uppsala, Sweden.

[118] Mendelsohn R. (2003). The challenge of conserving indigenous domestic animals. Ecological Economics 45: 501-510.

[119] Mtileni BJ, Muchadeyi FC, Maiwashe A, Phitsane PM, Halimani TE, Chimonyo M, Dzama K. (2009). Characterization of production systems for indigenous chicken genetic resources of South Africa. Applied Animal Husbandry for Rural Development 2: 18-22.

[120] Barker JS. (2001). Conservation and management of genetic diversity:

A domestic animal perspective. Canadian Journal of Forestry Research, 31: 588-595. Doi: 10.1139/x00-180.

[121] FAO (2006). Exchange, use and conservation of animal genetic resources: Policy and regulatory options. Centre for Genetic Resources, the Netherlands (CGN) Report, Rome, Italy

[122] Rodringuez LC, Herrero M, Baltenwoek I. (2011). Community-based intervention for the use and conservation of animal genetic resources: the case of indigenous scavenger chicken production in Benin. Tropical Animal Health and Production 43: 961-966.

[123] Simianer H. (2005). Decision making in livestock conservation. Ecological Economics 53: 559-572.

[124] Drucker AG, Hiemstra SJ, Louwaars N, Oldenbroek JK, Tvedt MW, Hoffmann I, Agwichew K, Abegaz Kebede S, Bhat PN, da Silva Mariante A. (2007). Back to the future: How scenarios of future globalization, biotechnology, disease and climate change can inform present animal genetic resources policy development. Animal Genetic Resources Information 2 (41): 75-89.

[125] Agbaje HA, Alabi OO. (2018). Application of GIS for biodiversity conservation of indigenous chickens in Ile-Ife, Nigeria. IOSR Journal of Agriculture and Veterinary Science 11 (12): 18-38.

[126] Gwaza DS, Dim NI, Momoh OM. (2018c). Effect of genetic drift versus natural selection on clutch traits in two populations of the Nigerian local chickens. Significances of Bioengineering and Biosciences 2 (1): 105-107.

[127] FAO (2003). Animal genetic resources conservation and development: The role of FAO. Arch. Zootec. 52: 185-192.

[128] FAO (2009). The use and exchange of animal genetic resources for food and agriculture. Background Study Paper 43.

[129] Mapiye C, Mwale M, Mupangwa JF, Chimonyo M, Foti R, Mutenje MJ. (2008). A research review of village chicken production constraints and opportunities in Zimbabwe. Asian-Australasian Journal of Animal Science 21 (11): 1680-1688.

[130] Naqvi AN. (2007). Application of molecular genetic technologies in livestock production: Potentials for developing countries. Advances in Biological Research 1 (3-4): 72-84.

[131] FAO (2013). In vivo conservation of animal genetic resources. FAO Animal Production and Health Guidelines 14, Rome, Italy.

[132] Kasso M, Balakrishnan M. (2013). Ex Situ Conservation of Biodiversity with Particular Emphasis to Ethiopia, Review Article, Hindawi Publishing Corporation. http://dx.doi.org/10.1155/2013/985037.

[133] Poudel D. (2015). On farm conservation of crop genetic resources: declining *de facto* diversity and optimal funding. Natural Resources 6: 196-207.

Chapter 9

The Reproductive Performance of Native Osmanabadi Goat of India

Monali Wakchaure, Mohammad Faheem Siddiqui and Akshay Sonawane

Abstract

Among the goat breeds of India, Osmanabadi goat breed is one of the most popular goat breed of the arid and semi-arid region of Maharashtra state. Historically this breed is known to exist on Deccan Plateau since decades. The name Osmanabadi is derived from its origin, i.e. Osmanabad district in Maharashtra state. The breeds is distributed mainly in 2 southern states of India viz. Western Telangana and North Eastern Karnataka state and are having largest contribution to meat production in Southern India as their meat is very tasty when compared with local breeds. The Osmanabadi breed is suited to all types of rearing systems, the most ideal being the semi-intensive system (grazing and closed enclosure) where higher production has been observed compared to extensive (grazing system) and intensive systems (zero grazing system). Osmanabadi goats reared in the Maharashtra, Karnataka and Telangana border region had been analysed with reproductive parameters and found that, the female kids attend puberty at the age of 349.8 ± 6.9 days with 17.45 ± 0.23 Kg body weight. The average gestation period found was 152.24 ± 0.24 days. The mean age at first kidding was found to be 494.4 ± 8.1 days. The average duration of post-partum anoestrus period was 67.34 ± 6.31 days which was responsible for short inter-kidding interval which shows high profile reproductive efficiency. The mean kidding interval recorded as 232.62 ± 5.45 days. Majority of kidding resulted in single births (87.27%) and with only 12.73% of multiple births. Breeding season and kidding season of Osmanabadi goats was observed from the month of June to September and November to February as a major.

Keywords: Osmanabadi goats, Reproductive performance

1. Introduction

India is one of the few countries in the world to make a rich contribution to the international pool of genes for livestock and to the improvement of animal production worldwide. India possesses an enormous goat population numbering 148.88 million [1], which is the second highest in the world after China and contribute about 27.80% of the total livestock population of India. As per the census report of the Department of Animal Husbandry, Dairying and Fisheries, Ministry of Agriculture, Government of India (2019), India is a rich repository of goat genetic capital with 23 well-recognised goat breeds and these breeds have evolved in relation to various geographical and climatic situations. In different agro-climate areas, these different breeds are of particular importance and have evolved unique adaptation characteristics over the years in order to survive and simultaneously produce sustained production under the agro-climate conditions of their habitats. In general, these breeds have been named after their place of origin and in some instances based on their prominent features.

Goats are one of the oldest domesticated species, and have been used for their milk, meat, hair and skins over much of the world. Ruminants are of great economic importance in livestock industry and small ruminants play very important role in the socio-economic status of the society. In fact, goat plays a significant economic role for the farming communities living in lowland, midland and highland provinces. Goat, being small sized and more prolific animal, requires minimum capital and maintenance costs with less risk in investment. Goats play an important role in the food and nutritional security of millions of rural people especially to the landless, marginal and small farmers. Further this is sturdy and adaptable animal and is known to provide sustainable source of income to more than 40% of below poverty line rural population. This is one of the indications of rural farmer preference to this animal for employment and income generation. The socio-economic value of goat rearing as compared to other livestock species, for poor farmers is immense. Goats and sheep are also among the main meat producing animals in India, whose meat is readily preferred. They also produce variety of other products, which are especially useful in the semi-arid and arid climatic conditions [2]. In India and other developing countries, the domestic goat (*Capra hircus*) is an important livestock species. It is popularly known as the "poor man's cow" because it provides a good source of meat, milk, fibre, and skin [3]. From very early times in human civilisation, goats have served agricultural, economic, cultural and even religious functions. Archaeological evidence suggests that at the start of the Neolithic period in the Fertile Crescent, the goat was one of the first animals to be domesticated by humans around 10,000 years ago [4, 5]. India posse's large number of goats which can be classified in to 23 different breeds adaptable to various climatic conditions. Among the various Indian goat breeds Osmanabadi goat breeds as one of the most popular goat breed of the arid and semi-arid region of Maharashtra state. Historically this breed is known to exist on Deccan plateau since decades. The name Osmanabadi is derived from its origin, i.e. Osmanabad district in Maharashtra state.

The goats are usually kept under extensive management and reared on natural vegetation, but due to shrinkage of grazing land and as is blamed for soil erosion and desertification, the maintenance of flocks under extensive system is threatened. However, Semi-intensive and intensive systems of goat rearing with small flocks are gaining momentum. In extensive system of management, the animals are reared on poor and degraded grazing lands resulting in low production and reproduction. The Osmanabadi breed is suited to all types of rearing systems, the most ideal being the semi-intensive system (grazing and closed enclosure) where higher production has been observed compared to extensive (grazing system) and intensive systems (Zero grazing system).

The knowledge of specific physical characteristics and production efficiency of this breed is the need of the hour to avoid indiscriminate breeding and to preserve sustainable productivity. Therefore, this review focuses on physical characteristics and production performance of Osmanabadi goat in India.

2. History and name of breeds

Osmanabadi goat mainly originated in Tuljapur taluka of Osmanabad District and Udgir taluka of Latur district of Maharashtra, both were earlier under the

Osmanabad district hence the name Osmanabadi became popular among the farmers. This goat is also known as Deccani. As these districts were earlier included in erst while Nizam State of Rule which was popularly known as Deccan State and hence the Osmanabadi breed was also synonymed as Deccani. The history of origin of Osmanabadi goat breed dates back to 150 years in the breeding tract of Marathwada region. The Osmanabad district being named after the Nizam Ruler of Hyderabad Deccan Estate Mir Osman Ali Khan the 7th Ruler of Nizam dynasty, hence the goat breed being named on the basis of breeding tract and origin from Osmanabad district as Osmanabadi (Seeri Hind and Tareekh-e-Khursheed, M. K. Shazli, 1968).

3. Distribution of breeds

The Osmanabadi breed distributed over greater part of central peninsular region, comprising the semi-arid areas or sub-tropical zones of Maharashtra, Andhra Pradesh and Karnataka states. It covers the major part of southern Maharashtra especially Osmanabad, Latur, Nanded, Parbhani, Hingoli, Beed, Jalna and Aurangabad district of Marathwada region and adjoining parts of Telangana and Karnataka State.

4. Breeding tract and climate

Osmanabad and Latur districts of the Marathwada region of Maharashtra state are the breeding areas for Osmanabadi goats [6]. The Osmanabadi breed has also been found to migrate to neighbouring areas. The breeding tract comprising Latur and Osmanabad districts is spread over 18^0-05′ to 18^0-07’ N Latitude and 73^0-25′ to 77^0-25′ E Longitudes and $17^0$35′ $18^0$40’ N Latitude and $75^0$16′ to $76^0$40′ E Longitude respectively in the deccean plateau [7]. Latur district is situated at 540 to 638 m height from mean sea level. The altitude of Osmanabad district is 600-611 m above mean sea level. The tract's agro-climate condition has been categorised as a subtropical zone and falls within a scarcity zone. Cereals, oilseeds, and pulses are the main crops grown in the Latur and Osmanabad districts, leading to harvests in the Kharif and Rabi seasons. The crops taken in Kharif are Jowar (*Sorghum bicolar*), udid or blackgram (*Phaseolus mungo*), tur or pigeonpea (*Cajanus cajan*), maize (*Zea mays*), post-monsoon sunflower (*Helianthus annuus*), Rabi jowar (*Sorghum bicolar*), wheat (*Triticum* sp.), bengal (*Cicer arientium*), and safflower (*Carthamus tinctorius* Linn.) in Rabi. The great bulk of the ration for ruminants is Jowar kadbi (*Sorghum bicolar*). In addition, mung or greengram (*Phaseolus aureus*), udid or blackgram (*Phaseolus mungo*), wheat (*Triticum sp.*), tur (*Cajanus cajan*) and groundnut (*Arachis hypogea*) crop residues are also used for animal feeding.

5. Management practices

Raskar *et al.* [8] was observed that 83.33% goat keepers provided housing only during night hours to protect them from the wild animals and theft, whereas 16.67% goat keepers provided day and night housing. The goats were kept in close housing (84.62%) as well as in open housing (15.38%). In case of close housing, the roofs were made up of locally available materials like tur straw, jowar straw, sugarcane trash, tree leaves, dry grasses (78.21%) and 21.79% goat houses were with tin sheds. In open houses the goats were kept under trees, open areas and fenced with thorny bushes etc. Majority of the goat keepers (98.72%) used kutcha type of floor in goat houses, while only 1.28% goat keepers provided pucca type of floor in shelter. In small flocks (3-4 animals) there was no separate housing and goats lived with the owner and shared the houses. Raskar *et al.* [8] was found that 73.72% farmers constructed the shelters separately, while 26.28% farmers maintained shelter as part of their houses to safeguard the animals during night time. The goat sheds were mostly half walled (77.56%) and few were full walled (22.44%) with 79.49% well ventilation mainly due to higher percentage of half walled structures. In few cases (20.51%) structures had poor ventilation (closed structure).

The goat house did not have well drained system for urine (98.72%) and only 1.28% had the proper drainage for urine, particularly noticed in pucca type of flooring structures. Shinde [9] reported 93.04% farmers provided housing for Osmanbadi goats during night hours only with 70.44% and 29.56% closed and open housing, respectively. It was further revealed that 98.27% had kutcha floors and only 1.28% pucca floor provided to the goats. Singh [10] and Gokhale *et al.* [11] reported that 66% of farmers maintained shelters as part of their residence.

6. Physical characteristics

6.1 Coat colour pattern

Osmanabadi goats have different coat colour patterns. According to the colour and presence or absence of horns, Osmanabadi goats were classified into five types. There is no specific name for these types, except Kali (Black), Morkani (White spotted ear), Hondi (Polled). The distribution of goats in surveyed area revealed that S1 (Kali) was 62.16%, S3 (Hondi) 17.12%, S2 (Morkhani) 10.68%, S4 (Hondi Morkani) 3.09% and proportion of remaining goats having different colour combinations i.e. S5 was 6.95%, [7].

Thus the majority of Osmanabadi goat population was comprised of breed sub-type S1 and S3 (79.28%); while the proportion of breed sub-types S2 and S4 was comparatively negligible than the first two subtypes. The breed sub-type S5 might be developed due to the admixture of different coat colours and breed combination in the population. The eyelids and hooves of Osmanabadi goat was 100% black for all categories of goats studied. Prakash and Balain [7] reported that the common colour of Osmanabadi goat was black and a mixture of white and black or red. Similarly, Ruben [12] reported that the coat colour of Osmanabadi goats was complete black or mixture of black and brown colour. Anonymous [13] recorded the distribution of Osmanabadi goats according to different breed sub-types in three districts and reported 82.60, 11.19, 2.67, 0.75 and 2.84 percent of the goats in the respective five breed sub-types.

The colour of muzzle was found black in 100% goats under survey. Deokar *et al.* [14], Verma *et al.* [15], Kumar *et al.* [16] and Kuralkar *et al.* [17] also observed the black colour of muzzle in Osmanabadi, Gohilwadi, Kutchi and Berari goats, respectively. It was observed that eyelids and hooves colour in Osmanabadi goat was 100% black for all categories of goat studied. Deokar *et al.* [14] reported black colour of eyelids and hooves in Osmanabadi goats, while, Deokar *et al.* [18] reported white colour of eyelids in Sangamneri goats and white (69.12%) and black (30.88%) hooves in Sangamneri goats.

6.2 Horn pattern

Horns were noticed in both the sexes in majority (79.54%) of Osmanabadi goats while 20.46% goats were polled, [7]. The 100% Osmanabadi goat had grey colour of horns. The length of horns in adult goats averaged to 7.48 + 0.52 cm. Osmanabadi goats had straight horn (64.40%), while, 35.60% had curved horn, [7]. The orientation of horn was mostly backward (48.55%) followed by upward (44.17%) and very less downward (7.28%) orientation. Presence of horns in both the sexes was also reported in Osmanabadi goat [14], Sangamneri goat [18], Jakhrana goat [8] and Surti goat [19]. The straight horn was reported by Motghare *et al.* [20] and Deokar *et al.* [14] in Osmanabadi as 67.17% and 98.57%, respectively, Deokar *et al.* [18] in Sangamneri (30.68%) and Deshpande *et al.* [21] in Surti (52.45%) goats. Kumar *et al.* [16] reported curved horn in Kutchi goats, while, Verma *et al.* [15] reported slightly twisted type horn in Gohilwadi goats. Motghare *et al.* [20] reported orientation of the horns in Osmanabadi goat was mostly upward (100%). Deokar *et al.* [14] reported maximum percentage had backward orientation (55.22%) followed by upward (36.40%) and only 8.38% had downward orientation.

6.3 Head, ear and tail patterns

The ear orientation of Osmanabadi goat was pendulous (drooping) with medium length (14.90 ± 0.26 cm), [7]. As regards the orientation of ears of Osmanabadi goats in the breeding tract, not a single case of erect ears was recorded. However, very few cases of horizontal ears were recorded. Majority of Osmanabadi goats had convex forehead (95.24%), [7]. The percentage of absence of wattle in Osmanabadi goats was 100%. Likewise the overall percentage of the goats not having beard was 100%. This clearly indicated that both wattle and beard characteristics were not the common feature of Osmanabadi goats. Deokar *et al.* [14], Deokar *et al.* [18] and Deshpande *et al.* [21] reported pendulous ear, convex forehead and absence of beard and wattle in majority of cases in Osmanabadi, Sangamneri and Surti goats, respectively. Tail pattern in Osmanbadi goat was observed as curved (96.27%) and only 3.73% goat has straight tail. Deokar *et al.* [14] reported 99.91% straight tail in Osmanabadi goats.

6.4 Body measurements and body weight

Raskar *et al.* [8] reported the least square means for height at withers, heart girth, body length, ear length, horn length and body weight for 3 months of age were 40.69 ± 0.55 cm, 37.61 ± 0.54 cm, 32.08 ± 0.55 cm, 11.69 ± 0.27 cm, 0.39 ± 0.15 cm and 6.29 ± 0.35 kg, respectively, in Osmanabadi goats. A survey on the Osmanabadi goat in its breeding tract was performed by Raskar *et al.* [8], i.e. The Latur and Osmanabad districts of Maharashtra's Marathwada region. Two blocks were selected from each district, namely Latur (B1) and Ausa (B2) blocks from the districts of Latur and Tuljapur (B3) and Osmanabad (B4) blocks from the districts of Osmanabad, and ten villages were considered from each block. The effect of location was significant source of variation for height at withers, chest girth, body length, ear length and body weight. DMRT (Duncan's multiple range test) showed that Block B1 goats had higher body weight, chest girth, body length, height at withers than goats of the other blocks. The effect of block was significant source of variation for body weight, chest girth, body length, height at withers and ear length. The least squares mean for height at withers, heart girth, body length, ear length, horn length, and body weight for 6 months of age were 55.04 ± 1.11 cm, 52.67 ± 1.11 cm, 46.36 ± 0.15 cm, 13.36 ± 0.44 cm, 2.91 ± 0.39 cm and 15.49 ± 0.57 kg in Osmanabadi goats, [7]. The effect of block was significant source of variation for height at withers, chest girth, body length and ear length. Block had significant source of variation for all the traits except horn length and body weight. The least square means for height at withers, heart girth, body length, ear length, horn length, and body weight for 12 months of age were 62.45 ± 0.87 cm, 60.03 ± 1.07 cm, 50.01 ± 1.08 cm, 15.58 ± 0.44 cm, 3.30 ± 0.67 cm and 19.56 ± 0.92 kg, respectively, in Osmanabadi goats, [7]. The effect of block was significant source of variation for chest girth, body length, ear length and body weight except horn length and height at wither. The effect of sex was significant source of variation for chest girth, body length and ear length. Raskar *et al.* [8] observed that the least square means for height at withers, heart girth, body length, ear length, horn length and body weight for 24 months of age were 71.00 ± 0.87 cm, 70.48 ± 0.91 cm, 58.34 ± 0.93 cm, 17.18 ± 0.37 cm, 6.51 ± 0.56 cm and 29.50 ± 0.64 kg, respectively, in Osmanabadi goats. The effect of block was significant source of variation for height at withers, chest girth body length and body weight except horn length and ear length. The effect of sex was significant source of variation for chest girth, body length and height at withers. The least square means for height at withers, heart girth, body length, ear length, horn length and body weight for 36 months of age and above were 73.12 ± 0.61 cm, 74.55 ± 0.67 cm, 62.10 ± 0.67 cm, 17.24 ± 0.24 cm, 7.48 ± 0.52 cm and 32.77 ± 0.60 kg, respectively, in Osmanabadi goats, [7]. The effect of sex was significant source of variation for all traits except ear length. Block had significant source of variation for all the traits except chest girth, horn length and ear length, whereas colour type had non-significant effect on all the traits in adult age group of Osmanabadi goats. Ravimurugan *et al.* [22] reported higher body weight, height at withers, body length, chest girth and horn length in Pallai Adu male goat than the female. The mean body height, length, girth and weight of Osmanabadi goats indicated that they belong to medium sized goat category. The Osmanabadi goat is medium sized meat breed thriving well in tropical wet and dry climate.

6.5 Performance

The average age at first kidding and kidding interval is 523 and 214 days respectively. Goats of this breed have very efficient reproduction and in well managed flocks [23], with 30% twining and 2% triplets. The daily milk yield ranged from 700 gm to 1500 gm under well managed village flocks with lactation length of 130-150 day. Raskar *et al.* [24] and Sahare *et al.* [8] found that in Osmanabadi goat maintained under farm condition, kidding percentage and twinning ability was 55.87% and 10.52% respectively.

6.6 Age at puberty

The mean age at puberty was recorded as 349.8 ± 6.9 days and ranged between 180 and 510 days. Kamble *et al.* [25] reported similar findings 335.3 ± 13.0 days while Lawer *et al.* [26] reported 219.34 ± 0.72 days pubertal age in Osmanabadi goats.

6.7 Weight at puberty

The weight at puberty in Osmanabadi goats occurred when the does attained an average body weight of 17.45 ± 0.23 kg, [27]. Smith [28] stated that, Angora goats should weigh 32-41 kg before being bred and recommended that breeding should be delayed until the animal has attained 60% or more of its adult body weight. It was also evident from the study that the age at puberty ranged between 6 and 17 months and it is highly probable that Osmanabadi goats attaining puberty at an early age had a better growth rate resulting in a better body weight cumulating in the onset of puberty.

6.8 Gestation period

The average gestation period in Osmanabadi goats was determined as 152.24 ± 0.24 days and it ranged from 137 to 158 days. The average duration of gestation in goats is generally reported as 147-155 days [4]. The gestation period was recorded as 150.08 days by Pathodiya *et al.* [4] and as 146.23 ± 0.49 days by Swami *et al.* [29] in Sirohi goats. In Ganjam goats gestation length was 148.26 ± 0.31 days, in Bengal type goats of Orissa it was 146.27 ± 0.37 days and in Ghumsur goats of Orissa it was 145.03 ± 0.48 days (Rao and Patro [30]). Mandakmale *et al.* [31] observed the gestation period to be 149.96 ± 0.82 days for Osmanabadi goats under field conditions and Bhusan and Rai [32] found it to be 151.33 ± 1.48 days in Jakhrana.

6.9 Age at first kidding

The first kidding in Osmanabadi goats occurred as early 330 days in some does and as late as 650 days in few others [27]. The mean age at first kidding was determined as 494.4 ± 8.1 days similar to the reports in other breeds [24]. Mandakmale *et al.* [31] observed the age at first kidding of 377.15 ± 2.67 days in Osmanabadi goats under field conditions. In Malabari it was observed to be 13.72 ± 0.10 months [33] and 531 days [34]. In Jamunapari goats first kidding was 700 ± 9.1 days (Rout *et al.* [35]), observed at 23 months (Rout *et al.* [36]).

6.10 Inter-kidding period

The mean kidding interval in Osmanabadi Does was recorded as 232.62 ± 5.45 days and ranged between 181 to 310 days, [27]. A similar kidding interval has been reported by Markendeya and Devanagare [37] in Osmanabadi goats. Mabari goats also appear to have a kidding interval similar to Osmanabadi does [38]. Raghavan *et al.* [33] reported the average kidding interval as 9.47 ± 0.11 months in Malabari goat breed of Kerala while Raghavan *et al.* [34] recorded kidding interval as 315 days (**Figures 1** and 2; **Table 1**).

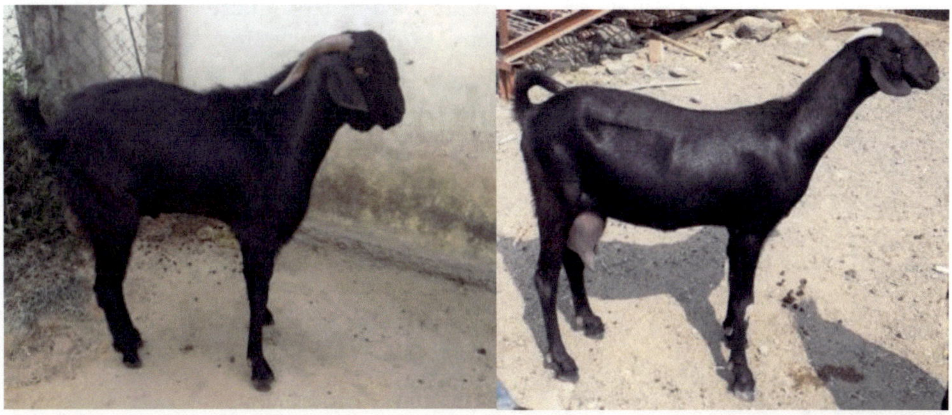

A B

Figure 1.
*Pure Osmanabadi breed (**A**) Osmanabadi Buck and (**B**) Osmanabadi doe.*

Figure 2.
Osmanabadi doe with its four kids.

Sr.No.	Traits	Average	Source
1	Age at Puberty	335.3 ± 13.0	Kamble *et al.* [25]
2	Age at First service	370 ± 14	Kamble *et al.* [25]
3	Age at First Conception	233.02 ± 0.89	Lawar *et al.* [26]
4	Age at First Kidding	0.081 ± 0.010	Patil *et al.* [38]
5	Kidding Interval	0.0317 ± 0.098	Patil *et al.* [38]
6	Gestation Period	0.292 ± 0.086	Patil *et al.* [38]

Table 1.
Average reproductive performance of Osmanabadi goats.

7. Conclusion

Generally, various factors such as genetic, development systems and management practices have influenced the productive and reproductive success of indigenous goats. The following recommendations have been forwarded from the aforementioned conclusion:

1. The ability to better adjust the climate for indigenous goat breeds; a regulated crossbreeding and selection policy should be established in line with the conservation of local adaptive characteristics of the breeds.

2. In particular, farmers should be provided with training and knowledge to increase the reproductive output of goats and farmers' livelihoods through improved management practises.

Acknowledgements

The author expresses his sincere thanks to those who have helped in preparation and enrichment of this article. The help received from **Dr. M. F. Siddiqui** and **Dr. A. S. Sonawane**, gratefully acknowledge.

Conflict of interest

Authors declare no conflict of interest.

Author details

Monali Wakchaure[1*], Mohammad Faheem Siddiqui[2] and Akshay Sonawane[3]

1 Department of LPM, G.M.D. Arts B.W. Commerce and Science College, Nashik, Maharashtra, India

2 Department of Livestock Production and Management, College of Veterinary and Animal Sciences, Parbhani, Maharashtra, India

3 Hester Bioscience Ltd., Ahmedabad, Gujarat, India

*Address all correspondence to: monali.wakchaure17792@gmail.com

References

[1] All India Report, 20th livestock census, 2019. Government of India, Ministry of Agriculture, Department of Animal Husbandry, Dairying and fisheries, New Delhi.

[2] Kumar S. and Roy M. M. 2013. Small ruminant's role in sustaining rural livelihoods in arid and semi-arid regions and their potential for commercialization. In: New Paradigms in livestock production from traditional to commercial farming and beyond. Agrotech publishing academy, Udaipur, pp. 57-80

[3] MacHugh, D. E. and Bradley D. G. 2001. Livestock genetic origins: goats buck the trend. Proc. Natl. Acad. Sci. USA 98:5382-5384.

[4] Pathodiya, O.P., Khadda, B.S., Gurjar, M.L. and Tailor, S.P., 2004. Some economic traits of Sirohi goats in field conditions, Indian Journal of Animal Sciences, 74 (1): 102-103.

[5] Prabhakaran, R. and Thirunavukkarasu M. 1992. Goat husbandry in Tamil Nadu (India). An economic appraisal V International Conference on Goats, Pre Conf. Abstracts. IGA Publication, New Delhi. (1): 33.

[6] Acharya R. M. 1982. Sheep and goat breeds of India. Animal Production and Health. Paper 30. Food and Agriculture Organization of United Nations, Rome.

[7] Prakash B. and Balain D. S. 1992. Conservation, evaluation and utilization of goat germplasm resources in India. Indian Journal of Animal Production and Management 8 (1 and 2): 1-22.

[8] Raskar, B. R., Chauhan, D. S., & Singerwad, P. S. 2018. Morphological Characterization of Osmanabadi Goat in Its Breeding Tract. An International Refereed, Peer Reviewed & Indexed Quarterly Journal in Science, Agriculture & Engineering, Vol. VII, Special Issue ICAAASTSD

[9] Shinde P. M. 2000. Studies on phenotypic characters of Osmanabadi goats in Rahuri tahsil of Ahmednagar district. M. Sc. (Agri.). Thesis submitted to M.P.K.V., Rahuri, Maharashtra.

[10] Singh M R. 2001. Comparative resource structure of goat keeping of landless labourers and marginal farmers in rural Mathura, Uttar Pradesh. Indian Journal of Small Ruminants 7 (1): 41– 44.

[11] Gokhale S B, Gokhale R B, Phadke N L and Desale R J. 2002. Status of village goat management practices in Maharashtra. Indian Journal of Animal Sciences 72(9): 810-814.

[12] Ruben D. 1997. Flock of our goats (Osmanabadi) Baliraja Magazine. 28 (3): 43.

[13] Anonymous. 1999. Final Report of Network Project on Survey of Osmanabadi goats submitted by M.P.K.V., Rahuri to the NBAGR, Karnal. 7: 66.

[14] Deokar D. K, Lawar V. S. and Ulmek B. R. 2006. Morphological characteristics of Osmanabadi goat. Indian Journal of Small Ruminants 12 (1): 213-215.

[15] Verma N. K, Dixit S. P, Dinesh Kumar, Aggarwal R. A. K. and Ahlawat S. P. S. 2007. Physical characteristics, performance status and genetic variation in Jakhrana breed of goat in its native tract. Indian Journal of Animal Sciences 77 (5): 390-394.

[16] Kumar A, Sushil K, Mishra A. K. and Singh V. K. 2006. Characteristics of

Kutchi goats of Gujarat. Indian Journal of Small Ruminants 12(2): 162-168.

[17] Kuralkar S. V, Verma N. K, Kjarkar Kranti and Kuralkar Prajakta 2013. Berari goats: Characterization, management, performance and population status. Indian Journal of Animal Sciences 83 (12): 1292-1298.

[18] Deokar, D. K., Lawar V. S, Pawar B. K. and Andhale R. R. 2007. Breed characteristics of Sangamneri goat. Indian Journal of Small Ruminants 13 (2): 13-15.

[19] Ekambaram B, Gupta B. R, Prakash M. G, Sudhaker K. and Reddy V. R. 2011. Housing, breeding and management practices of Mahabubnagar goats. Indian Journal of Animal Sciences 81(8):875-879.

[20] Motghare A. B., Ali S. Z. and Kuralkar S. V. 2005. Physical characteristics of Osmanabadi goats maintained in Vidarbha Region. Indian Journal of Small Ruminants. 11 (1): 75-76.

[21] Deshpande S. B, Desai P. M, Kharadi V. B and Sabapara G. P. 2009. Phenotypic and performance characteristics of Surti goats. Indian Journal of Small Ruminants 15 (1): 108-112.

[22] Ravimurugan T, Devendran P, Cauveri D. and Balachandran S. 2009. Performance of Indigenous goat (Pallai Adu) under field conditions. Tamilnadu Journal of Veterinary and Animal Science 5 (5): 203-207.

[23] Koratkar D. P., Bhoite Y. U. and Deshmukh A. K. 1998. Reproductive performance of Osmanabadi Goats, Indian Journal of Small Ruminants, 4:34-36.

[24] Sahare, M. G., Sawaimul, A. D., Ali, S. Z. and Kolte, B. R., 2009. Kidding percentage and twinning ability in Osmanabadi goat in Vidarbha climatic condition. Veterinary world, 2(2):60-61.

[25] Kamble et al. 2009. cited in AGTR, website (Animal Genetic Training Resource) ILRT (International Livestock Resarch Institute), SLV(Swedish University of Agricultural Sciences)

[26] Lawer et al. 2008 cited in AGTR, website (Animal Genetic Training Resource) ILRT (International Livestock Resarch Institute), SLV(Swedish University of Agricultural Sciences)

[27] Bijurkar, R.G., Krishnaswamy, A., Honnapa, T. G., Chandrashekhara Murthy, V. and Jayashankar, M. R. 2015. Reproductive Traits of Osmanabadi Goats in the Karnataka Maharashtra Border Region, Frontier J. Vet. Anim. Sci. Vol. 4, No. 2 (July-Dec.)

[28] Smith, M. C., 1997. Clinical reproductive anatomy and physiology of doe. Current therapy in large animal theriogenoloy: pp. 505-507.

[29] Swami, P.D., Barhat, N.K., Joshi, R.K., Murdia, C.K. and Vijay kumar, 2006 b. Reproductive performance of Sirohi goats and its crosses with Beetal in semi-arid condition of Rajasthan, Indian Journal of Animal Sciences, 76 (4): 346-348.

[30] Rao, P.K. and Patro, B.N., 2004. Goat genome diversity in Orissa, In: Proceedings of the seminar on Goat Genome, 5-6 April 2004, CIRG, Makhdoom, pp: 105-108.

[31] Mandakmale, S.D., Kamble, S.S., Dhage, S.A and Jagtap, D.Z., 2007. Post conception traits of Osmanabadi goats, In: Compendium of National symposium on role of Animal genetic resources in rural livelihood security, 8-9 February 2007, Ranchi (Jarkhand), pp: 271.

[32] Bhusan S, Rai B. 2008. Genetic evaluation and improvement of Jakhrana Breed of goats, In compendium of National symposium on Redefining role of indigenous animal genetic resources in rural development, 15-16 Feb, 2008 held at Bangalore, 2008, 180

[33] Raghavan, K.C., Raja, T.V. and Sasikanth, 2004. Malabari goats, In Proceedings of the seminar on Goat Genome, 5-6 April 2004, CIRG, Makhdoom, pp: 101-104.

[34] Raghavan, K.C., Sasikanth and Raja, T.V., 2007. Study on growth production and reproduction performance of Malabari goats under farm conditions, In:MCompendium of National symposium on role of Animal genetic resources in rural livelihood security, 8-9 February 2007, Ranchi (Jarkhand), pp: 281.

[35] Rout, P.K., Saxena, V.K., Khan, B.U., Roy, R., Mandal, A., Singh, S.K., and Singh, L.B., 2000. Characterization of Jamunapari goats in their home tract, Animal Genetic Resources Information (FAO), 27: 43-48

[36] Rout, P.K., Mandal, A., Singh, M.K., Roy, R., Sharma, N. and Haenlein, G.F.W. 2007. Jamunapari- A Dairy Goat Breed in India, Dairy Goat Journal, 82 (3):1-5.

[37] Markendeya, N.M. and Devanagare, A.A., 1997. XIV annual convention and National symposium on recent advances for enhancement of reproductive efficiency in farm animals. Nov. 14-16 held at Bidar (KS) pp. 94.

[38] Patil S. J., Mandakmale S. D., Walkunde T. R. and Kamble D. K. 2009. Heritability estimate of different traits in Osmanabadi goat under scarcity zone of Maharashtra, The Asian Journal of Animal Science (Dec., 2008 to May, 2009), Vol. 3 No. 2: (142-144)

Chapter 10

Reality of Mitogenome Investigation in Preservation of Native Domestic Sheep Breeds

András Gáspárdy

Abstract

This chapter deals with the study of extranuclear hereditary material and the possibilities of using it to maintain endangered animal breeds. The chapter characterizes mtDNA, presents its genes and their functions, while also emphasizing the hypervariable control region. It reports on the results of previous researches, referring to international publications. It sheds light on promising areas of mitogenomic research. It shows the maternal genetic background of local native varieties according to the results of the study of available country/geographical region. It deals with reasons for endangerment and the arguments for preservation of autochthonous breeds. In addition, it gives place to discuss some exciting professional concepts in rare breed preservation.

Keywords: mitogenome, sheep, genetic diversity, haplogroup, haplotype, breed preservation, maternal lineages, within-family selection

1. Introduction

For our domesticated animals, their domestication history has long preoccupied professionals. Substantially earlier evolution of domesticated species is also an area of research.

In the case of the horse, from the *Phenacodus* onwards, the last 60 million years have been exceptionally well known through a chain of transitional species, sometimes separations.

The phylogeny of family *Bovidae* (e.g., cattle, sheep, and goats) is less resolved. Here, the *Hypertragulidae* appeared as the first identifiable primitive ancestor around 50 million years ago (Mya) in Southeast Asia [1]. The complex, functional stomach developed about 40 million years ago. The molecular dating applied to cytochrome b gene, which is located in mitogenome showed that the separation of the sub-family *Caprinae* occurred 6.2 ± 0.4 million years ago, but there are proposed earlier radiation from about 14 Mya [2]. The *Myotragus*, which is basal to the *Ovis* clade within sub-family *Caprinea* stood out 5.35 million years ago.

In regard of evolutionary questions, besides nuclear microsatellites [3], SNPs [4], retrovirus integrations [5], and Y chromosomal mutations [6] mitochondrial DNA (mtDNA) represents a very informative genomic element. At the same time, this part of the hereditary material can also be efficiently used to better understand domestication. Nowadays, the study of mtDNA plays a role in the genetic characterization and differentiation of our animal species living with us. Looking to the future, we can believe that this will be essential for the conservation of genetic resources and preservation of endangered autochthonous animal breeds all over the world.

2. Mitochondrial hereditary material

2.1 Small circular genome

Majority of animal DNA as genetic information (about 3.3×10^9 base pairs) is stored in chromosomes within the cell nucleus. However, a minor part of DNA is located in chromosomes of mitochondria, outside the nucleus, in the cytosol. The circular mitochondrial genome is also built up of double-stranded DNA like nuclear genome, and consists of about 16,500 base-pairs. It is a semi-autonomous asexually reproducing genome in eukaryotic organisms [7]. Mitochondria are late descendants of free-living bacteria capable of metabolizing oxygen maintained by endosymbiosis in eukaryotic cells.

The mitochondrial DNA (mtDNA) which is not enveloped like nuclear DNA in chromosomes, is located in the mitochondrial matrix which can be found inside the inner mitochondrial membrane. The outer compartment of a mitochondrion is surrounded by the outer and inner membrane. The outer membrane contains porins through which smaller or larger proteins can enter the mitochondria. While, the inner mitochondrial membrane has all the elements of the electron transport system and the ATP synthase complex [8].

Considering that a cell has multiple mitochondria, and that a mitochondrion carries multiple copies of its own genome as opposed to the nuclear genome, the difference between the two remains significant.

2.2 High mutation rate

The mitochondrial genome or mitogenome mutates more frequently (approx. 100 times more often) than nuclear ones causing divergence in mtDNA at within-mitochondrion and between-mitochondrion level. Therefore, the mitogenome can be considered heterogeneous and heteroplasmic homoplasmic instead.

Little is known about the movement and segregation of mitochondrial DNA during mitotic growth or meiotic divisions. When a cell divides, mitochondria enter the progeny cells at random. If the DNA of the mitochondria of the dividing cell differs for several mitochondria, it is possible that the two daughter cells will receive the same genetic information, but it is also conceivable that they will not. Thus, it is hard to estimate the outcome of the transfer of genetic information, including defects. The random mutations that occurred further complicate that situation.

Because of the constantly frequent mutation rate of mitogenome, it has been widely used as a phylogenetic marker for both cladogram building and molecular dating. Brown et al. [9] first implemented the mitochondrial molecular clock in primates using fossil data. According to that work, scientists considered the 2% substitution rate per one million year as a reasonable reference in case of missing of relevant fossil data in vertebrates [10]. Since then, studies (e.g. [11]) have reported significant differences between species, it was found that because of the not fully clock-like evolution of the species the mtDNA mutation rate is of limited use in a comparison. The median (from 3 to 14.3 million years) of divergence dates between species are not related to body mass or generation interval [12]. However, Galtier et al. [13] found a proven correlation between mutation and longevity, which is closely related to the generation interval, and suggested a low (somatic) mutation rate could be responsible to achieve long life, in concordance with the mitochondrial theory of ageing. According to Song et al. [14] mitochondria may have a nucleotide imbalance that leads to higher mitochondrial DNA mutation rates. Their research suggests increased dGTP (deoxyguanosine triphosphate) level in free deoxynucleotide triphosphate (dNTP) pool which increases the rate of T to C substitutions.

2.3 Maternal inheritance

The offspring receive mostly mtDNA from the maternal ooplasm (sometimes this material from the sperm can also be included), but in the adult embryo only the maternal mtDNA remains functional. So, mitochondrial inheritance is considered as clonal or maternal, as one of the cases of uniparental inheritance. Paternal mtDNA, even if it enters the ovum, loses its function before (in crayfish [15]), during (in Ascidia [16]) or after (in mouse [17]) fertilization. That condition prevents an effective recombination. However, the paternal inheritance of mtDNA was displayed by Zhao et al. [18], what is to show the mtDNA patterns of progeny were identical to that of its male parent. Systematic surveys of within-species mtDNA data revealed departure from the clonality assumption in several species [19]. These prove that mitochondrial recombination is possible, and caution when in constructing and interpreting within-species mtDNA genealogies [20].

Thus, mtDNA testing can reveal the maternal background of individuals; which maternal lineage of mitochondria they belong to. But, of course, it can also be used to prove maternal kinship. Mitochondrial DNA sequencing has gained significance also in human rights cases [21].

It was observed that some characteristics (e.g. behavioral) do not follow Mendelian segregation. If a trait is such, it is either polygenic or extranuclear.

2.4 Perspectives of mtDNA research

Mitochondrial gene content is strongly conserved across animals, with very few duplications, no intron, and very short intergenic regions [22]. At the same time, mitogenome also contains a very limited presence of non-coding regions, approximately 3%, as opposed to nuclear DNA, where its proportion is 93%. These highly variable non-coding regions (e.g. the control region) are typically flanked by highly conserved ones (e.g. ribosomal DNA). The elevated mutation rate of highly variable regions creates the condition for monitoring the population history over relatively short time frames.

One of the perspectives of mitogenome research is therefore the discovery of mitochondrial genetic disorders (next to accelerated ageing, neurodegenerative disease, cancer, diabetes), and the study of their mechanism of action (e.g. [23]).

Another promising area is the mapping of evolutionary branches and the determination of the more precise taxonomic location and movement of different species (including humans [24]). Mutations will be passed over into all maternal progenies homoplasmic making individuals of a maternal lineages the same in mitogenome, especially, when they share entirely homoplasmic mitochondrial pool.

For the third time, it is worth mentioning the unfolding of the microevolutionary web of our domesticated animals and the knowledge of the origin (at the same time geographical) of the breeds that have developed today (e.g. [25]).

3. Non-coding and coding region on mtDNA

3.1 Control region

On the circle of the mtDNA, there is a specialized sequence, called control region (CR), and also called D (displacement)-loop because of its peculiar protrusion. The CR is made up of a triplex DNA structure at the site of origin of the heavy strand. This region is critical for the initiation of transcription and translation [7].

Zardoya et al. [26] determined the nucleotide sequence of the sheep mitochondrial DNA CR and its flanking tRNA genes. They found that several conserved motifs characteristic mammals have been identified along the 1189-bp sequence of the sheep control region: ten termination-associated sequences (TASs) and one conserved sequence block (CSB-1). CSB-2 and CSB-3, which are frequently determined in most species, are not present in the sheep CR, which shows instead a short direct repeat at their usual localization.

The CR contains hypervariable sites (mutational hotspots). This unstable segment gives the basis for dating estimation in many mammalian species. However, according to some authors (e.g. [27]) the high occurrence of recurrent mutations may bias dating estimates. Also within the sheep species, according to Pedrosa et al. [28], the time of separation of haplogroups is significantly earlier if hypervariable CR is taken into account. Researches on human mtDNA raise the concerns that focusing exclusively on CR can be inadequate [29].

3.2 Genes

A total of 37 genes of mtDNA are coding 2 ribosomal RNAs, 22 transfer RNAs, and 13 mitochondrial proteins as well in mammals. The latter, without exception, direct cells to produce protein subunits for enzyme complexes of the oxidative phosphorylation system. Mitogenome has a very similar conservative nucleotide sequence in all organisms.

A meta-analysis study [30] of over 1500 animal species revealed that the average within-species level of mtDNA diversity *per se* is remarkably similar across animal phyla. Reason for that is a recurrent selective sweeps which would affect mtDNA evolution in species causing frequent drops in diversity at the whole genome level. Based on that hypothesis and being high conservative gene dense, according to the report of Galtier et al. [31] mtDNA is by and large not the satisfactory marker of molecular diversity and representation of population history. After that, it may come as a surprise the most popular marker of molecular diversity in animals, a mitochondrial fragment, COX1, was recently elected as the standardized tool for molecular taxonomy and identification [32].

Some genes even overlap. In the mitochondrial genome, some triplet codons may be the final stage of one gene but also (in a functional overlapping) the initial stage of the next gene. Another feature is that the mtDNA is transcribed from several structural genes to the messenger RNA at the same time. Thus, large mitochondrial mRNAs contain instructions for the synthesis of different proteins. The inheritance of the extranuclear genes are independent of nuclear genes, but, they interact with each other in function. The mitochondrial genome is not able to produce all the proteins required for phosphorylation on its own, so mitochondria are highly demanding of gene products produced by the nuclear genome.

Investigation of Rocky Mountain bighorn sheep (*O. canadenesis,* [33]) revealed 16,466 bps, with about 40% GC content. Further on, it confirmed also the bighorn sheep mitochondrial genome has 22 tRNA genes, 2 rRNA genes (12S and 16S), and

13 respiratory genes (ATP6, ATP8, CYTB, COX1, COX2, COX3, ND1, ND2, ND3, ND4, ND4L, ND5, and ND6). Comparison of the genome of bighorn sheep with the genome sequence of other sheep showed 99.6% identity, indicating the separation of the bighorn 3 million years ago from the sheep living at that time.

Higher mutation rate of mtDNA will cause increased rate of genetic diseases of mitochondrial origin [34]. Therefore, such diseases are also all of maternal origin. Males can be carriers of a given genetic defect and can be affected by its manifestation, however, they are not responsible for transmitting that disease into their progenies.

Since mitochondria acts as the powerhouses of cells, tissues that have high energy demands (brain, retina, skeletal muscle, and cardiac muscle) are particularly vulnerable to the harmful consequences of mutation. As an inherent part of energy production, mitochondria create reactive oxygen species (ROS) as well which is seen to cause further mitochondrial mutations. Elevated levels of ROS negatively affect cellular metabolism, thereby accelerating the cell ageing process and increasing the likelihood of cell death. Symptoms of mitochondrial diseases in humans can usually include: poor growth; muscle- weakness, pain, low tone, exercise intolerance, and movement disorders; vision and/or hearing problems; learning disabilities and mental retardation; autism and autism-like features; heart-, liver- or kidney diseases; gastrointestinal disorders, swallowing difficulties, diarrhea or constipation, unexplained vomiting, and cramping, reflux; diabetes; increased risk of infection; neurological problems, seizures, migraines; strokes; thyroid problems; respiratory (breathing) problems; lactic acidosis; dementia [35]. Fortunately, next-generation sequencing techniques have substantially improved genetic diagnosis.

Individuals resulting from cloning procedures (*nuclear transplantation* or *somatic cell fusion,* [36]) are heteroplasmic. The initial heteroplasmic stage of chimeric offspring cell turns usually into homoplasmy with prevailing mitochondria of the host oocyte [37]. It is likely that heteroplasmy of mitochondrial genomes will be terminated by selective elimination of donor or recipient mitochondria by chemical or other means. Establishing biotechnical approaches allows women with mitochondrial diseases to have reproductive options. Recent advances in these including *in vitro* fertilization techniques with mitochondrial donation, will serve as a solution in the future [38].

Mutation in mtDNA-coded ribosomal RNA, called RNR1 indicates the presence of an environmental effect. That mutation causes deafness in children, but the clinical symptoms of the deafness are related to the administration of certain antibiotic type [39].

The percentual manifestation (penetrance) of a mitochondrial disease in males and females differing from the expectations points to the likely involvement of other (nuclear) genes and environmental factors. *Leber hereditary optic neuropathy* (LHON) is an example for mitochondria-associated disorders which is manifested in loss of vision [40].

First reported Pal et al. [41] the association of cytochrome b (Cyt b, CYTB) gene with disease traits in sheep. Mutations of Cyt b gene (non-synonymous substitutions: F33L and D171N) interferes with the site of heme binding domain and calcium binding essential for electron transport chain causing anemia, malfunctioning of most of the vital organs. This discovery raises the possibility that the sheep may come into play as a model of man.

Results of Reicher et al. [42] revealed ovine mitogenome genetic variation in protein- and tRNA coding genes (26 and 8 mutant sites, respectively) and emphasize that sequence variation is associated with ewe prolificacy.

Yüncü et al. [43] tested the restriction fragment length polymorphism (RFLP) method (applied to CR) and the single strand conformational polymorphism

(SSCP) method (applied to NADH dehydrogenase subunit 2 and 4) for reliability in haplogroup classification. Among these the SSCP analysis of NADH dehydrogenase subunit 2 exhibited the highest discrimination power among these. Starting with that, authors advice a stepwise screening, when whole sequencing is not easy available.

4. Mitochondrial investigation in sheep

4.1 Whole mitogenome

Over the half (58%) of the mitogenome was completely (included CR and CYTB, ND2, ND3, ND4L, COX3, and 12 tRNA genes, and the origin of L strand replication), and partially (12S and 16S rRNA and an additional six protein coding and six tRNA genes) analyzed by Hiendleder [44]. In that research, the CRs and the coding regions shown 4.34% and 0.44% divergence in the comparison of sheep haplogroup A and B, respectively.

Table 1 represents the complete mtDNA molecular sequencing of the domestic sheep (*Ovis aries*) achieved by Hiendleder et al. [45]. The length of the complete ovine mtDNA presented is 16,616 nucleotides (nt), which length is variable, due to heteroplasmy caused by the occurrence of different numbers of a 75-nt-long tandem repeat in the CR. The majority of domestic sheep contained four copies of that 75 bp repeat unit in work of Meadows et al. [46] resulting in a mitogenome of 16,616 or 16,620 bps.

Using 14 restriction enzymes Hiendleder et al. [47] evaluated haplotypes of restriction fragment length polymorphisms (RFLP) based on pairwise nucleotide sequence divergence between haplotypes, and proved that the domestic sheep come exclusively from *Ovis orientalis*.

Sanna et al. [48] reported the first complete mitogenome of the (*Ovis gmelini ophion*), and compared to the known five mitochondrial haplogroups. They suggest that the Cyprus Mouflon, a feral variant of domestic sheep diverged from urial (*O. vignei*) and argali (*O. ammon*) about 0.89 and 1.11 million years ago.

Lv et al. [49] performed a meta-analysis using complete and partial mitogenomic sequences. They suggest sheep individuals migrating east from Fertile Crescent on the Mongolian Plateau region may have formed another centre for the further spread of domestic sheep 3–5 thousand year B.C., at the same time extending the haplogroup C.

The complete mitochondrial genome provides complex information for knowledge of phylogeography and population genetics in sheep. The mitochondrial genomes of several breeds of domestic sheep (*Ovis aries*) have now been mapped by Davenport et al. [33].

4.2 Examination of fossils

The sheep (*Ovies aries*), together with the dog are the earliest domesticated animal species, and had remarkable role in the life of ancient societies. Sheep were domesticated around 7–9 thousand years B.C. in the area of the Fertile Crescent. Demirci et al. [50] found a time-dependent change in the incidence and proportions of haplotypes in Anatolia. Ancient samples showed the presence of haplogroup E (3%) in the Bronze Age and the presence of haplogroup C (6%) in the Hellenistic age, while haplogroups A and B were continuously present (with nearly 50–50 percent).

Feature	From	To	Size	Start codon	Stop codon[b]	3′ spacer
tRNA-Phe	1	68	68			
12S rRNA	69	1,026	958			
tRNA-Val	1,027	1,093	67			
16S rRNA	1,094	2,667	1,574			
tRNA-Leu (UUR)	2,668	2,742	75			AA
NADH1	2,745	3,700	956	ATG	TAa	
tRNA-Ile	3,701	3,769	69			
tRNA-Gln (L)	3,767	3,838	72			AT
tRNA-Met	3,841	3,909	69			
NADH2	3,910	4,951	1,042	ATA	Taa	
tRNA-Trp	4,952	5,018	67			A
tRNA-Ala (L)	5,020	5,088	69			A
tRNA-Asn (L)	5,090	5,162	73			
Origin of L-strand repl.	5,163	5,194	32			
tRNA-Cys (L)	5,195	5,262	68			
tRNA-Tyr (L)	5,263	5,330	68			C
COI	5,332	6,876	1,545	ATG	TAA	
tRNA-Ser (UCN) (L)	6,874	6,944	71			TAAAC
tRNA-Asp	6,950	7,017	68			T
COII	7,019	7,702	684	ATG	TAA	AAT
tRNA-Lys	7,706	7,773	68			T
ATPase8	7,775	7,975	201	ATG	TAA	
ATPase6	7,936	8,615	680	ATG	TAa	
COIII	8,616	9,399	784	ATG	Taa	
tRNA-Gly	9,400	9,468	69			
NADH3	9,469	9,815	347	ATA	TAa	
tRNA-Arg	9,816	9,884	69			
NADH4L	9,885	10,181	297	ATG	TAA	
NADH4	10,175	11,552	1,378	ATG	Taa	
tRNA-His	11,553	11,621	69			
tRNA-Ser (AGY)	11,622	11,681	60			A
tRNA-Leu (CUN)	11,683	11,753	71			
NADH5	11,754	13,574	1,821	ATA	TAA	
NADH6 (L)	13,558	14,085	528	ATG	TAA	
tRNA-Glu (L)	14,086	14,154	69			ACTA
Cyt b	14,159	15,298	1,140	ATG	AGA	CAA
tRNA-Thr	15,302	15,371	70			

Feature	From	To	Size	Start codon	Stop codon[b]	3′ spacer
tRNA-Pro (L)	15,371	15,436	66			
Control region	15,437	16,616	1,180			

[a]Nucleotide number 1 is the 5′ end of the tRNA-Phe-specifying gene. Anticodons for the two tRNA-Leu and the two tRNA-Ser are given in parentheses. (L) denotes light-strand sense. Positions include the 58 and 38 nt of each feature. ATPase6 and ATPase8, genes encoding subunits 6 and 8 of ATPase; COI-III, genes encoding subunits I–III of cytochrome c oxidase; Cyt b, gene encoding cytochrome b; NADH1–6, genes encoding subunits 1–6 of nicotinamide adenine dinucleotide dehydrogenase.
[b]Incomplete stop signals are denoted by lowercase letters.

Table 1.
Features of the Ovis aries *mitochondrial genome[a] [45].*

Dymova et al. [51] carried out archeological mitochondrial DNA D-loop fragment analysis based on about 4,000–1,000 years old sheep bone remains in Altai. They found all the previously determined haplogroups (A, B, C, D and E lineages). That richness of diversity led them to conclude that the Altai region had been a migratory area for many sheep and peoples in the past.

Study of Horsburgh and Rhines [52] evaluating sheep finds excavated in a South African Neolithic Age cave shown their assignation to haplogroup B.

In comparative study of samples dated primarily by archeological context and ranged from Late Bronze Age, through Iron Age to post-medieval period Rannamäe et al. [53] identified four novel ancient haplotypes specific to Estonia (H3, H4, H5 and H9), and haplogroups A and B in a ratio of one to two.

4.3 Displacement-loop

Mitochondrial displacement-loop (D-loop), called also as control region (CD) is a frequently investigated sequence in researches, also in sheep.

Divergence times estimated for types B and A (which was about 1.5–0.45 Mya) can be overestimated when it is based solely on hypervariable sequence of CR [54]. In regard of calibrating a molecular clock, the consideration of the codifying cytochrome b gene seems to be more accurate. To eliminate the distortive effect (known heteroplasmic behavior) of CR the repeat unit located within the CR region was removed before phylogenetic inference made by Meadows et al. [46].

The purpose of sequencing is, in addition to the genetic characterization of a given breed (population), to compare its genetic material to genetic material of other already sequenced breeds (GenBank sequences). Thus, we try to get an answer to the origin of the given breed and its genetic relatives. This is important when studied in the former natural range of the sheep species (primarily Asia, then Europe and Africa), but it is also important on the continents (America and Australia) where sheep individuals later entered with migrants.

For example, based on the CR, Annus et al. [55] confirmed the common origin of the Hungarian Tsigai with European sheep after finding that they are belonging to the haplogroup B (with the exception of 6% to the haplogroup A). An example of the latter case is the evaluation of the Mexican Creole sheep carried out by Alonso et al. [56], which revealed a narrow Iberian maternal origin.

Lancioni et al. [57] discovered relationships of the three Italian Merino-derived sheep breeds, and obtained that these are representatives of the predominant haplogroup B (99%). Since almost all the animals are carrying an own individual haplotype these are characterized with a diverse genetic background. On the other hand, this processing gives an example of how, despite upgrading (with Merino), the maternal background (Appeninica) is clearly discernible.

4.4 Cytochrome b gene

It was observed cytochrome b (Cyt b) gene is also quite mutable than other mtDNA coding regions. For this reason, phylogenetic studies have used that marker too to investigate the genetic relationships among breeds. By use of cytochrome b gene and displacement-loop together or individually, like before, five haplogroups of sheep can be distinguished from each other [46].

In the phylogenetic sequence analysis based on cytochrome b gene Bunch et al. [58] revealed an about 3.12 million yearlong evolution of true sheep (*Ovis*). On the course of its evolutionary history there have been three major genetic groups developed. Foremost, the Argaliforms (*Ovis ammon*, 2n = 56) and Moufloniforms (*Ovis musimon* or *O. orientalis*, 2n = 54, and Urial/*Ovis vignei* 2n = 58) diverged from the initial ancestral stock 2.3 million years ago and spread on Eurasian continent. The domestic sheep (*Ovis aries*, 2n = 54) descend solely from *O. orientalis*. Second, the snow sheep group (*Ovis nivicola*, 2n = 52) as a variant of Pachyceriforms took their own shape in Eastern Asia from about 1.96 million years ago, then the other variants of Pachyceriforms (*O. canadiensis*, 2n = 54; *O. dalli*, 2n = 54) separated from them at about 1.41 million years ago, and evolved further in North America. Argaliforms are represented by only one species, *O. ammon*. The ancestral karyotype had 2n = 60 chromosomes. During the evolutionary development of variants of the species, also acrocentric fusions of chromosomes are observed.

According to the initial studies (e.g. [59, 60]) haplogroup A predominates in Asian sheep, while haplogroup B predominates in European sheep. Nevertheless, haplogroup C seems to have a wide geographical distribution [61]. Pedrosa et al. [28] and Chen et al. [61] suggested the divergence time of haplogroup C from haplogroup A and B to be approximately 0.42–0.76 million year ago and approximately 0.45–0.75 Mya from the analysis of control region and Cyt b gene sequences, respectively. However, a study of Meadows et al. [46] using 12 protein-coding genes of mtDNA puts the separation between the haplogroups less early (0.59–1.17 Mya between A and B, and 0.26–0.09 Mya between C and E.

4.5 ND5 gene

Tserenbataa et al. [62] collected 71 argali sheep (*O. ammon*) samples from three main geographical regions of Mongolia, and additionally from Kazakhstan and Kyrgyzstan. Based on the sequenced 556 bp of the mitochondrial ND5 gene. They differentiated 17 haplotypes which were differed far from each other by transitional substitutions in most of the cases. Nucleotide diversity was low within the three regions from Mongolia (π = 0.0029) compared to Kazakhstan and Kyrgyzstan. While the variance occurred within populations was as much as 85.76%. Finally, the use of mitochondrial ND5 gene provided an opportunity to detect divergence between the Altai and Gobi, and Altai and Khangai populations at a significant level.

4.6 Haplogroup dispersion

For phylogenetic relationship between mitochondrial haplogroups of domestic sheep Meadows et al. [46] observed the greatest distance between B and C (nucleotide difference, D = 163.5), closely followed by B–E and C–D (D = 162.0, identically). The lowest number of nucleotide differences was 93.0 and 58.5 between A and B, and C and E, respectively. **Table 2** reveals the genetic distance between domestic sheep haplogroups in addition to Urial, Argali to us.

	HA	HB	HC	HD	HE	Mouflon	Urial	Argali
HA	—	0.57	0.93	0.75	0.90	0.58	2.19	2.53
HB	93	—	1.01	0.81	1.00	0.07	2.31	2.59
HC	150.5	163.5	—	1.00	0.36	1.00	2.33	2.65
HD	122.5	131.5	162	—	0.98	0.81	2.27	2.61
HE	147	162	58.5	159.5	—	0.98	2.30	2.63
Mouflon	94	11	162.5	131.5	160	—	2.31	2.60
Urial	357.7	377	380.3	370.5	375.7	377.3	—	2.32
Argali	413	423	433	425.5	429	424	379	—

The average number of nucleotide differences (D) is given below the diagonal and nucleotide substitutions per site (K, given as a percentage) are given above the diagonal for the full mitochondrial sequence after removal of both indels and the repetitive component of the control region.

Table 2.
Genetic diversity observed between domestic and wild sheep mitogenomes [46].

The distinct haplogroup diversity of sheep mtDNA is comparable with what is observed in goats and cattle, although the divergence of sheep haplogroups is less pronounced than the *taurine–zebu* divergence [63]. Also, sheep haplogroups show little association with the geographical origin, in contrast to bovine haplotypes. A given sheep haplogroup can assume several regions of origin, or the coexistence of several different maternal lineages in a domestication centre can be suspected.

4.6.1 Haplogroups A, B, C, D, and E in Asia

In a today phylogenetic study of Ganbold et al. [64] revealed three haplogroups (A, B, and C) in Mongolian native sheep. The Mongolian Plateau, as mentioned above played a determining role in the arrival of sheep in eastern Asia. And, as a consequence of it, they observed a small genetic differentiation between breeds from Mongolia and China.

The Moghani sheep of Iranian plateau was identified in haplogroup A [65].

Haplogroups D and E are the least frequent and have only been identified in samples from Turkey and the Caucasus [66, 67]. Slowly, haplogroup E was detected also in Iran [68]. In a paper of Liu et al. [69], the proportion of haplotypes of lineage D was 0.157% in Tibetan sheep, further demonstrating that lineage D is the rarest of the mtDNA lineages.

4.6.2 Haplogroups A, B, C, and D in Europe

Haplogroup B is scattered in numerous countries of Europe (e.g. [57, 70]. The haplogroup B seems to be expanded around 6,400 years ago and reached Western Europe before the haplogroup A [48].

Within Europe haplogroup C has been found, so far, only on the Iberian Peninsula (in Portugal [71] and in Spain [72]) and in the southern countries of the Balkan Peninsula (in Albania and Greece [73]. Haplogroup D is present in Italy in breeds Bergamasca and Laticauda [74].

4.6.3 Haplogroups B and C in Africa

Haplogroup B is also dominant in Africa as it was revealed in some today publications: in Benin in breed Djalonke [75], in Mauritania in breeds Peul and

Touareg [76], in Somalia and Kenya in Red Massai and Blackhead, respectively [77], and in Egypt in Barki and Ossimi breeds [78].

Ghernouti et al. [79] found thin-tailed Arabic breeds in Algeria belong to haplogroup B (87%) and C (13%). Authors believe the presence of haplogroup C in breed Ouled Djellal is a proof of the Middle Eastern origin of that breed. The haplotype C, also identified in Egyptian breeds is in agreement with the assumption of early spread in sheep [78]. Studying the Siroua sheep in Morocco two haplogroups (haplogroups B and C) were also identified by Kandoussi et al. [80] with a predominance of haplogroup B.

4.6.4 Haplogroup B in America

Analyzing mtDNA control region of 40 unrelated domestic sheep in Mexico, Campos et al. [81] revealed 31 different haplotypes with 74 polymorphic sites. The phylogenetic analysis identified all Mexican sheep as belonging to haplogroup B. Sheep from other American regions (Brasilia and Cuba) in that analysis made sure the high frequency of an ancestral haplotype (h15) in Ibero-American countries as well.

Revelo et al. [82] identified the Creole sheep as exclusive haplogroup B, and justified that the two seriously different types of Creole sheep (wooly and hairy) of Colombia descent from an Iberian and an African ancestor.

4.6.5 Haplogroup A and B in Australia

Hiendleder et al. [54] suggest that the high frequency of haplogroup A (beside B) in New Zealand resulted from early imports of fat-tailed Indian sheep (beside mouflon specimens) into Australia in accordance with the sheep stream hypothesis.

4.7 Haplotype diversity

With the third of the aforementioned goals of mitochondrial research, we can relate that subchapter the most. In characterizing the haplotypes of the breeds and comparing them with each other, CR may fulfill the expectations placed on it. Going further, CR can also gain ground in research into the genetic structure of the breed (sub-breed, variety, family).

D-loop of two Tibetan sheep breeds was analyzed by Wang et al. [83]. The length of the D-loop sequences varied considerably (between 1,107 bp and 1,259 bp) according to the copy numbers of a 75 bp tandem repeat located from 640 bp to 1,140 bp. That variability was most characteristic for haplogroup C, less so for A and B. Fu's test showed that the populations had not been expanded historically ($0.10 > p > 0.05$). Results are useful for the conservation and utilization of Chinese sheep genetic resources.

In randomly collected samples of four Nigerian breeds 96 haplotypes were observed with a high mean haplotype diversity of 0.899 ± 0.148. The high percent of variation (99.77) found by Agaviezor et al. [84] within populations indicates common origin of these breeds. However, the evolutionary divergence of the breeds (Yankasa, West African Dwarf, Balami, and Uda) based on mitochondrial DNA D-loop sequence may be coincident with their geographical distribution in Nigeria.

Arora et al. [85] compared 19 Iranian sheep breeds in their extended CR study. They confirmed the majority of the breeds belong to haplogroup A solely, and five breeds appear with of haplogroup B as well. Both haplogroups show unimodal patterns of mismatch distribution curves, and the significant minus F_S statistics values indicate population expansion in Indian sheep population.

The control region of mtDNA showed polymorphisms at 32 sites in the Hungarian Cikta evaluated by Kovács et al. [86]. However, herds shared 24 polymorphic sites, so the maternal background of the Cikta appears to be genetically uniform. The total number of haplotypes were 13, furthermore, most of the samples belonged to the haplogroup B of sheep. The average number of pairwise differences (k) and the average nucleotide diversity (π) were 6.863 and 5.95 × 10–3, respectively. The values of the Cikta population were not significant ($p < 0.10$) neither by the Tajima D-test (0.107) nor by Fu's Fs statistics (2.533), meaning that the greatly reduced population size of the breed known from the breed history did not cause genetic drift, it is in genetic equilibrium regarding its ancient families. The Cikta shown some degree of genetic narrowing based on Cyt b gene [87]. However, the average number of pairwise nucleotide differences is relatively high, which indicates different genetic characteristics of the families occurring in the farms.

Kusza et al. [88] investigated the two variants of Wallachian sheep by country sequencing 599 bps of the D-loop region. They isolated altogether, 42 haplotypes, of which 23 were common in both eco-types. Since they estimated a very low level of genetic differentiation between the Gyimesi Racka (in Hungary) and Turcana (in Rumania) breeds, therefore these are really two variants of one transboundary breed.

According to the haplotype diversity results Kirikci et al. [89] stated the Karayaka breed from Northern Anatolia cannot be categorized as a genetically homogeneous population. That breed not only has not suffered from a genetic bottle neck effect, but even has four different haplogroups (A, B, C, and E).

5. Animal genetic resources

Term animal genetic resources is defined shortly as a potential of domestic animals that is used for production of food and fiber [90]. Animal genetic resource management is necessary on a global scale and its improvement requires careful thinking. While the contribution of livestock sector to 43 percent of world's agricultural Gross Domestic Product, which in some developing countries accounts for about 30 percent of national agricultural GDP. Actual economic modeling estimates that for those rural populations, poverty is limiting, economic growth suggested to be critically low. The fate of poor people and their livestock is interlinked, so none should be overlooked in future food security efforts [91]. The World Bank forecasts that contribution of livestock sector to agricultural GDP in undeveloped regions will be necessary by about 80 percent between 2000 and 2030.

Sheep are very important in the socio-economic lives of the people. However, their potential is not realized under poor conditions because of low productivity resulting from high mortality and weak performance among others. That fact calls the attention to the environment of production. But a given loss of animal genetic resources concerns the loss of genetic diversity within improved, cosmopolitan breeds and not only the extinction of traditional breeds [92]. The first reason for loss, the uniformity with increasing homozygosity as consequence of enormous development of highly improved breeds has led to growing concerns about the erosion of genetic resources [93]. Lenstra et al. [94] give a detailed review about molecular tools and analytical approaches for the characterization of farm animal genetic diversity.

Integration of local breeds threatened by extinction but carrying appropriate alleles into the further refinement of breeds for mass production result in effective management of erosion of farm animal genetic resources (FAnGR, [95]). Therefore, the maintenance of old, local breeds is in any case justified by this requirement.

However, autochthonous breeds are the national treasure of a given country and, as such, their maintenance is the duty of that state. In addition, in my personal opinion, access to the benefits of this treasure needs to be regulated for one's own country, but especially for other countries.

6. Maintenance of endangered breeds

6.1 Reasons for endangerment/extinction

FAO [96] listed the broad categories of threats in three major groups: trends in livestock sector, disasters and emergencies, and animal disease epidemics and lack of control measures.

Considering the first one, the global reliance on a very limited number of international, specialized (single purpose) breeds suited to the needs of high input high output industrial agriculture can be mentioned. This expansion was accompanied by the grading-up of local breeds, by changes feeding-, housing, and reproduction technologies.

Under the second group of treats the lack of development interventions, appreciation, sustained breeding programmes, and loss of labour force (migration to urban areas in search of employment), traditional knowledge associated with livestock herding, further on changes in land use (destruction of native habitats), inappropriate management of climate change and natural disasters (floods, drought, famine). In many places, to this is added the local conflict (socio-political), and a range of political instability (civil strife, war).

The third means: inadequate control of disease epidemics, lack of disease control, preventive treatments, genetic control of inheritable defects, as well as lack of identification, transport, traceability, food chain controlled.

6.2 Arguments for preservation

The experts collect the following economic, scientific, human cultural, socio-economic, and environmental rationales for preservation beside the needs for development and sustainability mentioned formerly.

Genetic variation is the raw material for animal improvement. Prudent economy demands conservation. Lost flexibility will limit the ability of future generations to respond to changed markets and opportunities. Old breeds are of unique physiological or other traits. They can show specific adaptation ability, resistance to diseases. Biotechnology will need to reveal unique sequences of DNA. Based on microsatellites, Agaviezor et al. [97] concluded that these associated with unique ancestral alleles of certain functional genes may reflect a better adaptability in more agro-ecological zones. Firestone et al. [98] shown through simulations that with samples of at least 30–40 individuals found the correct ratio of private alleles in most cases can be. Due to the low frequencies of the private alleles in a study, the results should be interpreted cautiously and viewed more as a trend.

Animal husbandry is a special characteristic of human culture. It is comparable to other great reminders of man's past. Rare breeds are results of human creation (worth preserving and conserving as any other work of art, like monuments or buildings). They are kept for demonstration and showing of historical development of animal husbandry, and are of great advantage and value for physiological and genetic comparative studies. Some domestic animal breeds are historically closely linked to different farming cultures, environment and regions, traditional and regional. Livestock are part of life style in all the countries.

Impact of changed animal genetic diversity affects their whole life style. In integrated crossing programs with high yielding breeds it may be economical (crossed progenies for market and pure bred progenies for herd replacement!). Under unfavorable (harsh) conditions it can be bred with minimum input.

Our old, native breeds are multipurpose animals. However, we have become accustomed to using several benefits. Therefore, in addition to primary production (which is already known to be low), it is necessary to emphasize the use of animals in several ways during their lives (e.g. grazing/landscape care, traction, tourist attraction) and efficient processing of slaughtered animal parts through traditional handicrafts (e.g. fur and horned skull/trophy). What is very important, and what is realized exclusively in *in situ* gene conservation, is the regular use of animals. In the original, traditional environment, the genetic ability of the breed can be manifested. Raw materials of animal origin obtained directly or indirectly must be processed regularly and the product sold (although this is often seasonal).

Animal, plant, forest, fish, wildlife genetic resources are equal major components of global biological diversity and must be viewed together. The environmental implications of livestock are huge. Billions of livestock in human care moving over large areas of land where they affect the soil, water and vegetation, and interacting with other life. Therefore, the environment must be considered as a whole.

6.3 Breed history for preservation

In my opinion, maintaining a variety is more than finding and currently using a beneficial gene. The rationale for the preservation of old breeds is given by the history and former significance of the breed. Regarding the breed history, I would like to call the attention to the *in libro* concept of conservation worded by us formerly [99]. A longer version of the technical term is *in libro conservation in causa emoriendi*, namely the conscious preservation of an already extinct domestic animal breed or a lost characteristic entered. The meaning of the entered (in a book; booked) conservation in broader sense is the preservation of all the remaining knowledge, keepsake, documents and material heritage of a still living rare breed. As a reason for the *in libro* conservation the same arguments can be presented as for the *in vivo* and *in vitro* approaches of conservation. The keeping in life of an extinct breed does not crop up but the "keeping alive" its one-time presence in the common knowledge is an important role. The cradle of the breed should not be forgotten.

In contrast to mtDNA, Peter et al. [100] among others, reports the disadvantage of microsatellites in search for breed history, especially in open flocks. Although microsatellites on nuclear chromosomes describe the current genetic nature and genetic relatedness of the herds, the genetic character depends heavily on the genetic makeup of the rams being actually used. The breeds can no longer be clearly assigned to one of the breed groups, but can be found in the mixed populations.

6.4 Herd booking and pedigree analysis

The official, mandatory individual animal numbering is essential for everyday breeding work. The parentage control is also essential for reliable pedigree registration. Of course, among the pedigree data we also store the regularly recorded (in many cases seemingly redundant) production data. Most multi-purpose indigenous breeds are utilized solely through their meat production today, however, in order to preserve the valuable traits, not only the fattening and slaughter values, but also, for example, the wool production traits must be recorded and selected for.

Pedigree data are fundamental to the assessment of the demographic structure and risk status of livestock breeds. Careful investigation of herd book data will serve knowledge on pedigree completeness [101], effective population size [102], factual number and effective number of founders and ancestors, respectively (these two later explain the complete genetic diversity of a population [103]), and founder genome equivalents [104].

The length of the generation interval (defined as the average age of parents at the birth of their progeny kept for reproduction) is one of the keys to demanding breed maintenance. The four pathways (sire-son, sire-daughter, dam-daughter, and dam-son) of generation interval should be as long and similar as possible. The higher this is, the lower the possibility of an annual gain even in the case of a preserving section that may not be perfectly implemented.

Based on pedigree data the average relatedness coefficient of each individual is evaluated together with the inbreeding coefficient. Inbreeding coefficient (autozygosity) is influenced by the length and completeness of pedigree, the longer and more complete the available pedigree is the more reliable the estimated inbreeding coefficient.

6.5 Preserving selection

When maintaining endangered animal breeds, care must be taken not to preserve the name of the breed in the new, upgraded population, or to ensure that the critical herd size is maintained for generations. The purpose of breed maintenance is to preserve the original level of phenotypic (and underlying genetic) characters of the breed. The constancy and possible, unexpected change of the production level can be verified in historical comparative study (e.g. [105–107]. That is, in order to maintain within-breed diversity, it is necessary to leave individuals from all phenotype groups (production levels, ideally based on their breeding value) for further breeding. This is the aim of preserving selection. The principle of diversity-preserving selection is very far from conventional truncation selection. The selection limit is not determined by the performing-ability, but by the measure of the remount rate proportional to the Gaussian curve.

6.6 Founder sampling

The genetic composition of a given sheep breed is also likely to be mixed in terms of mtDNA haplotypes. As a breed has a remarkable higher number of different mutant sites overall it can be assumed that more founder animals were present in this population have to. The long-known history of this breed (sheep herd of the area) and the current genetic mapping can provide certainty, confirming the colorful background, possibly the contribution of several breeds or sheep subspecies to the gene pool.

From the current point of view, the essence of the pedigree study is the identification of female founder individuals and the families derived from them. In order to capture full diversity the DNA samples should be taken from the living descendants of the eldest families based on herd booking. In the study of mtDNA of two Hungarian native sheep breeds (Tsigai and Cikta), sample collection and analysis [55, 87] was preceded by the processing of pedigree data and the identification of ancient families [108, 109].

6.7 Maternal lineages

As previously described, reliable characterization of the breed for mtDNA should be performed on samples taken from representatives of maternal lineages. The genetic background and the current diversity thereby reliably discovering in any breeds. The number of haplotypes representing individuals in the sampled families or providing the sample is likely to be less than the number of samples. Unfortunately, it can also be predicted that the number of surviving founder families will decrease from generation to generation. Thus, it is recommended to leave breeding offspring from all families, from all haplotypes, to replenish the female herd. Since the number of individuals is higher than the number of haplotypes a rigid selection focussing on haplotype maintenance can lead to loss of many other genetic information (like in selection against scrapy genotypes).

Of course, it is a big question whether, in the case of outlier haplotypes, we choose to save or discard it in the endangered population of small size. Each cell contains many copies of mtDNA which, except in very rare cases of heteroplasmy, are identical and shared by all members of the maternal lineage. Another problem may then be the treatment of the heteroplasmy.

To my mind, the searching for ancient families is very crucial aspect of breed conservation. I consider it important to inform farmers about the family and haplotype of their sheep. In this respect, close long-term cooperation is needed between breeders and breeders' associations, and even with the breeding authority in obtaining state subsidies. In other words, the individual sheep genetic knowledge gained from the researches should be communicated to the animal owners.

6.8 Within-family selection

If differentiated families with their specific haplotype are already available, it is reasonable to select offspring for further breeding within these. Using a within-family selection, potential offspring are identified by sophisticated breeding software, while the breeder remains free to choose which one to actually stay for breeding.

At this point, I would like to draw attention to the differences between the professional work of the association and the ideas of the private breeder, and the necessary cooperation, or the antagonism in many places that the breeder is the owner of the animal but the breed must be preserved by the state.

7. Conclusions

During the phylogenetic investigation of sheep species, and characterization for improvement and conservation of sheep breeds, mtDNA diversity plays an important role. This is indeed noticeable as it is being applied more and more widely. However, opinions differ on the degree of success of the study in terms of its purpose. Here, too, it is true that any doubt encourages further appropriate work.

For comparison between domestic and wild sheep the less mutagen gene (Cyt b) sequences are advisable. But, for the exploration of haplogroup relationships among domestic sheep processing of a Cyt b gene dataset combined with CR is recommended. Then, for a precise differentiation of haplotypes a hypervariable sequence set of CR seems to be the most reliable. High levels of mutations observed in the control region (emphatically hypervariable sequence) may skew dating estimates for many mammalian species. Before drawing phylogenetic conclusions the removal or reduction of repetitive sequence elements located within the CR is to advice because of its known heteroplasmic behavior.

Coding regions are conservative, for function with sameness making them less useful for isolating species, breeds, and individuals. Investigation of these is important in screening for deleterious alleles.

At the same time, there is a need to consider more segments of the mitogenome which are used to outline the phylogeography of the species (since a single segregating locus does not adequately represent the origin of the entire genome), and the more of these, the more reliable they are. However, consideration of one or a few sequences in the context of species domestication may be advantageous in itself. In addition to the number of sequences investigated, their length (number of mutant sites) and the number of individuals sampled should not be overlooked in terms of processing reliability and comparability.

The fluent determination genetic diversity and its maintenance at a high level is indispensable on the course of preservation of our old, rare landraces. Indigenous breeds are national treasures and an aim of the saving of such breeds is not only to be prepared to compensate for the genetic erosion of another breed, but manufacture local products, preserve ancient knowledge and conserve natural landscape, furthermore transfer historical and cultural values continuously.

In order to map the broad genetic background biological samples should be taken from the descendants of the founders. During the future preservation work special attention should be paid to maternal families and representatives of ancient families should be preserved. A more intense focusing on the maternal side is motivated also by the fact that the females exceed in number the males, respectively they remain in breeding for a longer period of time, so they can at larger extent be the depositaries of realization and maintenance of genetic diversity.

Acknowledgements

This project is supported by the European Union and co-financed by the European Regional Development Fund (ERDF) (Grant Contract No: VEKOP-2.3.2.-16-2016-00012).

Conflict of interest

The Author declares no conflict of interest.

Author details

András Gáspárdy
University of Veterinary Medicine, Budapest, Hungary

*Address all correspondence to: gaspardy.andras@univet.hu

References

[1] Hackmann TJ, Spain JN. Mint: Invited review: Ruminant ecology and evolution: Perspectives useful to ruminant livestock research and production. Journal of Dairy Science. 2010;93(4):1320-1334. DOI: 10.3168/jds.2009-2071

[2] Lalueza-Fox C, Castresana J, SampietroL,Marquès-BonetT,AlcoverJA, Bertranpetit J. Mint: Molecular dating of caprines using ancient DNA sequences of Myotragus balearicus, an extinct endemic Balearic mammal. BMC Evolutionary Biology. 2005;5:70. DOI: 10.1186/1471-2148-5-70

[3] Joost S, Bonin A, Bruford MW, Després L, Conord C, Erhardt G, Taberlet P. Mint: A spatial analysis method (SAM) to detect candidate loci for selection: towards a landscape genomics approach to adaptation. Molecular Ecology. 2007;16(18):3955-3969. DOI: 10.1111/j.1365-294X.2007.03442.x

[4] Kijas JW, Townley D, D alrymple BP, Heaton MP, Maddox JF, McGrath A, Wilson P, Ingersoll RG, McCulloch R, McWilliam S, Tang D, McEwan J, Cockett N, Oddy VH, Nicholas FW, Raadsma H for the InternationalSheepGenomicsConsortium. Mint: A genome wide survey of SNP variation reveals the genetic structure of sheep breeds. PLoS ONE. 2009;4(3):e4668. DOI: 10.1371/journal.pone.0004668

[5] Chessa B, Pereira F, Arnaud F, Amorim A, Goyache F, Mainland I, Kao RR, Pemberton JM, Beraldi D, Stear M, Alberti A, Pittau M, Iannuzzi L, Banabazi MH, Kazwala R, Zhang Y-P, Arranz JJ, Ali BA, Wang Z, Uzun M, Dione M, Olsaker I, Holm L-E, Saarma U, Ahmad S, Marzanov N, Eythorsdottir E, Holland MJ, Ajmone-Marsan P, Bruford MW, Kantanen J, Spencer TE, Palmarini M. Mint: Revealing the history of sheep domestication using retrovirus integrations. Science. 2009;324(5926):532-536. DOI: 10.1126/science.1170587.

[6] Meadows JRS, Hawken RJ, Kijas JW. Mint: Nucleotide diversity on the ovine Y chromosome. Animal Genetics. 2004;35(5):379-385. DOI: 10.1111/j.1365-2052.2004.01180.x

[7] Cummins J. Mint: Mitochondrial DNA in mammalian reproduction. Reviews of reproduction. 1998;3(3):172-182. DOI: 10.1530/ror.0.0030172.

[8] Chial H, Craig J. Mint: mtDNA and mitochondrial diseases. Nature Education. 2008;1(1):217

[9] Brown WM, George M, Wilson AC. Mint: Rapid evolution of animal mitochondrial DNA. Proceedings of the National Academy of Sciences of the United States of America. 1979;76(4):1967-1971. DOI: 10.1073/pnas.76.4.1967

[10] Moritz C, Dowlin TE, Brown WM. Mint: Evolution of animal mitochondrial DNA: relevance for population biology and systematics. Annual Review of Ecology and Systematics, 1987;18(1):269-292. DOI: 10.1146/annurev.es.18.110187.001413

[11] Xu W, Jameson D, Tang B, Higgs PG. Mint: The relationship between the rate of molecular evolution and the rate of genome rearrangement in animal mitochondrial genomes. Journal of Molecular Evolution. 2006;63(3):375-392. DOI: 10.1007/s00239-005-0246-5

[12] Nabholz B, Glémin S, Galtier N. Mint: Strong variations of mitochondrial mutation rate across mammals – the longevity hypothesis. Molecular Biology and Evolution. 2008;25(1):120-130. DOI: 10.1093/molbev/msm248

[13] Galtier N, Jobson RW, Nabholz B, Glémin S, Blier PU. Mint: Mitochondrial whims: metabolic rate, longevity, and the rate of molecular evolution. Biology Letters. 2009;5(3):413-416. DOI: 10.1098/rsbl.2008.0662

[14] Song S, Pursell ZF, Copeland WC, Longley MJ, Kunkel TA, Mathews CK. Mint: DNA precursor asymmetries in mammalian tissue mitochondria and possible contribution to mutagenesis through reduced replication fidelity. Proceedings of the National Academy of Sciences of the United States of America. 2005;102(14):4990-4995. DOI: 10.1073/pnas.0500253102

[15] Moses MJ. Mint: Spermiogenesis in the crayfish (*Procambarus clarkii*) II. description of stages. The Journal of Biophysical and Biochemical Cytology. 1961;10:301-333. DOI: 10.1083/jcb.10.3.301

[16] Ursprung H, Schabtach E. Mint: Fertilization in tunicates: Loss of the paternal mitochondrion prior to sperm entry. Journal of Experimental Zoology. 1965;159(3):379-383. DOI: 10.1002/jez.1401590310

[17] Sutovsky P, Moreno RD, Ramalho-Santos J, Dominko T, Simerly C, Schatten G. Mint: Ubiquitin tag for sperm mitochondria. Nature. 1999;402(6760):371-372. DOI: 10.1038/46466

[18] Zhao X, Chu M, Li N, Wu C. Mint: Paternal inheritance of mitochondrial DNA in the sheep (Ovine aries). Science in China Series C: Life Sciences. 2001;44(3):321-326. DOI: 10.1007/BF02879339

[19] Ujvari B, Dowton M, Madsen T. Mint: Mitochondrial DNA recombination in a free-ranging Australian lizard. Biology Letters. 2007;3(2):189-192. DOI:10.1098/rsbl.2006.0587

[20] Hey J. Mint: Human mitochondrial DNA recombination: can it be true? Trends in Ecology and Evolution. 2000; 15(5):181-182. DOI: 10.1016/s0169-5347(00)01856-5

[21] Owens KN, Harvey-Blankenship M, King M-C. Mint: Genomic sequencing in the service of human rights. International Journal of Epidemiology. 2002;31(1):53-58. DOI: 10.1093/ije/31.1.53

[22] Gissi C, Iannelli F, Pesole G. Mint: Evolution of the mitochondrial genome of Metazoa as exemplified by comparison of congeneric species. Heredity. 2008;101(4):301-320. DOI: 10.1038/hdy.2008.62

[23] Polyak K, Li Y, Zhu H, Lengauer C, Willson JKV, Markowitz SD, Trush MA, Kinzler KW, Vogelstein B. Mint: Somatic mutations of the mitochondrial genome in human colorectal tumours. Nature Genetics. 1998;20:291-293. DOI: 10.1038/3108

[24] Ingman M, Kaessmann H, Pääbo S, Gyllensten U. Mint: Mitochondrial genome variation and the origin of modern humans. Nature. 2000;408(6813):708-713. DOI:10.1038/35047064

[25] Kim YS, Tseveen K, Batsukh B, Seong J, Kong HS. Mint: Origin-related study of genetic diversity and heteroplasmy of Mongolian sheep (*Ovis aries*) using mitochondrial DNA. Journal of Animal Reproduction and Biotechnology. 2020;35(2):198-206. DOI: 10.12750/JARB.35.2.198

[26] Zardoya R, Villalta M, López-Pérez MJ, Garrido-Pertierra A, Montoya J, Bautista JM. Mint: Nucleotide sequence of the sheep mitochondrial DNA D-loop and its flanking tRNA genes. Current Genetics. 1995;28(1):94-96. DOI: 10.1007/BF00311887.

[27] Achilli A, Bonfiglio S, Olivieri A, Malusà A, Pala M, Kashani BH, Perego UA, Ajmone-Marsan P, Liotta L, Semino O, Bandelt H-J, Ferretti L, Torroni A. Mint: The Multifaceted Origin of Taurine Cattle Reflected by the Mitochondrial Genome. PLoS ONE. 2009;4(6):e5753. DOI: 10.1371/journal.pone.0005753

[28] Pedrosa S, Uzun M, Arranz J-J, Gutiérrez-Gil B, Primitivo FS, Bayón Y. Mint: Evidence of three maternal lineages in near eastern sheep supporting multiple domestication events. Proceedings of the Royal Society of London, Series B: Biological Sciences. 2005;272(1577):2211-2217. DOI: 10.1098/rspb.2005.3204

[29] Torroni A, Achilli A, Macaulay V, Richards M, Bandelt H-J. Mint: Harvesting the fruit of the human mtDNA tree. Trends in Genetics. 2006;22(6):339-345. DOI: 10.1016/j.tig.2006.04.001

[30] Bazin E, Glémin S, Galtier N. Mint: Population size does not influence mitochondrial genetic diversity in animals. Science, 2006;312(5773):570-572. DOI: 10.1126/science.1122033

[31] Galtier N, Nabholz B, Glémin S, Hurst GDD. Mint: Mitochondrial DNA as a marker of molecular diversity: a reappraisal. Molecular Ecology. 2009;18:4541-4550. DOI: 10.1111/j.1365-294X.2009.04380.x

[32] Ratnasingham S, Hebert PDN. Mint: BOLD: The Barcode of Life Data System (http://www.barcodinglife.org). Molecular Ecology Notes. 2007;7(3):355-364. DOI: 10.1111/j.1471-8286.2006.01678.x

[33] Davenport KM, Duan M, Hunter SS, New DD, Fagnan MW, Highland MA, Murdoch BM. Mint: Complete mitochondrial genome sequence of bighorn sheep. Genome Announcements. 2018;6(23): e00464-18. DOI: 10.1128/genomeA.00464-18

[34] Taylor RW, Turnbull DM. Mint: Mitochondrial DNA mutations in human disease. Nature Reviews Genetics. 2005;6(5):389-402. DOI:10.1038/nrg1606

[35] United Mitochondrial Disease Foundation (UMDF) https://www.umdf.org/

[36] Wilmut I, Schnieke AE, McWhir J, Kind AJ, Campbell KHS. Mint: Viable offspring derived from fetal and adult mammalian cells. Nature. 1997;385:810-813. DOI: 10.1038/385810a0

[37] Evans MJ, Gurer C, Loike JD, Wilmut I, Schnieke AE, Schon EA. Mint: Mitochondrial DNA genotypes in nuclear transfer-derived cloned sheep. Nature Genetics. 1999;23(1):90-93. DOI: 10.1038/12696

[38] Gorman GS, Chinnery PF, DiMauro S, Hirano M, Koga Y, McFarland R, Suomalainen A, Thorburn DR, Zeviani M, Turnbull DM. Mint: Mitochondrial diseases. Nature reviews. Disease primers. 2016;2:16080. DOI: 10.1038/nrdp.2016.80

[39] Prezant TR, Agapian JV, Bohlman MC, Bu X, Öztas S, Qiu W-Q, Arnos KS, Cortopassi GA, Jaber L, Rotter JI, Shohat M, Fischel-Ghodsian N. Mint: Mitochondrial ribosomal RNA mutation associated with both antibiotic-induced and non-syndromic deafness. Nature Genetics. 1993;4:289-294. DOI: 10.1038/ng0793-289

[40] Yu-Wai-Man P, Griffiths PG, Brown DT, Howell N, Turnbull DM, Chinnery PF. Mint: The epidemiology of Leber hereditary optic neuropathy in the North East of England. American Journal of Human Genetics. 2003;72(2):333-339. DOI: 10.1086/346066

[41] Pal A, Pal A, Banerjee S, Batabyala S, Chatterjee PN. Mint: Mutation in Cytochrome B gene causes debility and adverse effects on health of sheep. Mitochondrion. 2019;46:393-404 DOI: 10.1016/j.mito.2018.10.003

[42] Reicher S, Seroussi E, Weller JI, Rosov A, Gootwine E. Mint: Ovine mitochondrial DNA sequence variation and its association with production and reproduction traits within an Afec-Assaf flock. Journal of Animal Science. 2012;90(7):2084-2091. DOI: 10.2527/jas.2011-4673

[43] Yüncü E, Demirci S, Baştanlar EK, Doğan ŞA, Taşdemir U, Togan I. Mint: Comparative study of three simple molecular approaches in search of mtDNA haplogroup identification of domestic sheep. Small Ruminant Research. 2013;114(1):64-71. DOI: 10.1016/j.smallrumres.2013.05.014

[44] Hiendleder S. Mint: A low rate of replacement substitutions in two major *Ovis aries* mitochondrial genomes. Animal Genetics. 1998;29(2):116-122. DOI: 10.1046/j.1365-2052.1998.00295.x

[45] Hiendleder S, Lewalski H, Wassmuth R, Janke A. Mint: The Complete Mitochondrial DNA Sequence of the Domestic Sheep (*Ovis aries*) and Comparison with the Other Major Ovine Haplotype. Journal of Molecular Evolution. 1998;47:441-448. DOI: 10.1007/PL00006401

[46] Meadows JRS, Hiendleder S, Kijas JW. Mint: Haplogroup relationships between domestic and wild sheep resolved using a mitogenome panel. Heredity. 2011;106(4):700-706. DOI: 10.1038/hdy.2010.122

[47] Hiendleder S, Mainz K, Plante Y, Lewalski H. Mint: Analysis of mitochondrial DNA indicates that domestic sheep are derived from two different ancestral maternal sources: no evidence for contributions from urial and argali sheep. Journal of Heredity. 1998;89(2):113-120. DOI: 10.1093/jhered/89.2.113

[48] Sanna D, Barbato M, Hadjisterkotis E, Cossu P, Decandia L, Trova S, Pirastru M, Leoni GG, Naitana S, Francalacci P, Masala B, Mereu P. Mint: The first mitogenome of the Cyprus Mouflon (Ovis gmelini ophion): New insights into the phylogeny of the genus Ovis. PloS ONE. 2015;10(12): e0144257. DOI: 10.1371/journal.pone.0144257

[49] Lv F-H, Peng W-F, Yang J, Zhao Y-X, Li W-R, Liu M-J, Ma Y-H, Zhao Q-J, Yang G-L, Wang F, Li J-Q, Liu Y-G, Shen Z-Q, Zhao S-G, Hehua E, Gorkhali NA, Vahidi SMF, Muladno M, Naqvi AN, Tabell J, Iso-Touru T, Bruford MW, Kantanen J, Han J-L, Li M-H. Mint: Mitogenomic meta-analysis identifies two phases of migration in the history of Eastern Eurasian sheep. Molecular Biology and Evolution. 2015;32(10):2515-2533. DOI: 10.1093/molbev/msv139

[50] Demirci S, Baştanlar EK, Dağtaş ND, Pişkin E, Engin A, Özer F, Yüncü E, Doğan ŞA, Togan I. Mint: Mitochondrial DNA diversity of modern, ancient and wild sheep (Ovis gmelinii anatolica) from Turkey: new insights on the evolutionary history of sheep. PLoS One. 2013;8(12):e81952. DOI: 10.1371/journal.pone.0081952

[51] Dymova MA, Zadorozhny AV, Mishukova OV, Khrapov EA, Druzhkova AS, Trifonov VA, Kichigin IG, Tishkin AA, Grushin SP, Filipenko ML. Mint: Mitochondrial DNA analysis of ancient sheep from Altai. Animal Genetics. 2017;48(5):615-618. DOI: 10.1111/age.12569

[52] Horsburgh KA, Rhines A. Mint: Genetic characterization of an archaeological sheep assemblage from South Africa's Western Cape. Journal of Archaeological Science.

2010;37(11):2906-2910. DOI: 10.1016/j.jas.2010.06.035

[53] Rannamäe E, Lougas L, Niemi M, Kantanen J, Maldre L, Kadorova N, Saarma U. Mint: Maternal and paternal genetic diversity of ancient sheep in Estonia from the Late Bronze Age to the post-medieval period and comparison with other regions in Eurasia. Animal Genetics. 2016;47:208-218. DOI: 10.1111/age.12407

[54] Hiendleder S, Kaupe B, Wassmuth R, Janke A. Mint: Molecular analysis of wild and domestic sheep questions current nomenclature and provides evidence for domestication from two different subspecies. Proceedings of the Royal Society of London, Series B: Biological Sciences. 2002;269:893-904. DOI: 0.1098/rspb.2002.1975.

[55] Annus K, Maróti-Agóts Á, Pásztor K, Vada E, Sáfár L, Gáspárdy A. Mint: Characterisation of Hungarian Tsigai variants based on control region of mtDNA (with English summary). Magyar Állatorvosok Lapja. 2015;137(10):625-631.

[56] Alonso R, Ulloa-Arvizu R, Gayosso-Vázquez A. Mint: Mitochondrial DNA sequence analysis of the Mexican Creole sheep (*Ovis aries*) reveals a narrow Iberian maternal origin. Mitochondrial DNA Part A. 2017;28(6):793-800. DOI: 10.1080/24701394.2016.1192613

[57] Lancioni H, Di Lorenzo P, Ceccobelli S, Perego UA, Miglio A, Landi V, Antognoni MT, Sarti FM, Lasagna E, Achilli A. Mint: Phylogenetic relationships of three Italian Merino-derived sheep breeds evaluated through a complete mitogenome analysis. PLoS ONE. 2013;8(9): e73712. DOI: 10.1371/journal.pone.0073712

[58] Bunch TD, Wu C, Zhang Y-P, Wang S. Mint: Phylogenetic Analysis of Snow Sheep (*Ovis nivicola*) and Closely Related Taxa. Journal of Heredity. 2006;97(1):21-30. DOI: 10.1093/jhered/esi127

[59] Guo J, Du LX, Ma YH, Guan WJ, Li HB, Zhao QJ, Li X, Rao SQ. Mint:, 2005. A novel maternal lineage revealed in sheep (*Ovis aries*). Animal Genetics. 2005;36(4):331-336. DOI: 10.1111/j.1365-2052.2005.01310.x

[60] Meadows JRS, Li K, Kantanen J, Tapio M, Sipos W, Pardeshi V, Gupta V, Calvo JH, Whan V, Norris B, Kijas JW. Mint: Mitochondrial sequence reveals high levels of gene flow between breeds of domestic sheep from Asia and Europe. Journal of Heredity. 2005;96(5):494-501. DOI: 10.1093/jhered/esi100

[61] Chen, S.Y., Duan, Z.Y., Sha, T., Xiangyu, J., Wu, S.F., Zhang, Y.P., 2006. Mint: Origin, genetic diversity, and population structure of Chinese domestic sheep. Gene. 2006;376(2):216-223. DOI: 10.1016/j.gene.2006.03.009

[62] Tserenbataa T, Ramey IIRR, Ryder OA, Quinn TW, Reading RP. Mint: A population genetic comparison of argali sheep (*Ovis ammon*) in Mongolia using the ND5 gene of mitochondrial DNA; implications for conservation. Molecular ecology. 2004;13(5):1333-1339. DOI: 10.1111/j.1365-294X.2004.02123.x

[63] Bruford MW, Bradley DG, Luikart G. Mint: DNA markers reveal the complexity of livestock domestication. Nature Reviews Genetics. 2003;4(11):900-910. DOI: 10.1038/nrg1203.

[64] Ganbold O, Lee S-H, Seo D, Paek WK, Manjula P, Munkhbayarc M, Lee JH. Mint: Genetic diversity and the origin of Mongolian native sheep. Livestock Science. 2019;220:17-25. DOI:10.1016/j.livsci.2018.12.007

[65] Mohammadhashemi A, Pirany N, Nassiri MR, Daloii TA, Kohnegroz BB. Mint: Studying the partially sequenced mtDNA hypervariable region1 (HVR1) of Iranian Moghani sheep. Annals of Biological Research. 2012;3(6): 2906-2910

[66] Tapio M, Marzanov N, Ozerov M, Ćinkulov M, Gonzarenko G, Kiselyova T, Murawski M, Viinalass H, Kantanen J. Mint: Sheep mitochondrial DNA variation in European, Caucasian, and Central Asian areas. Molecular Biology and Evolution. 2006;23(9):1776-1783. DOI: 10.1093/molbev/msl043

[67] Meadows JRS, Cemal I, Karaca O, Gootwine E, Kijas JW. Mint: Five Ovine Mitochondrial Lineages Identified From Sheep Breeds of the Near East. Genetics. 2007;175(3):1371-1379. DOI: 10.1534/genetics.106.068353

[68] Rafia P, Tarang A. Mint: Sequence Variations of Mitochondrial DNA Displacement-Loop in Iranian Indigenous Sheep Breeds. Iranian Journal of Applied Animal Science. 2016;6(2):363-368

[69] Liu J, Ding X, Zeng Y, Yue Y, Guo X, Guo T, Chu M, Wang F, Han J, Feng R, Sun X, Niu C, Yang B, Guo J, Yuan C. Mint: Genetic Diversity and Phylogenetic Evolution of Tibetan Sheep Based on mtDNA D-Loop Sequences. PLoS ONE. 2016;11(7): e0159308. DOI: 10.1371/journal.pone.0159308

[70] Ćinkulov M, Popovski Z, Porcu K, Tanaskovska B, Hodžić A, Bytyqi H, Mehmeti H, Margeta V, Djedović R, Hoda A, Trailović R, Brka M, Marković B, Važić B, Vegara M, Olsaker I, Kantanen J. Mint: Genetic diversity and structure of the West Balkan Pramenka sheep types as revealed by microsatellite and mitochondrial DNA analysis. Journal of Animal Breeding and Genetics. 2008;125(6):417-426. DOI: 10.1111/j.1439-0388.2008.00742.x

[71] Pereira F, Davis SJM, Pereira L, McEvoy B, Bradley DG, Amorim A. Mint: Genetic Signatures of a Mediterranean Influence in Iberian Peninsula Sheep Husbandry. Molecular Biology and Evolution. 2006;23(7):1420-1426. DOI: 10.1093/molbev/msl007

[72] Pedrosa S, Arranz J-J, Brito N, Molina A, Primitivo FS, Bayón Y. Mint: Mitochondrial diversity and the origin of Iberian sheep. Genetics Selection Evolution. 2007;39:91-103. DOI: 10.1051/gse:2006034

[73] Pariset L, Mariotti M, Gargani M, Joost S, Negrini R, Perez T, Bruford M, Marsan PA, Valentini A. Mint: Genetic Diversity of Sheep Breeds from Albania, Greece, and Italy Assessed by Mitochondrial DNA and Nuclear Polymorphisms (SNPs). The Scientific World Journal. 2011;11:1641-1659. DOI: 10.1100/2011/186342

[74] Mariotti M, Valentini A, Marsan PA, Pariset L. Mint: Mitochondrial DNA of seven Italian sheep breeds shows faint signatures of domestication and suggests recent breed formation. Mitochondrial DNA (The Journal of DNA Mapping, Sequencing, and Analysis). 2013;24(5):577-583. DOI: 10.3109/19401736.2013.770493

[75] Brahi OHD, Xiang H, Chen X, Farougou S, Zhao X. Mint: Mitogenome revealed multiple postdomestication genetic mixtures of West African sheep. Journal of Animal Breeding and Genetics. 2015;132(5):399-405. DOI: 10.1111/jbg.12144

[76] Álvarez I, Capote J, Traoré A, Fonseca N, Pérez K, Cuervo M, Fernández I, Goyache F. Mint: Mitochondrial analysis sheds light on the origin of hair sheep. Animal Genetics. 2013;44(3):344-347. DOI: 10.1111/j.1365-2052.2012.02398.x

[77] Resende A, Gonçalves J, Muigai AWT, Pereira F. Mint:

Mitochondrial DNA variation of domestic sheep (Ovisaries) in Kenya. Animal Genetics. 2016;47(3):337-381. DOI: 10.1111/age.12412

[78] Othman OE, Pariset L, Balabel EA, Marioti M. Mint: Genetic characterization of Egyptian and Italian sheep breeds using mitochondrial DNA. Journal of Genetic Engineering and Biotechnology. 2015;13(1):79-86. DOI: 10.1016/j.jgeb.2014.12.005

[79] Ghernouti N, Bodinier N, Ranebi M, Maftah D, Petit D, Gaouar SBS. Mint: Control Region of mtDNA identifies three migration events of sheep breeds in Algeria. Small Ruminant Research. 2017;155:66-71. DOI: 10.1016/j.smallrumres.2017.09.003

[80] Kandoussi A, Boujenane I, Piro M, Petit D. Mint: mtDNA genetic characterization of an isolated sheep breed in South of Moroccan Atlas. Small Ruminant Research. 2020;193:106250. DOI: 10.1016/j.smallrumres.2020.106250

[81] Campos E, Cuéllara J, Salvador O, García-Trejo EA, Pereira F. Mint: The genetic diversity and phylogeography of Mexican domestic sheep. Small Ruminant Research. 2020;187:106109. DOI: 10.1016/j.smallrumres.2020.106109

[82] Revelo HA, López-Alvarez D, Landi V, Rizzo L, Alvarez LA. Mint: Mitochondrial DNA Variations in Colombian Creole Sheep Confirm an Iberian Origin and Shed Light on the Dynamics of Introduction Events of African Genotypes. Animals. 2020;10:1594. DOI: 10.3390/ani10091594

[83] Wang X, Chen H, Lei CZ. Mint: Genetic diversity and phylogenetic analysis of the mtDNA D-loop region in Tibetan sheep. Asian-Australasian Journal of Animal Sciences. 2007;20(3)313-315. DOI: 10.5713/ajas.2007.313

[84] Agaviezor BO, Adefenwa MA, Peters SO, Yakubu A, Adebambo AO, Ozoje MO, Ikeobi CON, Ilori BM, Wheto M, Okpeku M, De Donato M, Imumorin IG. Mint: Mitochondrial D-loop genetic diversity of Nigerian indigenous sheep. Animal Genetic Resources. 2012;50:13-20. DOI: 10.1017/S2078633612000070

[85] Arora R, Yadav HS, Mishra BP. Mint: Mitochondrial DNA diversity in Indian sheep. Livestock Science. 2013;153(1-3):50-55. DOI: 10.1016/j.livsci.2013.02.006

[86] Kovács E, Maróti-Agóts Á, Harmat L, Annus K, Zenke P, Tempfli K, Sáfár L, Gáspárdy A. Mint: Characterisation of Hungarian Cikta sheep based on the control region of mtDNA (with English summary). Magyar Állatorvosok Lapja. 2020;142(7):421-428.

[87] Kovács E, Harmat L, Tempfli K, Sáfár L, Becskei Zs, Maróti-Agóts Á, Gáspárdy A. Mint: Ergebnisse der Sequenzanalyse des mitochondrialen Gens Cyt-b von Cikta Schafen (with English summary). Danubian Animal Genetic Resources. 2020;5(1):19-25.

[88] Kusza Sz, Zakar E, Budai Cs, Cziszter L, Padeanu I, Gavojdian D. Mint: Mitochondrial DNA variability in Gyimesi Racka and Turcana sheep breeds. Acta Biochimia Polonica. 2015,62(1):273-280. DOI: 10.18388/abp.2015_978

[89] Kirikci K, Noce A, Cam MA, Mercan L, Amills M. Mint: The analysis of mitochondrial data indicates the existence of population substructure in Karayaka sheep. Small Ruminant Research. 2018;162:25-29. DOI: 10.1016/j.smallrumres.2018.02.007

[90] FAO. The state of food and agriculture 2007. Rome. Agriculture Series No. 38

[91] Salmon GR, MacLeod M, Claxton JR,Pica Ciamarra U, Robinson T, Duncan A, Peters AR. Mint: Exploring the landscape of livestock 'Facts'. Global Food Security. 2020;25:100329. DOI: 10.1016/j.gfs.2019.100329

[92] Taberlet P, Coissac E, Pansu J, Pompanon F. Mint: Conservation genetics of cattle, sheep, and goats. Comptes Rendus Biologies. 2011;334(3):247-254. DOI: 10.1016/j.crvi.2010.12.007

[93] FAO. The State of the World's Animal Genetic Resources for Food and Agriculture. FAO, Rome. 2007

[94] Lenstra JA, Groeneveld LF, Eding H, Kantanen J, Williams JL, Taberlet P, Nicolazzi EL, Sölkner J, Simianer H, Ciani E, Garcia JF, BrufordMW,Ajmone-MarsanP,WeigendS. Mint: Molecular tools and analytical approaches for the characterization of farm animal genetic diversity. Animal Genetics. 2012;43(5):483-502. DOI: 10.1111/j.1365-2052.2011.02309.x

[95] Groeneveld LF, Lenstra JA, Eding H, Toro MA, Scherf B, Pilling D, Negrini R, Finlay EK, Jianlin H, Groeneveld E, Weigend S, GlobalDiv Consortium. Mint: Genetic diversity in farm animals – a review. Animal genetics. 2010;41(Suppl 1):6-31. DOI: 10.1111/j.1365-2052.2010.02038.x

[96] FAO. Molecular genetic characterization of animal genetic resources. FAO Animal Production and Health Guidelines. No. 9. Rome 2011 (available at http://www.fao.org/docrep/014/i2413e/i2413e00.htm). Google Scholar

[97] Agaviezor BO, Peters SO, Adefenwa MA, Yakubu A, Adebambo AO, Ozoje MO, Ikeobi CON, Wheto M, Ajayi OO, Amusan SA, Ekundayo OJ, Sanni TM, Okpeku M, Onasanya GO, De Donato M, Ilori BM, Kizilkaya K, Imumorin IG. Mint: Morphological and microsatellite DNA diversity of Nigerian indigenous sheep. Journal of Animal Science and Biotechnology. 2012b;3(1):38.

[98] Firestone KB, Houlden BA, Sherwin WB, Geffen E. Mint: Variability and differentiation of microsatellites in the genus Dasyurus and conservation implications for the large Australian carnivorous marsupials. Conservation Genetics. 2000;1(2):115-133. DOI: 10.1023/A:1026578821339

[99] Gáspárdy A, Csóri Zs, Daróczi-Szabó M. Legend of the four horned Racka or whether in the sight of what did Eugene of Savoy delight? Paper on Joint Annual Meeting of ÖNGENE and of DAGENE; 30-31 August 2012; Wels, Austria [Internet]. Available from: http://www.dagene.eu/docs/ongene_dagene_2012/legendo4_gaspardy_2012.pdf [Accessed: 2020-12-10]

[100] Peter C, Bruford M, Perez T, Dalamitra S, Hewitt G, Erhardt G, ECONOGENE Consortium. Mint: Genetic diversity and subdivision of 57 European and Middle-Eastern sheep breeds. Animal Genetics. 2007;38(1):37-44. DOI: 10.1111/j.1365-2052.2007.01561.x

[101] Boichard D, Maignel L, Verrier É. Mint: The value of using probabilities of gene origin to measure genetic variability in a population. Genetic Selection Evolution. 1997;29:5-23. DOI: 10.1186/1297-9686-29-1-5

[102] Gutiérrez JP, Cervantes I, Goyache F. Mint: Improving the estimation of realized effective population size in farm animals. Journal of Animal Breeding and Genetics. 2009;126(4):327-332. DOI: 10.1111/j.1439-0388.2009.00810.x.

[103] Huby M, Griffon L, Moureaux S, De Rochambeau H, Danchin-Burge C,

Verrier É. Mint: Genetic variability of six French meat sheep breeds in relation to their genetic management. Genetic Selection Evolution. 2003;35(7):637-655. DOI:10.1051/gse:2003044

[104] Caballero A, Toro MA. Mint: 2000 Interrelations between effective population size and other pedigree tools for the management of conserved populations. Genetics Research. 2000;75(3):331-343. DOI: 10.1017/s0016672399004449

[105] Gáspárdy A, Simon Cs, Andrásofszky E, Sáfár L, Kósa E. Mint: Historical evaluation of milk producing ability of the Hungarian native Tsigai sheep (with English summary). Állattenyésztés és Takarmányozás. 2016;65(1):24-36

[106] Mederle A, Maróti-Agóts Á, Matiuti M, Gáspárdy A. Mint: Challenges in conservation of Tyrolean Grey Cattle. Danubian Animal Genetic Resources. 2020;5(1):11-17

[107] Ledwith L, Kőrösi A, Daróczi-Szabó M, Gáspárdy A. Mint: Comparative skull geometry of recently lived Hungarian Rackas. Danubian Animal Genetic Resources. 2020;5(2):55-63

[108] Annus K, Arkenberg H, Prikoszovich M, Oláh J, Maróti-Agóts Á, Gáspárdy A. Mint: Characterisation of the female Tsigai population by use of Hungarian herd-book data. In: Hajas P, Gáspárdy A, editors. 25 years with DAGENE. Printed by Palatia Nyomda és Kiadó Kft. Győr, ISBN 978-963-12-3101-4, 2015;108-113

[109] Posta J, Kovács E, Tempfli K, Sáfár L, Gáspárdy A. Mint: Pedigree analysis of a population bottlenecked, the Cikta with special regard to its maternal lineages (with English summary). Magyar Állatorvosok Lapja. 2019;141(3):171-180.

Chapter 11

Domestic Pig Germplasms of Andaman and Nicobar Islands

Perumal Ponraj, Arun Kumar De and Debasis Bhattacharya

Abstract

Andaman and Nicobar Islands are endowed with immaculate flora and fauna biodiversity. Among the indigenous livestock species, pig occupies 27.26%. Andaman and Nicobar Islands have three different categories of domestic pig groups/breeds. Andaman Local pig is prevalent in Andaman group of Islands (South, Middle and North Andaman); Nicobari pig is in Nicobar group of Islands and long snouted Little Andaman wild pig (Schedule II animal under Forest Act, India). Other than the indigenous pigs, pure and crossbreds of Large White Yorkshire are available in Andaman and Nicobar Islands. Nicobari Pigs are reared exclusively by Nicobari tribes in Nicobar group of islands and create a well defined socio-economic-ecological status of their tribal society. Nicobari pig occupies a prominent place in custom, festivals and socio-economic status of Nicobari tribes. These Andaman local and Nicobari pigs are reared for meat purpose under free range or semi-intensive system. Nicobari pig is appeared as short, black/brownish in colour and living as a family. Andaman local pig is available in Andaman group of islands and body colour differs from rusty grey to black and brown. Neck and dorsal portion hair are long and thick whereas flank and sides hairs are shorter and thinner. Wild pig of Andaman (*Sus scrofa andamanensis*) is a most endangered porcine species of Andaman and Nicobar islands. Jarawa tribes in Andaman Islands prefer this wild pig as a good protein source. It is black in colour, short legged, small to medium sized and a prolific breeder. Litter size varies from 4 to 7 numbers. Another pig group is crossbred, cross between Large White Yorkshire and Andaman local or Nicobari pig. Crossbred pigs are light brown to complete white with different lines of blackish colour. This breed exhibits early maturity, high growth rate and fecundity. The Nicobari pig has high prolificacy as litter size is ranging from 8 to 10 numbers with good mothering ability and body weight of matured pig differs from 115 to 130 kg. Moreover, this crossbred is adapted highly to the local tropical humid environmental conditions and also can adjust with locally available feed resources on the different agricultural produces. This is highly suitable for commercial production of pork in this Andaman and Nicobar islands. However, the domestic pig breeds need to be protected and be conserved in this Andaman and Nicobar group of Islands.

Keywords: Andaman and Nicobar Islands, indigenous pigs, physico-morphological characters, haematological profiles, management, reproductive and productive profiles

1. Introduction

Andaman and Nicobar islands are one of the diversified unique ecosystems in the world. Being away from the main land and less population pressure, the area is still maintaining almost pollution free virgin environment, harbouring pure and rich germplasm resources. It is situated in the southern part of the Bay of Bengal between 92°12′ E and 93°57′ E longitude and between 6° 45′N and 13° 41′N latitude with 10°N channel dividing the Nicobar group of islands from Andaman group of Islands. Andaman and Nicobar is a group of 576 islands, islets and rocks covering a geographical area of 8293 km^2 and a population of 3.80 lakhs. Andaman and Nicobar islands share the same broad agro-ecological region as South East Asian countries. Majority of the 188 named islands are small in size. Thirty-six of these are inhabited. Only four islands namely North, Middle and South Andaman in the Andaman group and Great Nicobar in the southern group have an area greater than 1000 km^2. Little Andaman with an area of 731 km^2 is the next largest island. Among the rest, 32 islands exceed 10 km^2 while 96 are less than 1 km^2 in area. Of the inhabited islands, 12 have population exceeding 1000 persons. Andaman and Nicobar islands have the annual rainfall of average is 3070 mm covering the month from May to December. The period between January and April is the driest when the number of rainy days in each month hardly exceeds three. During these periods, agricultural crops often suffer severely. The mean temperature (24.3-30.5°C), relative humidity (82.5%) and wind speed (5.8 km/h) almost remains same throughout the year. The seasons were classified into rainy/wet (May to November) and dry/summer (December to April) in Andaman and Nicobar Islands. Average sun light hours per day differed significantly between rainy (4.28 ± 0.89) and dry summer (9.20 ± 0.74) seasons. From the month of December to April in Andaman and Nicobar Islands, the sun shines regularly, whereas the sky is often become cloudy from June to September. Average relative humidity (%) was differed significantly between rainy (84.21 ± 1.93) and dry summer (75.80 ± 2.06) seasons. Average temperature (°C) was differed significantly between rainy (29.71 ± 0.62) and dry summer (31.42 ± 0.80) seasons. Average rainfall (mm) was differed significantly between rainy (444.92 ± 13.62) and dry summer (89.04 ± 8.84) seasons. Average solar direct irradiance ($kWh/m^2/day$) was differed significantly between rainy (3.47 ± 0.95) and dry summer (6.24 ± 0.56) seasons. Average temperature humidity index (THI) was differed significantly between rainy (84.92 ± 1.59) and dry summer (85.59 ± 1.15) seasons. Average sea surface temperature (°C) was differed between rainy (27.97 ± 0.87) and dry summer (29.94 ± 1.30) seasons.

Livestock farming is considered to be a profitable enterprise in agriculture and constitutes an important activity for income enhancement. As per livestock census of 2012, the cattle, buffalo, goat, pig and poultry population including duck in the island is 45625, 7863, 65324, 35921 and 1165223, respectively. Livestock census, India revealed that the pig population was reduced (25.79%) significantly from 18th (2007: 48406 [1]) to 19th (2012: 35921 [2]) and then increased (5.98%) from 19th to 20th (2017: 40488[3]) livestock census, Government of India (**Tables 1** and **2**). Similarly, Tsunami, 2004 has significantly affected the Nicobari pig population in Nicobar group of Islands. There are four different genetic groups of pigs in the Islands, namely, Andaman local pig (ALP), long snouted Little Andaman wild pig (Schedule II animal under Forest Act, 1972, Govt. of India), Nicobari pig and pure and cross breeds of Large White Yorkshire (**Table 3**). Andaman and Nicobar group of Islands are endowed with immaculate flora and fauna biodiversity [5]. The indigenous livestock germplasm namely Nicobari, Andaman local and Andaman wild pigs, Teressa and Andaman local goats and Nicobari fowl are predominant in Andaman and Nicobar group of islands. Among the indigenous livestock, pig occupies 27.26% of the total livestock in Andaman and Nicobar Islands [6]. However, the Nicobari indigenous pig is under severe threat to endanger from the island, therefore immediate conservation effort is to be taken and its very much necessary [7]. Till 2012, this Nicobari breed received very little attention and no

District	Category	Male			Female			Grand total
		< 6 Months	>6 Months	Total	< 6 Months	>6 Months	Total	
Nicobar	Indigenous	4381	3928	8309	4442	3644	8086	16395
	Exotic/ crossbred	1892	1477	3369	1579	1438	3017	6386
North & Middle Andaman	Indigenous	1725	1513	3238	1604	1544	3148	6386
	Exotic/ crossbred	1004	951	1955	854	870	1724	3679
South Andaman	Indigenous	528	409	937	704	676	1380	2317
	Exotic/ crossbred	134	215	349	167	242	409	758

Table 1.
District-wise livestock census (19th) of pigs in Andaman and Nicobar Islands.

Tehsil wise	Exotic/Crossbred			Indigenous			Total
	Male	Female	Total	Male	Female	Total	
Diglipur	851	713	1564	1682	1823	3505	5069
Mayabunder	571	420	991	625	558	1183	2174
Rangat	553	591	1124	930	765	1695	2819
Middle and North Andaman Dist	**1955**	**1724**	**3679**	**3237**	**3146**	**6383**	**10062**
Ferrargunj	98	109	207	11	29	40	247
Port Blair (Rural)	134	200	334	9	15	24	358
Port Blair (Urban)	117	100	217	27	50	77	294
Little Andaman	0	0	0	890	1286	2176	2176
South Andaman Dist	**349**	**409**	**758**	**937**	**1380**	**2317**	**3075**
Car Nicobar	1	0	1	6369	6240	12609	12610
Nancowry	3056	2719	5775	1583	1528	3111	8886
Campbell Bay	312	298	610	358	320	678	1288
Nicobar Dist	3369	3017	6386	8310	8088	16398	22784
State total	**5673**	**5150**	**10823**	**12848**	**12614**	**25098**	**35921**

Table 2.
Tehsil-wise livestock census (19th) of pigs in Andaman and Nicobar Islands.

systematic documentation was made. This Nicobari pigs were considered as a recognised and distinguished pig breed of Indian Government (INDIA PIG-3300-NICOBARI09005 by NBAGR, Karnal). Genetic diversity was very high as compared to European breeds [7]. Nicobari pig breed is adapted well physiologically and anatomically and has high tolerable capacity to different humid tropical deleterious environmental conditions. Nicobari pigs are natural scavengers and size is from medium to large with low reproductive and growth performance. Nicobari pig breed is highly preferred among the tribals, it is a good source of protein supplement to them and also it helps to improve their family income.

The ALP is associated with the socio-culture-economic-tradition of tribals. Andaman local pig is in general as semi-feral in behaviour and is mostly reared in

Common name	Scientific name	Habitat	Status	Adaptation	Disease resistance	Management
Nicobari pig	*Sus scrofa nicobaricus*	Nicobar group of islands	Not-endangered	Adapted to hot and humid climate of Nicobar islands	Acquired Resistance to common pig diseases	Backyard pig production system
Andaman local pig	*Sus domesticus*	Andaman group of islands	Not-endangered	Adapted to hot and humid climate of Nicobar islands	Acquired Resistance to common pig diseases	Backyard and intensive pig production system
Andaman wild pig	*Sus scrofa andamanensis*	Andaman group of islands	Endangered	Adapted to hot and humid climate of Andaman islands	Acquired Resistance to common pig diseases	Completely scavenging Feral

Table 3.
Status of pigs in Andaman and Nicobar Islands [4].

extensive or free-range system with little amount of management. Mitogenome analysis revealed that this ALP can be evolved as an independent breed in Andaman and Nicobar Islands as merit for registration as a recognised pig breed [8]. This pig group is under the condition of endanger and immediate preservation, conservation and propagation effort is very much needed to save the local breed of pig from extinction [7]. Pig production system is highly economical due to high production potential, fast growth rate, short generation interval, prolific fecundity, highly efficient carcass yield and higher adaptability to the different micro and macro environmental as well as the climatological conditions [9]. ALP is very well adapted and tolerable to the different tropical humid harsh environmental conditions with higher relative humidity, higher temperature as well as higher temperature humidity index. Further, these local indigenous Andaman local pigs are scavengers and also semi-wild in their behaviour or character. Andaman local pigs have very good maternal ability and are aggressive when farrowing or delivery. Although the ALPs have lower in their growth rate and reproductive and productive performances, it is highly liked by the rural tribal communities for supplementation of sufficient protein and income for the family. Pork production is essential especially in the Nicobar Islands than in the other part of the Andaman and Nicobar Islands. People of Nicobar group of Islands consume 70% of the pork produced in Andaman and Nicobar Islands while the rest of the islanders consume 30% of pork [2]. Wild pig of Andaman (*Sus scrofa andamanensis*) is a threatened endangered porcine germplasm of Andaman and Nicobar islands. Jarawa tribes prefer this wild pig as a protein source. It is black in colour, short legged, small to medium sized and a prolific breeder. Another pig group is crossbred, cross between Large White Yorkshire and Andaman local or Nicobari pig. This Nicobari breed exhibits early maturity, high growth rate and fecundity than other pig breeds. The Nicobari pig has high prolificacy as litter size is ranging from 8 to 10 numbers with good mothering ability and body weight of matured pig differs from 115 to 130 kg. This is highly suitable for commercial production of pork in this Andaman and Nicobar islands. However, the domestic pig breeds need to be protected in this Andaman and Nicobar Islands. Reorganisation and rearrangement of these pig breeds is significant for its conservation, preservation and propagation. Efforts have to be made to conserve this breed outside its breeding tract with different managemental condition. The chapter describes the different aspects of pigs namely, Nicobar pig, Andaman local pig,

Andaman wild pig and crossbred pigs, which are available in Andaman and Nicobar Islands.

2. Nicobari pig

Nicobari pig (*Sus scrofa nicobaricus*), locally known as Ha-un and is reared by Nicobari tribes in Nicobar group of Islands. Majority of the pigs in the Nicobar group of islands are Nicobari pigs. The total area of Nicobar group of islands is 1841 square km which lies between 6° and 10° North latitude and it comprises of 19 islands, of which, important islands are Car Nicobar, Chowra, Teressa, Nancowrie, Little Nicobar and Great Nicobar. A small population of Nicobari pig are reared by Nicobari tribes on Little Andaman (Nicobari settlement area at Harminder Bay) Island [10–12]. The registered unique Nicobari indigenous pig is considered as a sign of integrity and wellbeing of the Nicobari tribes in this Andaman and Nicobar island territories. Nicobari pig is considered as an endemic in the island region; however, this pig is still in the domestication process as witnessed by phenotypic expression and are generally believed as it has originated from Eurasian wild boar (*Sus scrofa*) and also believed that *Nicobaricus* as a Nicobar regional specific sub-species and are called generally as "Nicobari Pig" [11]. These pigs are available since immemorial time with primitive tribes of these Andaman and Nicobar islands. Nicobari means 'eating pork'. Pig growing is very common, preferred and custom within the Nicobari tribes and which provides as an essential source of animal rich protein [13]. Nicobari Pigs are reared exclusively by Nicobari tribes in Nicobar group of islands and create a well defined socio-economic-ecological status of their tribal society. Pig rearing has always been an integral part of the rich cultural traditions of the Nicobarese Tribes [14]. Pigs are treated as an asset and bring prestige to the joint family of Nicobarese. In Nicobari society, religion, custom, festivals and social status, the pig occupies a prominent place. Of which, pigs constitute the major portion of their economy. The economic prosperity of a family, village and its position in the island is judged by the number of pigs as they have in a village or lineage. As such *Nicobarese* maintain a large number of pigs that are freely roaming in their settlement as well as in mature coconut plantation. Nicobari pigs have well adapted to the tropical humid island ecosystem in physiological and anatomical over the long period of the times and express their potential very well under the integrated farming (plantation based) production system in Andaman and Nicobar Islands [12]. In overview, the male Nicobari pig is very much temperament (nervousness) than his counterpart of female pigs and to catch the male pigs at least ten people are needed. However, the Nicobari sows are very calm and could be managed easily with the Nicobari tribal women. Ill-treatment or misbehaviour of the same is treated as serious offence. In case, any Nicobari or outsider hit or beat the pig, then their tribal village Council deals the matter sternly and the same treatment was given to the person concerned and impose handsome amount of fine to prevent repeat of the act [15].

2.1 Phenotypic characterisation

Nicobari pig is a registered descriptive domesticated pig breed of India (INDIA_PIG_3300_NICOBARI_09005). Molecular characterisation with use of microsatellite markers on local pig breeds revealed that the Nicobari pig has mean observed heterozygosity of 0.70 ± 0.09 and Andaman local pig has 0.72 ± 0.07 and both were has significantly higher mean observed heterozygosity than in the Large White Yorkshire as 0.56 ± 0.07, the present study result indicated that Andaman

Local pig as well as Nicobari pigs are genetically different from the exotic pig breeds of LWY as well as from other Indian indigenous pig breeds such as Gahuri, North Indian desi and Ankamali [16]. Nicobari pig population was highest in Car Nicobar followed by Chowra, Teressa, Nancowry and Katchal. At present, this indigenous pig breed is under the endangered and threatened category and immediate conservation effort is necessary. In Nicobar and other parts of Islands where the Nicobar tribes are living, there is no commercial pig rearing system or commercial pork production system among the Nicobari tribal community. The pigs are also exchanged as gift between families and islands, bartering within their communities. This porter or exchange practice has significantly reduced the inbreeding among the pigs in the villages. Nicobari pigs are short in stature with compact body, black/brownish or creamy white or reddish brown or blackish brown coat, light brown or pinky and strong muzzle, light brownish creamy-white or light blackish white hooves and small eyelids with brown or creamy white and short, coarse, straight ears and attached with close to the head or body. In some pockets of islands, piglets were shown with dark brownish red stripes in the dorsal part of the body and this appearance of striped piglets is an essential indicator of primitive pig type or marker of origination of this pig from wild group of pigs. The majority of the pigs appears small to medium sized, short legged, short with a long body and their skin colour includes black, grey, brown and blackish brown. Sometime, the ventral side (belly region) were coloured with cream or white and in some pigs the pattern of colouring has extended throughout the whole body. The bristles are dense, coarse with black or brown or creamy in colour. This pig has a marked bristle crest or mane on the dorsal part of the pig which is extending to tail base from mid head/shoulder. Slight downward curvature or arch of the back/low back is considered as the most common feature in this pig breed. There are no facial warts in pigs. These breeds are sturdy and short compared to other desi breeds. Head is short with a strong slightly curved (downward) snout and large jowl. Some pigs inside the jungle are reported with long big head and strong lengthy snout with aggressive indicates wildness. Neck is short, clean and heavy. The shoulders are light, firm and free from coarseness, medium width and well attached to body. The body is medium length, slightly arched (downwards) at back, no uniform breadth/sides, well sprung ribs, strong and slightly wider loin and back; slightly broad hams, well-filled but not up to hocks. Nicobari pigs have large, capacious/heavy and moderately pot-bellied abdomen. The fascial profiles of pigs vary from flat to concave giving a docile nature and rooting behaviour. The legs of pig are short, strong, smooth pattern with or without wrinkles and they are fast runners. The legs are square with body. Tail is generally medium to long in size and the characteristic feature of the tail is that no curling observed and it is straight extending beyond hock. Uncastrated pigs live inside the jungle are heavy weight with well grown tusk, ferocious in nature and attempts to attack the strange people, those enter inside the jungle. This indigenous pig breed is healthy, very active, alert and fast runner and well adapted to the local environment of Nicobar. These parameters mark that the Nicobari pigs are originated or descendent from wild boar i.e. *Sus scrofa*, however, they are still in domestication process [7, 10–12].

2.2 Feeding practice

The pig is managed under open grazing, free range systems in the coconut plantation and inside the dense forest. These pigs have the natural habitats include rain forest, mountain forest and plantation area. During day time, pigs roam freely in jungles in search of food and in evening, they return to their respective owners or remain in the forest. Feed and feeding practices reveal that none of the farmers provide pigs with commercial feed. Feeding the pigs, both in morning and evening is the important routine of the day and these pigs are very active both in very early morning and late evening and move in batches of 4 to 20 to eat feeds [17]. No feed (ration) is prepared separately for the pigs. The pigs are grown and fattened using locally available feed resources and without any concentrate ration. Nicobari pigs are omnivorous, though largely vegetarian; are opportunists and most will eat a wide range of food of animal origin. The pigs are fed with copra, coconut and its water, ripe pandanus fruit, bread fruit, Nicobari aalu, root crops, both fresh and cooked fish, poor quality fish waste, crab, coconut beetles, kitchen and vegetable waste and other commonly found arthropods. Pigs are also fond of dehusking ripe coconuts and breaking the hard nut to enable them to eat the coconut kernel. Some of these pigs are also found on the sea shore, especially during the low tide, scavenging for snails, shellfish and other sea creatures. This Nicobari pig breed has very good behaviour on rooting and gets sufficient nutrient rice feeds especially on wild palm root and also eats crabs, small insects and also other sea wastes which are in sea shore areas. Four local or indigenously available feed materials such as pandanus, coconut, and Nicobari aalu and bread fruit are commonly fed to pigs by the tribals. Coconut is the main feed for Nicobari pigs and almost one third of the total coconut produce is reserved for feeding their pigs. Pigs are fattened mainly on coconut feeding. In Teressa Island, pigs are fed with poor quality fish, snails and meaty portions of seashells [13]. Although higher market value of coconuts, this Nicobari pigs are fed continuously with coconuts as these pigs are placed in a high position in the minds of the Nicobarese. At the time of feeding, the tribes have very different and distinct ways of calling the pigs, for example, by beating bamboo, producing different sounds by shouting at a peculiar high peak or singing particular songs. These animals soon respond to their owners and come to their respective place of feeding. It is observed that only the pigs of the concerned *tuhet* are turn up after listening such tuned call from his master. The calling the pigs for various purpose is different from one family to another one varying from mild vocal sweet tunes to heavy beating of bamboo pieces in a serially particular sound rhythm [15]. All the pigs gather at the place where the tribal man or woman breaks the coconut, remove the coconut with use of a special instrument from its outer shell and place the coconut with coconut water in the feeding trough or feeder which is locally known as “naam” in their Nicobari language, which is generally made from wood in various sizes as length is from 30 to 100 cm and width is from 15 to 20 cm. All the *Nicobarese* maintain one wooden stilt platform in their horticulture plantation to feed the pig. A lengthy hardwood with the size of 5 feet is removed the central portion till it forms in the shape of food container. It is used to keep sliced raw (*kutcha)* coconut to feed the pigs twice a day on regular basis [15]. After feeding, the animals wander back to the jungle, only to return to the village/household premises during night hours. Stem of big size bamboo is cut into two halves in middle with different length for feeding purposes. Hallow empty space is commonly used for feeding purpose. Other materials viz. old cans, shells of Giant clams (*Tridacna* spp.) and aluminium plates are used for feeding and watering. The pig feeder is most commonly prepared from thick wood, sea shell, bamboo or un-utilisable plastic drums. The quantity of feed provided to the pigs is not measured; however, the farmers reported that it differs with presently availability as well as age of pigs. For each adult pig, approximately two to three small raw coconuts (weighing approximately 0.5 kg) are provided daily. The Nicobari tribes lay down on a raised or height of the wooden platform to supply feed to the pigs. Normally they feed 3-5 coconuts to each pig. Both men and women are equally involved in the feeding and management of the pigs. Women pay special attention to pregnant and nursing pigs. Moreover, oil extracted coconut powder, Nicobari aalu, pandanus fruit (locally known as kevri); tapioca (malayal aalu) and also fish waste are fed in addition to feeding of raw coconut. The tribals do not cook or prepare any rationed feed separately to feed the pigs. However, pigs are still deficiency of balanced nutrition (energy, protein and minerals) and therefore, it is an urgent need to improve the knowledge and skill of technical know-how on feed resource management for the tribal people to enhance the pig/pork production system in Andaman and Nicobar Islands. It is also observed that some pigs never came to the residential area and just lives in the forest. The respective farmer regularly goes to the forest in the evening time to feed their pigs. It is interesting to observe that some tribes carried incense (locally available) sticks while feeding in the forest. By smelling the smoke, the roaming pigs recognised their owners and knew they are going to be fed [7, 10–12].

2.3 Genetic characterisation

The allele size range, observed and effective number of alleles, observed and expected heterozygosity and polymorphic information content (PIC) at 23 loci in Nicobari pig are studied. The allele size range varies from 86 to 116 bp at locus SW936 to 280–296 at locus IGFI. The total number of alleles ranges between 5 (S0178, SW951, SW24 and S0386) and 11 (S0355). The effective number of alleles ranges from 2.97 (SW24) to 7.9 (S0355). The mean observed number of alleles for all 23 loci in Nicobari pigs is 6.96 ± 0.31. The observed heterozygosities are lower than the expected values at all the 23-studied loci in Nicobari pig. The mean expected and observed heterozygosities are 0.75 ± 0.01 and 0.655 ± 0.02, respectively. The mean PIC for all the 23-studied loci is 0.74 ± 0.01 [7].

Herd composition of individual Nicobari pig family is 15.56 ± 2.59, 2.33 ± 0.33, 2.00 ± 0.48, 2.70 ± 0.90 and 1.83 ± 0.31 for herd size, sows, boars, growers and piglets respectively. They can survive with a very low level of management [4, 18]. The herd statistics reveals that the pig herd size per household ranges from 7.5 to 10.0 with a mean of 8.9. The herd size of Nicobari pig in every individual family in Nancowry, Teressa and Car Nicobar varies between 10 and 15 and is higher than on other islands. The overall herd size of the Nicobari pig is 12.46. It is recorded that 33.2% of farmers kept less than 5 pigs, 47.1% of farmers kept more than 10 pigs and 19.7% of farmers kept 5–10 pigs in their house. The herd composition reveals that the adult female population ranged between 9 and 20 percent. The adult breeding populations are important to further propagate the germplasm and there is an immediate need to increase breedable population of female pigs in Nicobar group of islands. 97.2% of household are rearing indigenous Nicobari pigs whereas remaining 2.8% are rearing Large White Yorkshire crossbreeds. Pigs for fattening purpose are reared by 84% of farmers. Black-coloured pigs are preferred by 86.7% of farmers, 6.7% liked white ones and 6.6% had no colour preference. Husbandry practices reveal that the tribal farmers did not rear pigs as a source of income. All the animals are used for domestic consumption during weddings and other festivals [7, 10–12].

2.4 Husbandry practices

Pigs in Nicobar do not have separate house/shelter or sty. Pigs are mainly resting underneath the tribal's hut/shelter. The Nicobari shelters/hut are prepared in sufficient height from the ground floor with approximately 2-3 m to assure a sufficient space/place for pig and piglets to get sufficient rest and also to protect from heavy rainfall in Nicobar islands [7, 10–12]. No separate pig house or sty or any housing pattern is constructed for the pigs. A separate enclosure/shelters for piglet are made using locally available material usually bamboos by the tribes of all the islands. Two types of night shelter are provided for pigs. In 58% of households, the pigs are housed underneath the tribe's hut. The huts are made in appropriate height from the floor to protect the pigs from heavy rainfall and other inclement weather. Seventy-six percent of shelters have concrete floors and they are cleaned regularly. On the other hand, in 42% of household, separate indigenous pig sties made of bamboo or other indigenous plant material are provided. The size of the sty varies according to the size of the pig and population. The pigs marked for slaughter and feral pigs caught from jungles were kept in separate wooden enclosures with roofs made of wild leaves or long grasses. The shelters are generally made of pieces of wooden planks, tree branches and the roofing is made using leaves/grasses [7, 10–12]. The shelters are made in different sizes depending on the population. Nicobari tribes in Teressa Island are using sea sand as the suitable bedding material for sows as well as old rags; cloths and/or dry big size leaves are used for new born piglets as bedding materials [13].

2.5 Slaughter practices

Adult body weight of Nicobari pig is 175-200 kg. Dressing percentage, live weight at slaughter (kg) and average age at slaughter (months) was reported as 70-80, 112.82 $\pm$ 14.26 and 12.76 $\pm$ 1.07, respectively. Both growers and adults are slaughtered. Pig slaughter and pork consumption pattern revealed that there is no commercial system of pig rearing or sale of pork prevailed among the tribal people. They rear pigs mainly for consumption during different festivals, ceremonies, village functions and inter village sports. During the functions, all the tribal people of a particular village assemble and take part in a fight between pigs and the tribe popularly known as "pig fight". The tribe members fight the pigs either alone or in a group. After defeating a pig, it is subjected to slaughter. Slaughtering of the pigs is carried out in different locations including the pig farmer's own premise, as there is no organised slaughterhouse. The slaughter procedure is done very systematically. Pigs are killed by direct cardiac puncture using a sharp-ended stick and the entire pig is roasted in a fire for scalding and cut off parts for consumption. Dressing percentage is found high; varied from 70 to 80%. The pig fat is smeared over the meat for long-term storage. Most of the festivals and ceremonies are cantered on pig and the pig festival (Cana-haun in Nicobari language). Mostly, male/boars are preferred for slaughter [7, 10–12].

2.6 Reproductive performances

Pigs are allowed for open range feeding and breeding occurs in the forest area. Reproductive performance reveals that natural mating is occurred in the jungle as in free-range systems of farming. The reproduction of this Nicobari pig stock is very high in comparison to other livestock. In general, the breeding male resides in jungle and also it is difficult to see or collect. The adult mature female pig goes to inside the jungle to cross with the adult boar at the breeding cycle. Reproductive performances of pigs are as age at first farrowing (months), litter size (number) and farrowing interval was 10.91 $\pm$ 0.85, 8.06 $\pm$ 0.33 piglets and 17.91 $\pm$ 0.33, respectively. The age at first farrowing is 10 to 12 months, the litter size is normally 6 to 10, farrowing interval is 8-10 months and the method of mating is natural. It is observed that before farrowing, the pregnant sow goes to jungles and prepares nests with wild leaves, grasses and some other plant materials and this indicates that the Nicobari pig has inherent behaviour for fashioning or building their nest. Pregnant sows and nursing sows are cared by the tribal women with utmost important which is same as traditional pig rearing practices which is followed at Kebar and Manokwari where pregnant females get top prior attention on feeding and management and they are kept as close to the tribal house and also supplied good quality food, water and shelter to them. It is also reported that sows in their last stage of pregnancy go to the jungle, farrow there and do not return to tribal shelter until 2 to 3 weeks, later bringing along with piglets. As farrowing occurred in the forest, exact litter size and piglet mortality are not known. Based on tribal farmers assumption, it is revealed that the mean age at first farrowing (months), litter size (number) and farrowing interval (months) were 10.8 ± 0.8, 6.8 ± 0.4 and 8.3 ± 0.4, respectively. It is reported that the teat number in Nicobari sows is varied from 5 to 6 pairs indicated that it has higher fecundity. The pigs are observed with good mothering characteristics. No weaning practice is followed. There is no information about the pre or post weaning mortality in pigs [7, 10–12].

The tribes have the knowledge and benefits of castration. It was believed that the castration might improve the body weight gain of pigs and makes the male pig more docile. Castration practices revealed that it was found that 94% of the farmers used to castrate their male pigs at the age of 3–4 months. Among the male piglets, the piglets with better vigour, body weight and health are not castrated; they are kept for breeding purpose. These practices indicate that the tribes have the knowledge of selection of good boar for breeding. Castration is performed in the dry season. The farmers use a surgical method of castration [7, 10–12].

One study was conducted to assess the effect of intensive and extensive system on different reproductive parameters. Results revealed that age at first oestrus (160.10 ± 6.83 vs. 173.6 ± 2.91 days), oestrus duration (66.00 ± 0.44 vs. 88.56 ± 3.57 hrs), age at first mating (160.00 ± 5.77 vs. 188.10 ± 2.41 days), gestation period (114.64 ± 0.23 vs. 116.12 ± 0.11 days), age at first farrowing (301.70 ± 2.4 vs. 319.20 ± 4.25 days), farrowing interval (226.00 ± 6.20 vs. 242.40 ± 4.84 days), litter size at farrowing (6.50 ± 0.34 vs. 7.19 ± 0.18), stillbirth (0.20 ± 0.01 vs. 0.59 ± 0.04 number per sow) and mortality (0.22 ± 0.08 vs. 0.68 ± 0.02 number per sow) are significantly lower in intensive system than free range system in female animals. Similarly oestrus cycle duration (26.09 ± 0.22 vs. 21.01 ± 0.20 days), litter weight at birth (0.83 ± 0.29 vs. 0.79 ± 0.71 kg), litter size at weaning (5.33 ± 0.33 vs. 5.23 ± 0.14), litter weight at weaning (31.28 ± 3.19 vs. 24.52 ± 3.15 kg) and litter weight at weaning (31.28 ± 3.19 vs. 24.52 ± 3.15 kg) are higher in intensive than in extensive system of rearing. In male, age at first mating (156.30 ± 2.08 vs. 143.1 ± 2.11 days) was significantly higher in intensive than in the extensive rearing system [19].

Body measurements (cm) such as chest girth (84.45 ± 3.01 vs. 93.77 ± 3.87), body length (84.88 ± 4.08 vs. 78.56 ± 2.77), height at withers (56.11 ± 2.44 vs. 60.65 ± 2.68) and neck girth (78.10 ± 3.40 vs. 67.64 ± 3.86) were differed between male and female Nicobari pigs in Nicobar group of Islands [20]. Similarly, body weights (kg) were significantly higher in intensive system than in extensive system in male and female animals at birth (0.86 ± 0.05 vs. 0.81 ± 0.06 and 0.81 ± 0.09 vs. 0.79 ± 0.07), weaning (6.56 ± 0.27 vs. 4.95 ± 0.15 and 5.17 ± 0.12 vs. 4.42 ± 0.13), 3 months (8.32 ± 0.14 vs. 6.47 ± 0.10 and 7.17 ± 0.17 vs. 6.15 ± 0.15), 6 months (50.00 ± 0.20 vs. 28.39 ± 0.30 and 42.27 ± 0.32 vs. 26.47 ± 0.22), 9 months (64.00 ± 0.27 vs. 38.39 ± 0.34 and 54.60 ± 1.07 vs. 36.57 ± 0.54) and 12 months (77.50 ± 0.29 vs. 43.06 ± 0.74 and 66.90 ± 1.08 vs. 40.95 ± 0.78). It is concluded that growth and reproductive performances of Nicobari pigs reared under intensive system has significantly higher beneficial than in free-range system [19].

2.7 Identification of pigs

Nicobari tribes identify their pigs in the systematic methods as they create identical cuts on the piglets' ears in such a way to easily identify or differentiate the piglets from one lineage or joint family to another lineage or family. Generally, markings resemble the symbols of claw and eyes of the crab, circle, half moon and similar identification cuts. It is a locally formatted act which is done by experienced tribal people of that specific lineage or family. In case, the identification cuts are wrong, the particular pig is killed and slaughtered and the particular concern person who marked wrongly is forced to eat the whole amount of pork without dividing with the other fellow *Nicobarese*. Such of different categories of ear cuts are visible in animals of the *Nicobarese* to know or identify the specific pig owner. In case no such identifications are seen on the pig ears, indicated that this is considered as wild boar and any person can hunt it for personnel purpose or consumption. In case mistakenly hunted the domestic one in the forest, it is given back to the concerned family by identifying its symbol on its ear. Pig slaughtering during ceremonial or any other domestic purposes, *Nicobarese* first remove the elongated piece of pork right from the earmarks to tail and displayed in front of the concerned house to prove its identity. Otherwise it is believed that others may mistake of its authenticity [15].

2.8 Pig trapping techniques and tools

Nicobarese generally uses the technique of *hinkuoñn* for catching the pig. In this technique, a tight rope is placed at different locations in the soil dibbled strong sticks and a round rope trap is located on ground close to its feeding trough or drinking water points. Finally it is attached to stick which is in the custody of hunter who hides in the nearby bush or tree, whenever it entangles the prey it tightly pull and caught by the waiting hunting party [15]. Hinkuoñn is a kind of trap used to catch pigs in the forest. An elongated rope is tied to a stick and a knot is made intermittently to facilitate easy tie of the rope to the leg of the pig. It is kept nearer to the regular feeding place and calls their hogs. When the pigs turn up for feeding are caught into trap [15].

2.9 Disease management practices

Lack of feed is found to be the biggest constraint. The main disease constraints are swine fever, parasitic diseases and respiratory problems. Swine fever as the main disease constraint in Nicobar pigs in Nicobar group of Islands. It was reported in the last decade, there was an outbreak of swine fever which caused mortality in young and adult pigs. A vaccination program against swine fever had been implemented by the Veterinary Department of Andaman and Nicobar Administration. Higher prevalence of parasitic diseases has also been observed. Worm infection is diagnosed by discerning of abdomen (67%), unthriftiness (43%) and poor appetite (41%). Only 33% of farmers practiced deworming. Other than diseases, the pigs were killed by the predators like python in the forest and street dogs [7]. There are various conditions such as natural calamities (tsunami and earthquake), predators (Reticulated pythons), outbreak of disease (swine fever) and non-availability of scientific breeding and farming practices leads to severe threat to the Nicobari pigs. Upto date there is three swine fever outbreaks have been observed from these Andaman and Nicobar islands. Nicobari pig breed which is available in Nicobar Islands are very much susceptible to this swine fever disease. The sero prevalence of swine fever was 41.75%, of which Lapathy in Car-Nicobar showed highest Sero-prevalence of 21.87% followed by Diglipur (18.75%), Nancowry (14.28%) and Tamaloo (Car Nicobar, 3.13%). The prevalence of Ascariasis, infestation with tape worm and abnormal nutritional deficiencies were reported in pigs in Andaman and

Nicobar Islands. A mild outbreak of Foot and Mouth Disease (FMD-type O) in pig was reported just after the episode of tsunami, 2004 [5].

2.10 Complete uterine prolapse-a case report

A Nicobari sows aged 2 years with complete prolapsed uterus was presented for treatment. History revealed that farrowing was normal and the hanging of prolapsed uterus was unnoticed for long period in the night time after farrowing. Everted uterine horns were protruded from vulva in clinical examination. The uterine masses which prolapsed were severely congested as well as oedematous. The values of body temperature, pulse and respiration rate were 102°F, 95/minute and 17/minute respectively. The prolapsed mass was cleaned with cold potassium permanganate (1:1000) solution. Ice packs were applied to reduce oedema. The rear part of the animals was elevated by placing gunny bags. After that attempts were given to replace the everted organ by gently pushing to its original position. However, the sow died due to prolonged exposure of complete prolapse in Nicobari pig [21].

2.11 Conservation of Nicobari pigs

Nicobari pig breed is believed as a local/indigenous pig germplasm belongs to this Andaman and Nicobar island territories. The external/phenotypic parameters revealed that this Nicobari pig breed is indigenous/ ethnic to these bay islands and their presence was reported since many decades. Nicobari pig revealed higher prolificacy as litter size varies from 8 to 10 numbers as well as lower preweaning mortality prevailed. Castrated boar and adult sow revealed significantly higher body weight (110-160 kg). The pigs are reared and considered as family asset among the tribal. No commercial farms or sale of meat is practiced. However, most of the pigs are slaughtered mostly during festive seasons or family/community ceremony. Awareness programme on conservation of indigenous pig germplasm and training on scientific pig farming is given for the Tribal and island farmers [11, 12]. Tribal families were identified for maintaining /conserving the pig germplasm.

3. Andaman local pig

3.1 Physical characterisation

Andaman local pig (ALP) has been introduced by settlers in these Islands from mainland India. This ALP is one of the indigenous pig breeds of Andaman group of islands and is mostly found in Baratang and Mayabunder area of Andaman. Its body coat colour differs from rusty grey to black or brown. Neck and back portion hair are very thick as well as long whereas hair on the sides and flank regions are shorter and thinner relatively. The adult male body weight varies from 75 to 80 kg female body weight varies from 60 to 70 kg. Age at first farrowing is about 300 days with litter size of about 7-8. They maintain good health with low plane of nutrition [22].

ALP is very well adapted and tolerable to the different tropical humid harsh environmental conditions with higher relative humidity, higher temperature as well as higher temperature humidity index. Further, these local indigenous Andaman local pigs are scavengers and also semi-wild in their behaviour and character. The Andaman local pigs have good mother caring ability and are more aggressive at farrowing or delivery time. Although the ALPs have lower in their growth rate and reproductive and productive performances, it is highly liked by the rural tribal communities for supplementation of sufficient protein and income for the family. The ALP is associated with the socio-culture-economic-tradition of tribals. Andaman local pig is in general as semi-feral in behaviour and is mostly reared in extensive or free-range system with little amount of management. Mitogenome analysis revealed that this ALP can be evolved as an independent breed in Andaman and Nicobar Islands as merit for registration as a recognised pig breed [8]. This indigenous local pig is under the endangered position and immediate preservation, conservation propagation effort is need to be taken to safeguard the indigenous pig breed from disappearance [7].

3.2 Genetic characterisation

Microsatellite markers have been used widely for the genetic characterisation of animal breeds including pig [23–26]. The microsatellites are used to assess the genetic diversity at higher level among the large genetic resource pools of pigs throughout the world [25]. Andaman local pig was characterised by 23 FAO recommended microsatellite markers. The allele size range, observed and effective number of alleles, observed and expected heterozygosity and polymorphic information content (PIC) at 23 loci in Andaman local pig is explained. The allele size range varies from 86 to 116 bp at locus SW936 to 280–296 at locus IGFI. The alleles' total number is ranged between 5 (SW122, S0228, SW951, S0178 and SW24) and 12 (S0355). The effective number of alleles ranges from 3.14 (SW24) to 8.1 (S0355). The mean expected and observed number of alleles for all the different 23 loci in Desi pigs of Andaman are 5.09 ± 0.20 and 7.04 ± 0.37, respectively [18]. The mean value of effective number of alleles for Andaman local pig is found higher than South-African pig breeds, Mozambique (8.45), Kolbroek (6.18) and Kune-Kune (5.97) but is lower than Duroc (3.98) [27]. The mean effective number of alleles of the Indian pig breeds Desi, Gahuri and Ankamali are 5.00, 5.33 and 5.34 respectively [28]. Higher allele numbers in India populations than in European breeds indicated that isolation and selection effects of these pig populations have been mild or minimum. Andaman local pig has the observed heterozygosities is lower than the expected value at the 22nd loci in S0005. The mean expected and observed heterozygosities are 0.77 ± 0.01 and 0.69 ± 0.01, respectively. The mean PIC for all the 23 studied loci is 0.74 ± 0.01. The genetic diversity in Andaman local pig is higher than the European pig breeds. The PIC is higher than Large White Yorkshire but comparable with other Indian pig breeds like Desi, Gahuri and Ankamali [28]. PIC values of all the microsatellite loci are above 0.5 which indicates that the microsatellite loci are suitable for detection of genetic diversity in Andaman local pig. Mean observed and expected heterozygosities of 23 microsatellite loci of Andaman local pig are found high indicating high genetic diversity of this pig breed. From the microsatellite data, it is also found that this pig breed is distinguishable from other pig breeds. As the pig breed is under the threat of extinction due to extensive cross breeding, serious effort must be initiated to conserve this breed in its breeding tract [18].

3.3 Reproductive profiles

Reproductive parameters such as litter size at birth (no.), total and individual litter weight at birth (kg), litter size at weaning (no.), total and individual litter weight at weaning (kg) and pre and post-weaning mortality (%) are recorded. Growth parameters such as body weights (kg) from month 1 to 9 are recorded. Dressing percentage, fat thickness, percentage of lean, fat, meat: bone ratio and also bone are recorded for Andaman local pigs in separately for male and female pigs [20].

Significantly higher body weights are observed from month 1 to 9 under intensive system in male than in female pigs. The rate of body weight growth in different months revealed that the rate between first and second months was 37.10% and this rate has been increased from second (15.39%), third (17.79%) and fifth months (22.81%) and then decreased from fifth (13.42%), sixth (12.45%), seventh (2.87%), eighth (3.14%) to ninth months (2.34%) in male pigs. Similar trend is also observed in female pigs as in month 1 (37.39%), month 2 (15.96%), month 3 (18.28%), month 4 (21.18%), month 5 (16.56%), month 6 (9.93%), month 7 (4.09%), month 8 (3.21%) and month 9 (1.64%). In overview, the body weight of female pig is significantly lower than the male pig (47.87 vs. 52.12%) with an average of 42.66 and 46.18 kg, respectively for female and male pigs [20].

Weight of total litter size, litter size at birth, and individual at birth, litter size at weaning, weight of total litter size and individual at weaning and pre and post weaning mortality differs significantly between male and female pigs at the rate of 17.72, 9.94, 7.79, 3.41, 5.18, 34.19 and 23.46%, respectively. The male has significantly higher value than in female with respect to all these parameters except the pre and post weaning mortality which are significantly higher in female than in male. However, these values are within the normal range of pigs of indigenous population [20].

One study was conducted to assess the reproductive parameters in Andaman local pigs. Results revealed that these reproductive parameters such as litter size at birth (3.87 ± 0.16 vs. 3.17 ± 0.12), average individual weight at birth (1.66 ± 0.02 vs. 1.42 ± 0.02 kg), litter weight at birth (6.41 ± 0.27 vs. 4.48 ± 0.17 kg), litter size at weaning (3.33 ± 0.13 vs. 3.11 ± 0.11), average individual weight at weaning (10.55 ± 0.09 vs. 9.51 ± 0.06 kg), litter weight at weaning (35.08 ± 0.31 vs. 29.56 ± 0.19 kg), pre-weaning mortality (8.87 ± 0.12 vs. 4.35 ± 0.08%) and post weaning mortality (3.42 ± 0.11 vs. 2.12 ± 0.03%) were significantly higher in male than in female animals. Similarly body weight (kg) at 1st Month (6.67 ± 0.15 vs. 5.96 ± 0.20), 2nd month (14.51 ± 0.18 vs. 13.08 ± 0.18), 3rd month (19.79 ± 0.22 vs. 18.05 ± 0.19), 4th month (28.36 ± 0.24 vs. 26.13 ± 0.29), 5th month (45.13 ± 0.17 vs. 40.77 ± 0.27), 6th month (59.13 ± 0.30 vs. 56.96 ± 0.27), 7th month (75.96 ± 0.29 vs. 69.53 ± 0.39), 8th month (80.45 ± 0.14 vs. 75.47 ± 0.22) and 9th month (85.67 ± 0.23 vs. 78.00 ± 0.37) was significantly higher in male than in female animals in Andaman local pigs [20].

3.4 Semen collection and artificial insemination

Cross breeding with the use of Artificial insemination (AI) can be a tool to upgrade genetically inferior local pigs and avoid inbreeding that usually happens with less number of available breeding boars or small pig population. The purpose of semen preservation for AI is to maximise the use of superior germplasm with extended sperm viability but without much effect on the sperm fertility essential for successful breeding. With the aforesaid vision, semen collection was attempted in Andaman local pigs using gloved hand technique. This is for the first time to be reported in Andaman local pigs.

Preliminary study indicated that Andaman local pig has released total semen volume, gel free semen and gel in semen volume was 220, 30 and 190 ml, respectively, and pH of semen was found to be 7.5. Objective assessment of total and progressive sperm motility was done which were 80 and 75%, respectively. Sperm concentration was estimated with use of haemocytometer chamber and count is 210×10^6/ml. Morphometric measurements of pig spermatozoa with software enabled microscope were performed. Average head length, head width, tail length and full sperm length was observed to be 9.42, 5.24, 43.93 and 52.37 μm, respectively. The preserved liquid semen was used for artificial insemination purpose in the organised pig breeding farm, ICAR-CIARI, Port Blair, India and the sow was conceived and farrowed 5 piglets in the year 2019 [29].

3.5 Carcass characteristics

Carcass characters such as dressing percentage, meat: bone ratio and fat thickness are not significantly different between male and female pigs whereas other parameters such as percentage of lean, fat and bone differs significantly between them. Percentage of fat (10.10%) in female and lean meat percentage (4.80%) and bone percentage (7.20%) in male are significantly higher than those in the other sex. Carcass characteristics such as dressing percentage (76.54 ± 0.31 vs. 75.52 ± 0.41), meat: bone ratio (5.53 ± 0.15 vs. 5.69 ± 0.15), fat thickness (5.55 ± 0.18 vs. 5.61 ± 0.16 cm), lean meat percentage (58.79 ± 0.36 vs. 53.4 ± 0.41; $p < 0.05$), fat percentage (30.37 ± 0.25 vs. 37.2 ± 0.20; $p < 0.05$) and bone percentage (10.86 ± 0.24 vs. 9.4 ± 0.23; $p < 0.05$) differed between male and female Andaman local pigs [5].

3.6 Feeding practices

Locally available feed resources such as rice bran, maize, wheat, coconut, taro (*Colocasia esculenta* and *Colocasia antiquorum*), tapioca, kitchen/ hotel waste, vegetable waste and poultry offal are fed to the indigenous local pigs in Andaman and Nicobar Islands [4]. In general, feed, fodder and soil of these Islands are deficient in minerals particularly Zn and is limiting factor for the growth of the pig. Age at puberty, age at first conception, age at first furrowing, litter size at birth, individual and total litter weight at birth, litter size at weaning, individual and total litter weight at weaning and weaning percentage are found significantly increased in pigs treated with 80 ppm zinc as zinc sulphate in Andaman local pig and its crossbred. Similarly, the fortnightly body weight gain (kg), total weight gain (kg) and the average daily weight gains (ADWG) are significantly higher in Zn supplemented Andaman local pigs and its crossbred [6].

3.7 Mastitis-Metritis-Agalactia (MMA) syndrome

Mastitis-Metritis-Agalactia (MMA) syndrome causes huge economical losses in the swine industry. Andaman local sow aged 3 years with the history of farrowing 18 days ago and complaint of anorexia, restlessness and inattentive towards her piglets, agalactia and lameness was presented with the elevated rectal temperature, congested mucus membrane, swollen painful mammary glands with foul smelling muco-purulent vulval discharge. Based on the visible clinical signs, sow was tentatively diagnosed as suffered from mastitis-metritis-agalactia syndrome. The affected sow was treated with ice fomentation, cleaning with liquid soap, application of Lugol's iodine solution and antiseptic ointment on the udder, injection of gentamicin, streptopenicillin, non-steroidal anti-inflammatory drug, prostaglandin F2α, intra-uterine infusion of normal saline followed by Lugol's iodine solution along with supportive therapy with multivitamin and hydrotherapy in water bath. The pig was fed with boiled chicken eggs for supports to her health. The piglets were fed with toned cow milk during the treatment regimen along with creep feed. On day 3rd post treatment, the sow was recovered and allowed the piglets to suckle. Thus the quick diagnosis and prompt treatment saved the pigs from the life threatening syndrome along with eliminating the pre-weaning piglet mortality. The MMA prevalence could be reduced through optimization of husbandry, feeding and managemental practices.

This is first report of MMA syndrome in Andaman local pig in Andaman and Nicobar Islands that too affected after 18 days of farrowing [30].

3.8 Foster mother behaviour

Piglet movement from one sow to another is known as fostering which is frequently observed when the number of piglets a sow gives birth to do not match her rearing ability. This practice is very common in Andaman local pigs. Andaman local sow aged 3 years farrowed 6 piglets with good health condition. At the near farrowing room, another Andaman local pig farrowed 8 piglets with good health condition. The second Andaman local sow died due to complete uterine prolapse. These orphan piglets were allowed to suck in another normal sow. The unaffected sow accepted and fostered till the weaning age.

3.9 Coprophagia behaviour

The coprophagy was observed in Andaman local pig is autocoprophagy (eating its own faeces). This may possibly to rebalance their microbiome or to ingest missing nutrients. Coprophagy is thought to be a source of vitamins B and K, produced by gut bacteria.

3.10 Placentophagy behaviour

An Andaman local sow aged 2 years was observed to eat her own placenta after 1-2 hours of farrowing. History revealed that the farrowing was normal with 6 piglets, sow was late attended and the placenta was eaten by dam. Body temperature, pulse and respiration rate were observed within the range. It is advised to attend sow after farrowing along with feeding the pig with good balanced diet enriched with vitamin and mineral supplements.

3.11 Complete mitochondrial genome sequence of indigenous pig germplasm

The complete mitochondrial DNA sequences of Nicobari pig and Andaman local pig were submitted to GenBank with the accession numbers MK248681 and MK248682, respectively. Both the Nicobari as well as Andaman local pigs have the length of the mitogenome of 16,613 bp and are have 37 encoded genes which include protein coding genes (13 PCGs), two rRNAs and 22 tRNAs. In addition, one AþT rich region (D-loop) was present. The orientation and order of the genes was same as to the mitogenomes of similar vertebrate species. Protein coding genes were located on heavy strand except ND6. Start codons for 13 protein coding genes were having ATN codon followed by truncated/ abbreviated stop codon was found in ND1, COX3, ND2, ND4 and ND3. From the phylogenetic tree, it was found that Nicobari pig has close phylogenetic relationship with Banna mini and Breed I pig, whereas Andaman local pig is close to Mong Cai and Jinhua pig. Mitogenome analysis on local indigenous pig breeds revealed that the analysis will be useful to format conservation strategy of the swine breeds in Andaman and Nicobar islands [8].

4. Andaman wild pig (*Sus scrofa andamanensis*)

Long snouted Little Andaman wild pig (Schedule II animal under Forest Act, India) is a threatened and endangered in Andaman and Nicobar Islands. Andaman wild pig is preferred by the local people as a meat source. Wild pig of Andaman is commonly spread at the Jarawa as well as Onge tribal forest reserve areas of Andaman group of islands. The Jarawa tribes also prefer this wild pig. They are being poached by the primitive Jarawa tribes and are the main source of protein for them from time immemorial. Due to unauthorised poaching, the number of this wild pig is reducing day by day, which needs attention for its conservation. Andaman wild pig was once found all over in the forests of the Andaman group of Islands, but have become extremely rare and currently the last strong holds are the Jarawa Reserve forest area, Rutland and Little Andaman Islands. It is a scheduled animal, black in colour, short legged, small to medium sized and a prolific breeder. As per the literature, Andaman wild pig has the litter size (number) from 4 to 7; however, due to unavailability of food and water and illegal hunting, their numbers has been decreasing very fast in Andaman and Nicobar Islands. Presently, this pig comes under schedule I Part I of the Wild Life (Protection) Act of India, 1972. These wild pigs of Andaman are well adapted physiologically and anatomically to this island ecosystem over the many centuries as they are native of these islands. They are black in colour, short legged, small to medium sized animal and very active, alert and fast runner [22].

4.1 Physical profiles

Body height of Andaman wild pig was measured as 20 inch at the level of shoulder with the compact body. The pig is very active, wild expression and a fast running animal. The RBC concentration, PCV and Haemoglobin concentration were found very high [22]. The phenotypic characters of male pig (in inches) are presented as body length (from shoulder to base of tail): 23, body height at shoulder level: 20, neck width: 15.5, ear length: 3, ear width: 3, Leg length; 9, hoof circumference: 2.5, tail length: 4, abdomen width: 20.5, chest width: 21.5, testis length: 2.5, testis width: 2 and body weight based on chest girth: 16 kg [31]. Boden Kloss [14] observed that the pigs (*Sus scrofa andamanensis*) in Andaman islands appeared were diminutive in stature and the fully grown boar was only 20 inches high at the shoulder.

4.2 Haematological profiles

The blood profiles of Andaman wild male pig revealed that the RBC: 9.72 x 10^6/μL, MCV: 63.1 f1, PCV: 61.3%, MCH: 17.77 pg., MCHC: 28.17 g/dl, Hgb: 17.27 g/dL. The leucocytic parameters of Andaman wild male pig revealed that the WBC: 35.12 10^3/μL, lymphocyte: 62.80%, monocyte: 8.37%, neutrophils: 4.80%, eosinophils: 21.37% and basophils: 0.70%. The thrombocytic parameters of Andaman wild male pig revealed that the platelet: 696.00 10^3/μL, MPV: 6.83 f1, Pct: 0.43% and PDW: 11.90 [22]. The WBC is also found high in Andaman wild pig. WBC count of Andaman wild pig is higher than that reported in wild boar of Croatia [32].

4.3 Comparision study among the wild, indigenous and exotic pigs

This comparision study was conducted in Andaman and Nicobar islands. Blood indices revealed that PCV, RBC and Hgb were significantly higher in Andaman wild pig than in other all pig breeds (Nicobari pig, Large White Yorkshire and Andaman local pig). The RBC, PCV and Hgb of LWY were significantly higher as compared to Andaman local pig and Nicobari pig. No significant differences in RBC, PCV and Hgb were found between Andaman local pig and Nicobari pig. Andaman wild pig has significantly higher Hgb, RBC and PCV indicates that a higher level of oxygen is required for wild pig as it is a fast running animal

Andaman wild and Nicobari pig are not differed significantly in their MCV, however, both these pigs were had significantly higher MCV than in Andaman local pig as well as LWY in Andaman and Nicobar Islands [31]. Higher values of MCV in wild pigs impute an enhanced need for oxygen [33]. Nicobari pigs are too growing in open grazing or free range systems; which also fast running animal. The increased blood profile is due to environmental effect on haematological traits as haematological and biochemical values may be affected by a wide range of factors, including environment, season, diet, age and stress [32]. Whereas, MCH of Nicobari pig was found significantly higher in comparison to all the other pig breeds, the value was lowest in LWY. However, MCHC did not show significant differences within the pig breeds. Wild pig of Andaman is well adapted anatomically and physiologically in the humid tropical climate of Andaman and Nicobar Islands [31].

The blood leukocyte indices revealed that the WBC was significantly higher in Andaman wild pig in comparison to Nicobari pigs and was lowest in LWY. Similarly lymphocyte concentration was significantly lower in LWY than in all the other pig breeds; however, there was non-significant difference between the Andaman wild pig, Andaman local pig and Nicobari pig in Andaman and Nicobar Islands. A significantly higher monocyte was found in LWY as compared to all the other pig breeds. Wild pig of Andaman had lower neutrophils significantly as compared to other all pig breeds available in Andaman and Nicobar Islands; LWY has highest value. Eosinophil was highest in Andaman wild pig followed by Andaman local pig, Nicobari pig and LWY. No significant differences were found in basophils among all the pig breeds studied. The neutrophil and lymphocyte ratio was lowest in Andaman wild pig and was highest in LWY. The MCV of Andaman wild pig was also significantly higher in comparison to Andaman local pig and LWY [31].

Blood thrombocytic values in Andaman wild pigs revealed that no significant ($p < 0.05$) difference was found in PLT between Andaman wild pig and LWY but the values were significantly higher in comparison to Andaman local pig and Nicobari pig. MPV value of LWY was significantly lower in comparison to all the other pig breeds studied. PCT of Andaman wild pig was significantly higher than Nicobari pig and LWY but did not differ significantly with Andaman local pig. PWD of LWY was lowest among all the breeds [31].

The reports on Andaman wild pigs revealed that based on the physical appearance, phenotypic characters and haematological profiles, these pigs are native to these islands and are well adapted to this island ecosystem over the centuries. Extensive survey on population status and studies on characterisation (*in situ* and *ex situ*) measures to protect this protected breed and scientific breeding methods should be implemented [22].

5. Andaman pig crossbred

Andaman cross breed is a cross between Large White Yorkshire and Andaman local or Nicobari pig. They are dark brown to slight white with different lines of black colour. This crossbred pigs exhibit high growth rate, fecundity and early maturity. It has high prolificacy (litter size 8–10 nos.), maternal care and the average body weight of matured animal varies from 110 to 125 kg. Moreover, this crossbred is adapted highly to the local tropical humid environmental conditions and also can adjust with locally available feed resources on the different agricultural produces. This is highly suitable for commercial production of swine meat in the island [16].

Different reproductive parameters like age at puberty (221.67 ± 3.99 days), age at first conception (245.50 ± 3.94 days), age at first farrowing (357.00 ± 4.07 days) and various litter traits like litter size at birth (6.17 ± 0.48), total litter weight at birth (7.26 ± 0.87 kg), individual litter weight at birth (1.18 ± 0.12 kg), litter size at weaning (5.17 ± 0.48), litter weight at weaning (30.46 ± 1.98 kg), individual litter weight at weaning (6.11 ± 0.60 kg) and weaning percentage (84.7 ± 5.51) were reported in Large White Yorkshire x Andaman local crossbred pigs [6].

6. Conclusion

Andaman and Nicobar Islands are completely packed with rich biodiversity. Porcine species occupies 27.26% of total livestock in these islands, of which, 70% pork consumption in Nicobar group of islands. There are three different groups of pig groups/breeds in ANI. Andaman Local is in Andaman group of Islands, Nicobari is in Nicobar group of Islands and Andaman wild pig in Andaman and Nicobar islands. Besides, crossbreds of LWY are prevalent in this ANIs. Nicobari pig plays significant roles in custom, festivals and socio-economic status of Nicobari tribes. Andaman local and Nicobari pigs are reared for meat purpose under free range or semi-intensive system. Andaman wild pig is an endangered pig germplasm of ANI. Another pig group is crossbred of LWY with Andaman local or Nicobari pig. This crossbreed exhibits high growth rate, early maturity and fecundity. In addition, it is highly adapted to the local environmental conditions and can be reared with locally available feed resources. This is highly suitable for commercial pork production in ANI. However, these domestic pig breeds need to be protected and be conserved in this Andaman and Nicobar group of Islands.

7. Outlook work for future work

7.1 Conservation and propagation strategy

The indigenous porcine population is decreasing gradually in ANI. Therefore conservation and propagation strategy needs to be established for domestic indigenous pig breeds/groups with formation of nucleus of elite flocks in farms [5]. This can be performed as follow as

- Survey of natural population: this can be performed by scholars or the people of local community. Survey training would be performed by ICAR-Central Island Agricultural Research Institute, Port Blair, Andaman and Nicobar Islands, India

- Establish shelters and farms for producing young ones for distribution.

- Analysis of recording of the breed performance in different conditions and locations on and off the farm.

- Establishing farms are: With the help of the Tribal Council in Car Nicobar and Hut Bay, Kamorta, ICAR-CIARI, Mayabunder, Diglipur and Port Blair, help from Department of Animal Husbandry and Veterinary Science, Andaman and Nicobar Administration. Creation of multiple centres can help that if one centre is severely affected by a natural calamity or disaster or disease outbreak, other remaining centres could help to restore.

- Caution in conservation of the Nicobari pig: In the Andaman Islands, the Nicobari pigs do not place near the areas where Andaman wild pigs are reported.

7.2 Feeding practices

The pigs are fed with locally available feed resources and agricultural, horticultural and marine by-products or waste without analysing the chemical composition or ration. Therefore, it is needed to survey the available local feed resources and formulate the ration suitable for the pigs for better growth rate and better production and reproduction performances.

7.3 Breeding programme and artificial insemination

The indigenous pigs are reared in free range or semi-intensive system and breeding has been occurred in the jungle or forest with dominant boars. This will create more inbreeding lines which inturn affect the growth rate, reproduction and production performances. This needs to be addressed with formation of suitable breeding strategy and artificial breeding programme with use of elite porcine germplasm.

7.4 Control of diseases

Prevalence of various diseases has been reported in pigs. These diseases should be analysed in season wise, island wise, age group and sex wise and are to be treated accordingly. Time schedule for deworming and vaccination protocol are need to be implemented in different islands of Andaman and Nicobar. As there are many diseases are zoonotic in nature in pork eating community, therefore, prevention and control of the diseases in pigs is very important.

7.5 Slaughtering procedure

Pig slaughtering is need to be modernised and hygienic handling of the pork is need to be improved. There is need to be established the modernised slaughter house, quality control lab, pork processing and preservation chamber in Andaman and Nicobar Islands. Carcass characters, meat quality, chemical composition of pork of different breeds of pigs need to be studied.

7.6 Housing management

Housing facilities need to be improved. Different houses like piglet, grower and adult pen are to be established. Clean and hygienic house and its surrounding are to be maintained.

7.7 Identification of pigs

Identification of pigs is done with cuts on the ears in Andaman and Nicobar Islands. This can be improved with the help of tags or electronic chip method.

7.8 Comparision study

Comparision study needs to be conducted between different indigenous and cross or pure exotic breeds of pigs in island ecosystem on different growth, production and reproduction parameters. Further detailed study needs to be conducted on effect of free range, intensive and semi-intensive system on different growth, production and reproductive parameters in different breeds of pigs.

7.9 Abnormal behaviour/condition

Abnormal behaviours/conditions like Coprophagia behaviour, placentophagy behaviour and mastitis-metritis-agalactia syndrome are need to be studied thoroughly about the aetiology, pathophysiology, prevention, treatment and control in pigs.

Author details

Perumal Ponraj*, Arun Kumar De and Debasis Bhattacharya
ICAR-Central Island Agricultural Research Institute, Port Blair, Andaman and Nicobar Islands, India

*Address all correspondence to: perumalponraj@gmail.com

References

[1] Livestock Census of India (18th). Department of Animal Husbandry, Dairying and Fisheries, Ministry of Agriculture, Government of India, New Delhi, India, 2007.

[2] Livestock Census of India (19th). Department of Animal Husbandry, Dairying and Fisheries, Ministry of Agriculture, Government of India, New Delhi, India, 2012.

[3] Livestock Census of India (20th). Department of Animal Husbandry, Dairying and Fisheries, Ministry of Agriculture, Government of India, New Delhi, India, 2012.

[4] Jeyakumar S, Kundu A, Yadav SP, Sunder J, Balakrishnan M, Kundu MS, Sujatha T, Verma SK, Srivastava RC. Diversity and conservation of farm animal genetic resources (FAnGR) of Andaman and Nicobar Islands, Ecology of faunal Communities on the Andaman and Nicobar Islands, (Springer Publication), 2012.

[5] Kundu A, Sunder J, Jeyakumar S, Verma SK, Kundu MS, De AK, Srivastava RC. Livestock and poultry production policy for Andaman and Nicobar Islands: a scientific perspective. Published by Director, ICAR-CARI, Port Blair, India, 2010; pp: 1-48.

[6] Kundu MS, Sunder J, Kundu A, De AK, Sujatha T. Reproductive and productive performances of crossbred Andaman local pigs under small holder production system at Bay Islands, India. Indian Journal of Animal Research. 2017; **51**(2): 377-381.

[7] De AK, Jeyakumar S, Kundu MS, Kundu A, Jai Sunder, Ramachandran M. Farming practices and genetic characterization of Nicobari pig, an indigenous pig germplasm of Nicobar group of islands, India. Tropical Animal Health and Production. 2014; DOI 10.1007/s11250-014-0547-z

[8] De AK, Muthiyan R, George Z, Perumal P, Sunder J, Kundu A, Malakar D, Bhattacharya D, Kundu MS, Muniswamy K. Mitochondrial landscape of indigenous pig germplasm of Andaman and Nicobar Islands. Mitochondrial DNA Part B: Resources. 2019; **4**: 2, 2808-2810.

[9] Holness DH. The tropical agriculturist (Pigs). CTA, Wageningen, 1991; pp. 1-29.

[10] Jeyakumar S, Sunder J. Conservation and Characterization of Nicobari Pig. AP Cess Project Report, Central Agricultural Research Institute, Port Blair, Andaman and Nicobar Islands, 2009.

[11] Jeyakumar S, Sunder J, Kundu MS, Kundu A, Swapna TP. Traditional pig rearing practices among the Nicobari tribes of Nicobar group of Islands, India. IOSR Journal of Agriculture and Veterinary Science. 2014a; 7(12): 35-41.

[12] Jeyakumar S, Sunder J, Kundu A, Balakrishnan P, Kundu MS, Srivastava RC. Nicobari pig: an indigenous pig germplasm of the Nicobar group of Islands, India. Animal Genetic Resources. 2014b; **55**: 77–86.

[13] Srivastava N, Ahlawat SPS, Chatterjee RN, Roy MM, Choudhuri NC, Saha SK. Backyard swine rearing practices among Nicobari tribes of Andaman and Nicobar Islands. Indian Journal of Animal Health. 2002; **41**(1): 9-12.

[14] Boden Kloss C. In the Andamans and Nicobars: The Narrative of a Cruise in the Schooner "Terrapin", with notices of the Islands, their Fauna, Ethnology, etc. John Murray, Albemarle Street, W., London, 1903.

[15] Prasad DV. Livestock management among the Nicobarese of Katchal Island. International Journal of Multidisciplinary Research Review. 2016; **1**(17): 147-153.

[16] Sujatha T, Kannan A, Jeyakumar S, Kundu A, Velmurugan A, Sunder J, Swarnam TP, De AK. Livestock and People – The Intimate Relation Under Threat (Chapter 15). In: Biodiversity a nd Climate Change Adaptation in Tropical Islands. Sivaperuman C, Velmurugan A, Singh AK and Jaisankar I (eds). Academic Press, 2008; pp. 433-457.

[17] Tikader BK, Das AK. Glimpses of Animal Life of Andaman and Nicobar Islands, Zoological Survey of India, Calcutta, 1985.

[18] De AK, Jeyakumar S, Kundu A, Kundu MS, Sunder J, Ramachandran M. Genetic characterization of Andaman Desi pig, an indigenous pig germplasm of Andaman and Nicobar group of islands, India by microsatellite markers. Veterinary World. 2013a; **6**(10): 750-753.

[19] Kundu MS, Perumal P, Ravi SK, Sawhney S, Bhattacharya D, Kundu A, Sunder J, Muniswamy K, De AK. Evaluation of reproductive and production performance of Nicobari pig under humid tropical island ecosystem. Indian Journal of animal Sciences. 2019; **89**(3): 73-78.

[20] Kundu MS, Perumal P, Ravi SK, Bhattacharya D, Kundu A, Sunder J, Muniswamy K, Sawhney S, De AK. Reproductive and production performance of Andaman Local Pig of Andaman and Nicobar Islands, India under intensive system of rearing. International Journal of Bio – Resource and Stress Management. 2020; **11**(1): 20-26.

[21] Ravi SK, Perumal P, De AK, Alyethodi RR, Kumari S, Bhattacharya D. Postpartum uterine prolapse in Nicobari sow-a case report. Journal of Andaman Science Association. 2019; **24**(1): 148-149.

[22] De AK, Jeyakumar S, Kundu MS, Kundu A, Sunder J. Andaman wild pig (*Sus scrofa andamanensis*): A preliminary report on phenotypic and haematological characteristics. Zoo's Print 2013b; **XXVIII**(9): 9-11.

[23] Amigues Y, Boitard S, Bertrand C, Sancristobal M, Rocha D. Genetic characterization of the Blonde d'Aquitaine cattle breed using microsatellite markers and relationship with three other French cattle populations. J. Anim. Breed. Genet. 2011; **128**(3): 201-208.

[24] Tamara AJ, Choumane W, Hmeshe M. Characterization and Estimation of Genetic Diversity in Two Syrian Chicken Phenotypes Using Molecular Markers. International Journal of Poultry Science. 2012; **11**(1): 16-22.

[25] Nidup K, Moran C. Genetic diversity of domestic pigs as revealed by microsatellites: a mini review. Genomics and Quantitative Genetics. 2011; **2**: 5-18.

[26] Sollero BP, Paiva SR, Faria DA, Guimarães SEF, Castro STR, Egito AA, Albuquerque MSM, Piovezan U, Bertani GR, da S. Mariante A. Genetic diversity of Brazilian pig breeds evidenced by microsatellite markers. Livestock Science. 2009; **123**(1): 8-15.

[27] Swart H, Kotze A, Olivier PAS, Grobler JP. Microsatellite-based characterization of Southern African domestic pigs (*Sus scrofa domestica*). South African Journal of Animal Science. 2010; **40**(2): 121-132.

[28] Behl R, Sheoran N, Behl J, Vijh RK. Genetic analysis of Ankamali pigs of India using microsatellite markers and their comparison with other

domesticated Indian pig types. Journal of Animal Breeding Genetics. 2006; **123**: 131-135.

[29] Ravi SK, Perumal P, Kundu MS, Bhattacharya D, Jai Sunder, De AK, Alyethodi RR, Kundu A. Physical, biochemical and molecular characterization of semen in pigs of bay islands vis-a-vis study on feasibility of artificial insemination. Annual Report (2018-19), ICAR-CIARI, Port Blair, Andaman and Nicobar Islands, 2019.

[30] Perumal P, Ravi SK, Sarkar G, De AK, Bhattacharya D, Sawhney S, Kundu A. Mastitis-Metritis-Agalactia Syndrome in Andaman Local Pig-First Case Report. Journal of Andaman Science Association. 2020; **24**(1): 14-18.

[31] De AK, Kundu A, Kundu MS, Sunder J, Jeyakumar S. Comparative study on haematological traits of endangered Andaman wild pig and other indigenous pig breeds available at Andaman and Nicobar Islands, India. Veterinary World. 2013c; **6**(10): 794-798.

[32] Harapin I, Bedrica L, Hahn V, Sostaric B, Gracner D. Haematological and biochemical values in blood of wild boar (*Sus scrofa ferus*). Veterinarski Arhiv. 2003; **73**(6): 333-343.

[33] Tusek T, Mihelic D, First L, Janicki Z, Opancar D. Komprativni prikaz crvene krvne slike divljei domace europske svinje. Vet. Stanica. 1994; **25**: 81-84.